Preserving archaeological remains *in situ*?

PROCEEDINGS OF THE
2ND CONFERENCE
12–14 SEPTEMBER 2001

Published by the Museum of London Archaeology Service

A CIP catalogue record for this book is available from the British Library

Production and series design by Tracy Wellman
Typesetting and design by Sue Cawood
Technical editing by David Bowsher
Copy editing by Ann Clark
Series editing by Sue Hirst/Susan M Wright

Printed by the Lavenham Press

Front cover: one of the large timber beams from London's Roman amphitheatre arena entranceway being winched off site

Preserving archaeological remains *in situ*?

PROCEEDINGS OF THE 2ND CONFERENCE 12–14 SEPTEMBER 2001

edited by Taryn Nixon

MUSEUM OF LONDON ARCHAEOLOGY SERVICE

Contents

Summary

This volume presents the papers and posters presented at the second Preserving Archaeological Remains *in situ* (PARIS2) conference. Five years after the first conference, which focused on the processes at work in buried archaeological deposits, this second conference set out to address three main themes: to review recent research; to examine the relative successes and consequences of decisions (especially those taken in the last decade or so); to preserve particular sites *in situ*; and to try to identify strategic directions for future research into the protection of our cultural heritage. The 36 papers in this volume comprise 32 presentations and 4 detailed poster contributions.

A great deal of research has taken place in the last five years which will inform how we treat and protect archaeological remains, whether those remains are terrestrial, coastal or submerged, and whether in urban or rural settings. The selection of papers here represents a range of studies which examine evidence of and key indicators for biological, chemical and physical change, both to artefacts and to soils. The research presented here covers topics such as the decay of biological remains in waterlogged soils (paper 2), the potential to use bone as an indicator of degradation of deposits (3), the acoustic characteristics of wood (4), the deterioration of iron and bronze (5 and 9), geochemical modelling to study post-depositional changes in the burial environment (7), and physical pressures on inclusions in sand through compression (6) and the geotechnical properties of archaeological sediments and artefacts as they respond to the sorts of loading and unloading scenarios commonly encountered in construction work (8).

Although the principles of preservation *in situ* have only been embraced in policy relatively recently, preservation decisions are required everyday; the second theme examines the consequences and efficacy of a range of decisions, giving case studies and details of particular monitoring regimes. 'Design solutions' at four development sites in London (16) and a site in Gloucester (18) describe the balance achieved between preservation *in situ*, development and archaeological excavation. Issues of ethics, principle and pragmatism are explored, with reference to the discovery of the Bronze Age Seahenge found in Norfolk (20), and the Zeumga project in advance of the construction of the Birecik dam in Turkey is outlined (21). Further case studies include an examination of biodegradation at Wareham experimental earthwork and its application to *in situ* preservation (10); the *in situ* preservation of waterlogged archaeological remains from the Tower of London (17); and the damage and survival of the earthwork at the Cawthorn Camps, North Yorkshire (23). The monitoring regimes at sites where archaeological remains preserved *in situ* are described and assessed at Trondheim (11) and Tønsberg (14) both in Norway; an Early Anglo-Saxon cemetery in Croydon, London (16); at Chedworth Roman villa, Gloucestershire (24); and at 12 sites in Voorne-Putten, Holland (22). Conservation management and *in situ* preservation is described in a range of different archaeological situations, including waterlogged remains at wetland sites at Sutton Common, South Yorkshire (13), and Nydam Mose, Denmark (15), historic shipwreck sites such as Scapa Flow (12) and arable landscapes (25). Methods of protection, such as the use of calcite precipitation systems to protect masonry remains (19) are presented.

There is a clear and growing focus within the historic environment community towards crafting a coherent and sustainable strategy for research and policy implementation. The requirements for our evolving conservation strategy are addressed in the third theme. Policy is examined in the role of archaeology in the conservation planning process (27), coastal management schemes (28) and the National Trust soil policy (29). The processes involved with mitigation of construction impacts on archaeological remains (30) are described and the preservation, conservation and display of London's Roman amphitheatre (26) illustrate the complex nature of such endeavours. The differing requirements of archaeological preservation and other environmental issues are seen in examples such as the groundwater control programme to control rising water levels in Old Cairo, Egypt (31), and at five projects in Oxfordshire (33) where the challenge of cleaning-up contaminated sites and preserving archaeological remains is addressed. The effects of peat wastage on waterlogged archaeological remains is examined (32). The rationale for conserving the historic environment and its role in urban and brownfield regeneration is examined both in Sheffield (34) and in London (36). The nature of a sustainable historic environment and archaeological remains as record, resource and heritage is illustrated with examples from St Albans, Hertfordshire, Annapolis, United States of America, and Sedgeford, Norfolk (35).

These conference proceedings represent a comprehensive collection of some of the key issues facing researchers and historic environment managers today, and are an important reference work for the protection of the historic environment tomorrow.

Acknowledgements

The main conference organisers were MoLAS – the Museum of London Archaeology Service – and English Heritage, with much appreciated support from the University of Bradford and the Institute of Field Archaeologists.

Without any agreement that there would not be a PARIS2, it is therefore important to commend Dana Goodburn-Brown for her timely, enthusiastic and insistent reminders to Mike Corfield, Taryn Nixon and Mark Pollard that a second conference was overdue! The conference was the product of a great deal of hard work by the steering group – led by Mike Corfield, Taryn Nixon, Jane Sidell and Ian Panter – and by the secretariat co-ordinated by Fiona MacDonell and Taryn Nixon. Considerable thanks are due to the other members of the steering group, who helped to plan the conference during the course of the preceding year and who organised and chaired sessions: Peter Hinton, Richard Hughes, Peter Murphy and Mark Pollard. Between them, the steering group provided the inspiration, the glue and the energy to deliver the event.

Fiona MacDonell deserves thanks and congratulations for doing a sterling job of organising and communicating with delegates over the many months before and since the event. MoLAS was grateful for English Heritage's immediate commitment to the conference by agreeing to underwrite the early administrative costs. Thanks are also due to the individuals who contributed to administration and events over the course of the three days: Rosie Joynson, Ruth Peacock and her team at London Wall, and Lindy Casson and Kath Edwards for demonstrating the interactive exhibition, 'The Dig', during one of the evening receptions.

It was undoubtedly the very high quality of the speakers and delegates, however, that ensured the great value of the conference. The conference organisers wish to thank all the speakers for their energy and excellent papers, and also the individuals – Chris Caple, Henk Kars, John Maloney, Vanessa Straker and Bob Sydes – who kindly provided commentary and incisive cross-examination, as rapporteurs, during the conference.

The publication of the *Proceedings* has been funded generously by English Heritage.

1

Introduction

MIKE CORFIELD

and TARYN NIXON

The impetus for the first PARIS conference held in 1996 was concern that, under the English Planning Policy Guidance Note 16 (PPG16, DoE 1990) and similar guidance in Scotland, Wales and Northern Ireland, many archaeological sites were being reburied without due understanding of the physical, chemical and biological processes that influence the preservation of archaeological evidence in the ground. The site protection measures, in other words, were unproven. Although PPG16 had promoted the concept of mitigation and preservation *in situ*, little research had been undertaken to ensure that the evidence contained within sites would be able to survive the changes resulting from construction. PARIS1 (Corfield et al nd) therefore concentrated primarily on the basics: what we mean by preservation *in situ*; the physical, chemical and biological characteristics of the burial environment, and the processes of change constantly at work within it; the engineering issues; the relationship between developers and archaeology and the principles at stake; the workings of the planning process; and some examples of recent research – mostly concerned with waterlogged environments.

In the introduction to PARIS1 it was noted that the conference was born of frustration and optimism (Nixon nd). The frustration came from the knowledge that it may be many years before the success of a site that was preserved *in situ* could be tested, and the optimism from the hope that professionals might work together in multi-disciplinary research programmes to achieve a better understanding of the interactions between archaeological deposits and their burial environments.

It was the need to review progress – to assess the impact of PARIS1 – that led to the planning of PARIS2.

In PARIS2 we have sought contributions to three main themes: recent research; the consequences and effectiveness of decisions where archaeological remains had been preserved *in situ*; and a strategy for future research and policy implementation. The papers, unlike those in PARIS1, are closely focused on actual archaeological sites or archaeological problems, and are predominantly presented by archaeologists and scientists working in the discipline. This shows the growing strength of the profession in tackling the fundamental questions of concern, but it also exposes the weakness of the research base and the need for considerably greater investment in order to build confidence in mitigation strategies. The range of focused research reported in these proceedings is much broader than at PARIS1 as is the geographical spread of participants, extending from Norway in the north to Egypt in the south. A number of important topics were also covered in poster contributions which were displayed throughout the conference, and some of these have been incorporated into the *Proceedings* as short papers.

So, did PARIS2 achieve its objectives? The answer is a qualified yes. It suggested that PARIS1 had indeed stimulated some of the research that was felt to be lacking, and that the research directions were extending beyond the predominance of wetland archaeology seen previously. In drawing in colleagues from Europe and beyond it was shown that the

Mike Corfield
(formerly Chief Scientist, English Heritage)
Heritage Science and Conservation Consultant
Northwood, Stafford Road
Gnosall, Staffordshire ST20 0ES, UK

Taryn Nixon
Managing Director
Museum of London Archaeology Service
Mortimer Wheeler House, 46 Eagle Wharf Road
London N1 7ED, UK

question marks still surrounding preservation *in situ* are many, and are of wide concern. There was some cross-examination of the rationale behind the policy principle of seeking to preserve a 'representative sample'. The importance of local rather than national values was underlined, and the role of end-users was emphasised time and again throughout the conference. No one argued that excavation must always be the response to threats of development, the costs are too high, the volume of material is beyond the capacity of museums to curate, and we have a duty to the future as well as the past to preserve evidence of human development. However, the requirement is to ensure that the evidence survives, and that the changes taking place in the ever-dynamic burial environment do not compromise the quality of the resource itself.

The qualification lies in the apparently severe limitations on available funding. The European Commission has funded research into the preservation and decay of metals and bone in the ground, though not in the particular context of development and mitigation. At least two applications for funding for major research focused on preservation *in situ* have been unsuccessful, although whether this was due to policy or the quality of the application was unclear. Of greater concern was the refusal of the Commission to include the cultural heritage sector overtly within the 6th European Union Framework Programme for Research and Technological Development on the grounds that it is a low priority (although, no doubt, niches will be found within other research areas in the 6th Framework to pursue this important topic). The UK Research Councils appear, similarly, to have been reluctant to fund research proposals around preservation *in situ*. Much of the research needed is derivative rather than basic, although well-constructed projects with clear scientific objectives can be successful and useful work has been done, particularly to doctorate level. English Heritage has been supporting a widening range of research through its Archaeology Commissions, often picking up on smaller projects that have not been funded by research councils. Recently too there has been collaboration with the Environment Agency in funding research into the relationship between contaminated land remediation and archaeology, and on the effects of piling on perched groundwater tables. There remains however the need to promote original research proposals with high scientific objectives that fit into the broad objectives of research councils: for NERC with good earth systems science; for EPSRC with a sound engineering focus; and for AHRB with sound archaeological objectives. The conference was unanimous in its call for multi-disciplinary research teams, and the need to ensure that research embraces and confirms a set of standardised key indicators to enable sites that have been preserved to be monitored effectively.

A further problem that has to be addressed is that of the timescales involved. It takes time to see evidence of change in burial environments unless they are brought about by catastrophic events. It also takes time to disentangle deterioration that has taken place before site formation, and that which has occurred in the ground. During the time since sites were formed there may have been many small changes to the ground environment that individually may have little impact, but which collectively may be significant. We have to understand how to recognise these changes so that we can separate them from the more recent impacts of modern development. Laboratory simulations may be one way of doing this, another is through earthwork experiments. The outcome of the 1996 opening of the Wareham experimental earthwork is reported here (paper 10). When the earthwork at Wareham and that at Overton Down were planned in the early 1960s it was primarily to examine the processes of weathering and site formation. It was only later that it was realised that there were opportunities to examine the impact of burial on archaeological materials. It is timely to think of setting up new very long term experimental sites where processes of site formation can be linked closely to a monitored environment, and where the environment of sections of the site can be artificially altered to simulate the sort of changes seen at actual sites.

The major challenge emerging from PARIS2 is to develop a coherent research framework for preservation *in situ*. Research frameworks are being developed in many areas of archaeology. Regional frameworks were the primary objective, but frameworks for archaeological periods and for the study of particular materials have been developed.

Preservation *in situ* is a fundamental plank of archaeological and planning policy. There are still questions about its viability, and as long as these remain there will still be concerns when important sites are affected by developments. Mitigation has to be designed to reflect the needs of the entire range of archaeological evidence and the risk to it of the processes involved in development (Davis et al 2004). Is it really sufficient to backfill significant sites with sand and pebbles as was recently reported to be the case at a major Roman site in East London? This may preserve the structural remains, but what will the effect be on artefacts and ecofacts of the increased water flows through the site? PARIS3 should be the time to present a new research framework to the community and plan the actions needed to implement its recommendations.

Bibliography

Corfield, M, Hinton, P, Nixon, T, and Pollard, M (eds), nd [1998] *Preserving archaeological remains in situ: proceedings of the conference of 1st–3rd April 1996*, London

Davis M J, Gdaniec, M, and White, L, 2004 *Study of the mitigation of construction impact on archaeological remains*, London

DoE, 1990 Department of the Environment, *Planning and policy guidance note 16: archaeology and planning*, London

Nixon, T, nd [1998] Practically preserved: observations on the impact of construction on urban archaeological deposits, in Corfield et al nd, 39–46, London

I

RECENT RESEARCH

2

Actively decaying or just poorly preserved? Can we tell when plant and invertebrate remains in urban archaeological deposits decayed?

HARRY KENWARD

and ALLAN HALL

Harry Kenward
Senior Research Fellow

Allan Hall
Research Fellow

Department of Archaeology
University of York
York YO10 7EP, UK

Abstract

We have recently argued that poorly preserved delicate macrofossil remains of plants and invertebrates in near-surface deposits in York are in active decay, rather than being preserved in stasis, part-way down the decay trajectory. Observations of both archaeological and modern deposits suggest empirically that remains either survive for a long period (if conditions are conducive) or decay rapidly (if they are not). The hypothesis that very gradual decay has led to large numbers of deposits containing remains in a similar state appears illogical. It is more likely that, where poorly preserved biological remains are found, they either decayed in the past and then were stabilised when ground conditions became anoxic, or are currently in decay. Long-term patterns of decay cannot easily be investigated experimentally, but arguments concerning patterns and rates of decay can be. Apart from the question of in-ground preservation, understanding patterns of decay will allow us to address a range of taphonomic problems fundamental to drawing archaeological information from these remains.

Introduction

Biological remains – including organic artefacts – preserved in archaeological deposits by anoxic waterlogging are hugely important as sources of information about the past. It is thus essential to prevent, as far as possible, loss or degradation of this resource. Here we consider easily-decayed plant and animal remains (and obviously artefacts made from such organisms), defined as 'delicate' by Kenward and Hall (2000): material such as plant cell walls and insect cuticles, generally preserved in a largely unchanged state only under conditions of anoxia. Other kinds of preservation, including permineralisation, are a completely different issue.

Our concern here is with urban archaeological deposits (often deeply stratified, as in parts of York and London). Relevant aspects of anoxic preservation have been considered by Caple (nd, see also references therein), although urban occupation deposits appear to have special characteristics rarely found in the mainly rural sites he discusses. In particular, we perceive that anoxia is maintained in many urban deposits by the water-retaining ('sponge') effect of large concentrations of organic matter, holding water above the general water table and making them exceptionally vulnerable to change.

The immediate stimulus leading to the authors' involvement in this line of enquiry was observations of remains from (and the organic matrix of) deposits of medieval date at the Marks and Spencer site, Parliament Street, York (referred to by Oxley nd), where there were biological remains which showed very odd decay, believed to be ascribable to recent changes in ground conditions (Kenward and Hall 2000; Davis et al 2001). In the course of writing about this site we realised that it was probably not very typical of the kind of degradation being undergone by buried organic matter, but we began to suspect that there was a much more general problem: widespread recent decay of organics in the top 1.0–2.0m of York's archaeology. In samples from these deposits rather uniform poor preservation can be seen throughout the 'fossils' and (where present) the organic matrix: in visual terms there is generally some reddening of remains, although the invertebrates sometimes tend to yellow. This in turn led to a broader consideration of the decay of the fossils with which we had dealt on a regular basis for 25 years or more. Our interest centred on how the rate and timing of decay might be determined, factors as crucial to the interpretation of suites of biological remains as to studies of in-ground decay.

In this paper we pose questions relevant to both these areas of study, and attempt to arrive at research methodologies by which they may be addressed in the shorter and longer terms. The questions may be grouped into four related areas: what patterns of decay, within tissues and among species and higher taxa, do assemblages of delicate biological remains acquire under different depositional conditions? In particular (and this is important in arguments about recent decay), can uniform poor preservation come about during the initial process of deposition? What is the relationship between long-term ground conditions and the preservation or decay of the various kinds of delicate biological remains within a given deposit? Given constant conditions, can decay of delicate remains occur extremely slowly (over centuries), or do most of them either decay rapidly or remain in good condition, the transition from good preservation to rapid decay being a catastrophe threshold? Can we determine whether the observed uniform decay of some organic deposits occurred in a past episode, or is a symptom of ongoing degradation?

Placing an academic value on bioarchaeological remains

Before considering these matters, we wish to address a question which we do not think has previously been posed explicitly: how do we measure the potential usefulness to archaeology (and other areas of palaeoscience) of assemblages of delicate biological remains? This is important when judging the seriousness of in-ground decay. We are not concerned here with the questions about relevance to research agendas which are asked during assessment. What matters for the present purpose is: can a degree of degradation be tolerated without significant loss of information which may be considered important in future?

We have attempted to summarise our thoughts in Fig 2.1a. Although, if numerous assemblages are examined, there is a continuum in overall decay states, we suspect that the distribution of quality of assemblage preservation (rather than that of individual fossils) would be found to be neither normal nor unimodal. We have suggested elsewhere that in our experience delicate remains are either well (or fairly well) preserved, or absent, or represented by what appear to be differentially preserved assemblages of the tougher remains (Fig 2.1b; Kenward and Hall 2000).

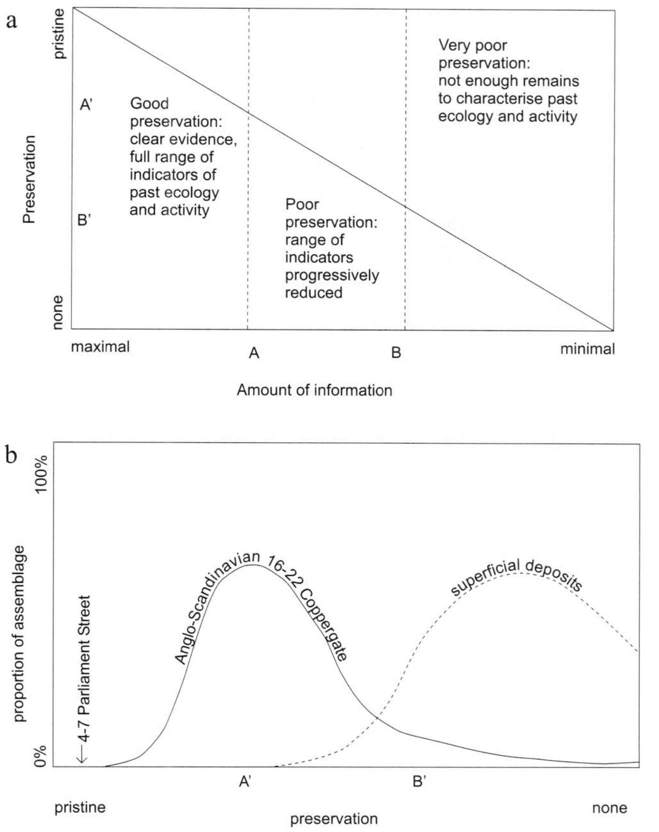

Fig 2.1 a – quality of preservation and information content of assemblages of 'delicate' biological remains in archaeological deposits (scales are arbitrary), note that this diagram does not include information from robust remains such as charred plant remains and bone; b – distribution of overall quality of preservation of 'delicate' plant and invertebrate remains in Anglo-Scandinavian deposits at 16–22 Coppergate, and in superficial deposits generally, in York. Quality of preservation at 4–7 Parliament Street was superior to that in almost all deposits at Coppergate. A' and B' correspond to the intercepts of A and B in Fig 2.1a. These curves are subjective estimates only: early methods of recording preservation and deliberate selection of 'rich' deposits for sampling and analysis mean that there is no overall objective record

With exceptions, these categories correspond to the potential archaeological value of the remains. Good preservation provides clear evidence, with a full range of indicators of past environments, materials and activity. Assemblages with more limited preservation ('poor' in Fig 2.1a) have a reduced potential, and may only provide crude information. A pit fill, for example, may be recognisable as rich in faecal matter by degraded parasite eggs or food remains, but it will not be possible to reconstruct ecological conditions in the pit or its surroundings, or resolve the sources of materials, the full range of foods, or even the species producing the faeces. Where there is very poor preservation there will no longer be enough remains to characterise past ecology or activity at all. We would suggest that in-ground decay is liable to shift assemblages of remains down these categories, and in particular from 'good' (and interpretatively potentially very useful) to 'poor' (and often of limited value). Thus if the information from biological remains is considered at all important, all in-ground decay is to be avoided. (Note, however, that the vertical broken lines in Fig 2.1a cannot strictly be equally positioned for both preservation and information obtainable: the latter will vary with the distinctiveness of what is being interpreted.)

When did decay occur?

We have observed deposits containing delicate remains which showed a substantial degree of rather unusual uniform decay at the Parliament Street site, and showing generally poor preservation in the top metre or so of numerous other sites in York. The former site is probably atypical, and it is primarily with the latter phenomenon that we are concerned here. From an initial assumption that this general degradation had occurred over a long period, because deposits were 'not ideal for preservation', we came round to a very strong suspicion that these layers may be currently in decay as a result of changing groundwater content and loss of anoxia (Kenward and Hall 2000). The problem was that we could not be *sure* that this decay was recent, and not just a different kind of stable preservation, some way down the decay trajectory. Our arguments concerning this decay require elaboration, and demand a wider consideration of decay of delicate remains. We are concerned with phenomena in medium to high rainfall areas (ie as experienced in north-west Europe), with temperate climates.

Decay of these remains may have occurred during deposition or subsequently, either episodically or very gradually, in the ground. How and when did decay of delicate remains usually happen? Clearly decay will have occurred at different stages in different deposits depending on how the deposit formed, and for different kinds of remains according to their pre-burial history and chemical nature. Over 25 years the authors have examined plant and animal remains in archaeological deposits of all dates (but mostly Roman and later and especially from urban contexts), examining samples from many hundreds of sites, involving analysis at some level of perhaps ten times as many samples. In doing this, we have formed strong (and it must be emphasised, subjective) impressions of the way preservation and decay of delicate biological remains has occurred. We need to stress, shamelessly at this stage of the development of the subject, that we have not carried out experiments to investigate the decay phenomena. The following arguments are thus informed speculation rather than 'hard science'. Our purpose is to arrive at suggestions for practicable investigations which can be carried out on realistic timescales.

The taphonomy of delicate biological remains

The biggest single problem in studying in-ground decay is determining when the observed degradation occurred. We will briefly consider the process of decay from the death of the organism through to the present. Decay can occur: before final deposition; immediately after deposition; in stable deposits subsequent to this; and as a result of changes in deposits. Likely patterns of decay under various preservational regimes are presented in Fig 2.2.

Decay before final deposition

This may have considerable interpretative significance: it may be the result of processing of raw materials (eg fermentation of woad, degradation of comminuted bark in a tanning pit, or passage through mammalian guts) or through decay on surfaces prior to burial (eg in insect 'house fauna' in floor sweepings dumped into pits).

Decay during and immediately after deposition

This will obviously be strongly influenced by the depositional environment (Fig 2.2a and b). It may consist only of fermentation of the more labile components of organisms, for example the food stores of seeds, or muscles of animals. But in many environments there will be fiercer decay which leads to varying degrees of destruction of most plant tissue and non-calcareous animal remains. Following this, decay may be arrested by an aquatic environment, by a moist substratum and the 'sponge' effect, by further deposition, or by other special circumstances which promote anoxia. We suspect that normally the point at which decay ceases depends entirely on the rate at which anoxia is established (there will be exceptions). Remains in anoxic deposits may thus be in pristine condition or be at any stage from this to having completely disappeared. Not only this, but (of course) different remains will decay at different rates, so that quite different suites of organisms may survive to the point where anoxia is established. Which materials decay will depend to an extent on the depositional environment. However, we would argue that decay before, during, and immediately following deposition will typically lead to assemblages of remains showing heterogeneous degrees of decay, and differential preservation. In brief, we would expect a skew towards tougher remains and within any one category of remains, a range of preservation from good (last in) to bad (first in). The importance of this is considered below.

After burial

After burial there may be: effectively no further decay (we beg the question of *very* long-term processes which belong in the

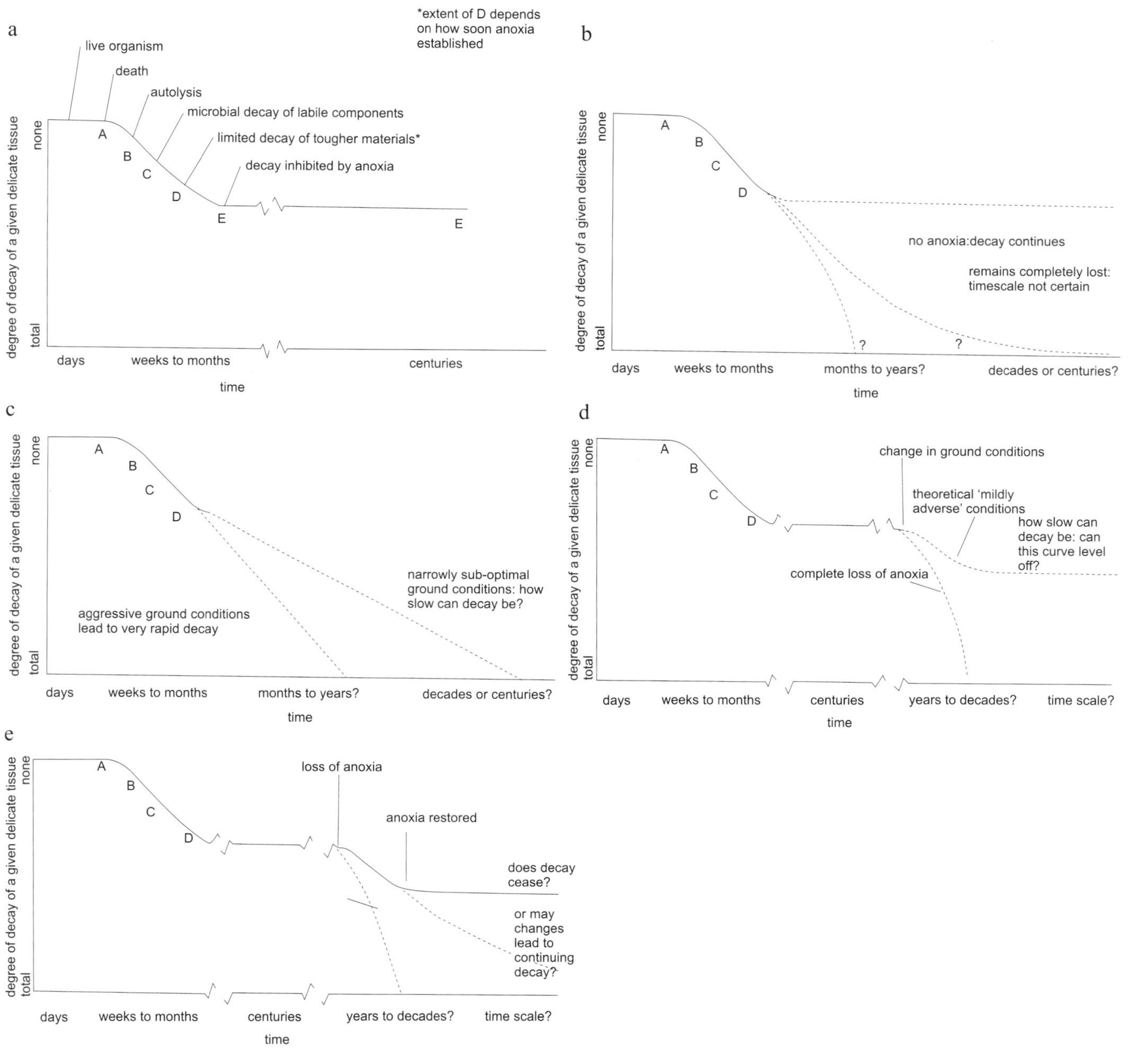

Fig 2.2 Decay of delicate biological remains: a – stages of decay of a given tissue (eg wood or insect cuticle) prior to anoxic preservation; b – decay of delicate tissue where anoxia is not established; c – decay of delicate tissue under aggressive and theoretical 'suboptimal' ground conditions; d – effect of loss of anoxia on delicate tissues; e – what happens to delicate remains when anoxia is restored following an adverse episode?

realm of geology); slow to very rapid degradation leading to loss of most remains on the scale of a year to a few decades; or theoretically (but to us improbably) very slow decay right up to the present and beyond (Fig 2.2c). If the deposit is close to an unstable water table there may be a longer or shorter annual (or occasional) season of decay as the water table falls.

Recent dewatering versus slow decay

The question of how slow decay can normally be, whether it is 'gradual' or incremental (annual or even more intermittent), is crucial in judging when badly preserved remains underwent decay. We think that in general, and under stable conditions, delicate remains are either preserved more or less indefinitely or decay rather rapidly. We thus think that the uniformly poorly preserved remains in superficial deposits are currently in decay as a result of recent dewatering or other changes in ground conditions, and may not last much longer unless conditions are restored (and perhaps not even then?).

The outcome of dewatering of deposits with anoxic preservation will in *general* terms be the same as that of sub-optimal preservational conditions from initial deposition: fossils will decay and eventually disappear (Fig 2.2d). Complete dewatering will generally lead to complete decay: it will mirror rapid post-burial decay.

The effect of seasonal dewatering of formerly waterlogged deposits caused by development (or climatic change) will resemble that of incremental seasonal decay and the same problem of determining the slowest likely rate of decay exists. Even if this kind of seasonal decay is rapid (eg most fossils destroyed in decades), it will leave us with observable part decayed fossils if the onset was fairly recent. We argued in our recent paper in *Antiquity* (Kenward and Hall 2000) that part decayed fossils in near-surface deposits in York probably fell in this category – we still think that this is the most sensible working hypothesis in the absence of experimental evidence.

Can we determine objectively whether decay is ancient or modern?

How might we distinguish the result of decay during deposition, incremental seasonal decay (providing it can lead to long-term survival of at least some delicate remains), and recent-onset decay caused by changing ground conditions? We cannot answer this question objectively and doubt whether anyone else can at present, but we can at least offer some thoughts on the matter. One obvious line of approach is long-term monitoring (providing that a reliable objective scheme for recording the preservational condition of fossils is established): over a decade or so we might see appreciable further degradation. It is essential that such monitoring is established in many places as soon as possible. However, there may be other approaches which can be adopted in the meantime, and tested in the shorter term. One such approach is to consider likely patterns of decay which might have occurred: during initial deposition and the first burst of post-burial decay; and, recently in dewatered sediments which were formerly anoxic.

Decay during deposition

During deposition, many different species of fungi and bacteria (and some higher organisms) will be attacking different materials at different rates. This will lead to the typically heterogeneous preservation seen in most anoxic deposits – only extremely rarely in our experience are conditions such that pretty much everything seems to survive in excellent condition (eg Littlewoods Store, 4–7 Parliament Street, York, Hall and Kenward 2000, with green leek leaf tissue and insects in superb preservation). A further source of heterogeneity of decay is the different histories of the remains: some may have been processed before burial and others may have lain on the surface for a long period (in each case leading to decay, perhaps characteristic), and others may have been sealed very soon after death (and so be very well preserved). Remains which have lain in a biologically active 'compost heap' environment during the burial process may show patchy preservation across individual fossils as a result of localised biological attack. This is the kind of pattern of decay generally to be observed in 'well preserved' assemblages of delicate biological remains, for example, from many contexts at 16–22 Coppergate, or in the Skeldergate well (Kenward and Hall 1995; Hall et al 1980).

What will in-ground decay look like?

We think this logically predictable pattern of decay is important, for the decay which occurs as a result of changing ground conditions is likely to occur in a very different way. We would argue that the range of organisms present deep in the ground will be limited and that decay will result from the overall conditions of the sediment rather than the varied and lively ecological mêlée of the fresh deposit. We suggest that this will lead to broadly uniform decay, effectively by chemical oxidation whether or not driven by micro-organisms, of all delicate remains and of the amorphous organic material in the matrix. Individual remains and the matrix may show zonal decay as a result of enhanced decomposition along the cracks which form in sediment as it loses water. This is exactly the kind of decay that can be seen in many relatively shallow archaeological deposits and we believe that where it is observed we should work on the strong presumption that recent decay has taken place; this leads to a conclusion horribly relevant to *in situ* preservation: large areas of York's archaeology may be rapidly losing the organic preservation which makes the city internationally important in studying the human past.

How may these phenomena be investigated?

Patterns of decay: can uniform poor preservation arise during initial deposition?

It is essential that we obtain information about how different kinds of remains decay under various depositional regimes (and as a result of different craft, industrial, domestic and natural processes): this includes patterns among tissues and among species. Do assemblages normally show an internal range of decay according to the history of individual remains and the small-scale variations inevitable in most forming deposits?

This can be approached as a piece of experimental science by setting up numerous replicates with a wide range of remains in different sedimentation regimes (mimicking pond silts, house floors, pit fills, etc). (The experimental earthworks projects have done this for one kind of burial environment, but unfortunately they are relevant to contexts with challenging ground conditions rather than the anoxic ones in which we are interested here.) This will be time-consuming and probably require long-term experiments, but would allow experimental control, measurement of sediment parameters, and investigation of the microbiology and biochemistry of the phenomena. However, the same questions can be attacked empirically using observations of suites of remains from modern sediments. Casual observations of the condition of such remains during research into the relationship between death assemblages and local ecology (eg the studies reported by Kenward 1978, and numerous others, unpublished) suggest that for insect remains, at least, heterogeneity of preservation is normal, not least because deposits cover a timespan during which the earliest remains to arrive have had the opportunity to decay before burial. What pattern of decay is normal in active soils, in peats, in anoxic and oxygenated ditch fills, or in piles of decaying vegetation?

This may be the most productive area of research in the short term, and is amenable to work on the timescale funded by research councils, or even in undergraduate projects. The results will also be important in attempts to investigate the pathways followed by delicate remains to archaeological deposits (for example, in a cesspit in which plant remains and 'house fauna' insects from floor sweepings, showing a heterogeneous pattern of decay, are mixed with food plants from faeces and insects which would have lived in the pit, showing excellent preservation).

The relationship between long-term ground conditions and state of preservation

It is almost impossible to establish objectively the relationship between stable long-term ground conditions and the preservational condition of remains, because measuring current sediment characteristics does not tell us what was happening in the sediment in the past. It may sometimes be reasonable to assume steady conditions have existed, for example in deep deposits where there is no obvious mechanism for changing groundwater status. (Even then, changes in rainwater chemistry may have had an effect.) Studies of sediment and fossil micromorphology may give hints that conditions have been stable or have changed substantially. Very long-term experiments in which sediment conditions are monitored in parallel with examination of fossils may begin to build up a body of relevant evidence, but the timescale will need to be multi-decadal.

Can delicate remains decay extremely slowly?

We have argued elsewhere that very slow decay is not likely to be normal (Kenward and Hall 2000), although it may occur under exceptional conditions. We see no reason to alter our opinion, unless experimental work shows otherwise (although it is hard to see how rates of decay so slow as to allow remains to survive for centuries could in practice be measured, or how they could be deduced *post hoc* by examining fossils and sediments).

Can we distinguish past and ongoing uniform decay?

If current ground conditions are good (ie clearly permanently, and not just seasonally, anoxic), then it is reasonable to suppose that where general decay is observed across the biota, then that decay occurred in the past, either during deposition or in some adverse episode. However, we do not know objectively what ground conditions are required to maintain excellent preservation of the full range of remains seen in the 'best' deposits (eg at the Littlewoods Store site, 4–7 Pavement, York, Hall and Kenward 2000). So poor preservation may be the result of current decay even where deposits *appear* suitable for preservation, yet in fact are in some way hostile.

Much research is needed here. It can include an empirical component, characterising in an objective way those deposits where preservation is still excellent, and where it is less good. But in-ground studies and laboratory work on rates of decay of different materials under controlled conditions are essential: we need to relate decay to the biochemical environment. The timescale for such investigations of gradual phenomena may be a problem, but some results should be obtainable over the periods of study appropriate to doctorate research.

There are thus practical approaches to the related problems of the taphonomy of delicate remains and of in-ground decay which can be carried out with modest means. However, the large-scale long-term investigation of *in situ* deposits remains an inescapable priority if we really believe the information locked up in these deposits with anoxic preservation is worth preserving for future research, or even just as a non-renewable component of our archaeological heritage.

Does decay stop when deposits are re-watered?

There appears to be a general assumption that even if it is established that organic remains in particular archaeological deposits are in active decay, the process can be halted by raising the water table and re-establishing anoxia (Fig 2.2e). This may be true, but an argument can be made that in some cases deposits will have been modified by the decay episode in a way which makes them more vulnerable to further damage. It is possible that the loss of organic matter will have two effects: the dense texture typical of richly organic deposits may be lost, and the reduction in organic content may decrease the buffering effect that it has against oxidation (Table 2.1). Aspects of groundwater microbiology and chemistry are emphasised, respectively, by Caple (nd) and Pollard (nd). We wonder whether the decay which follows loss of anoxia is in fact primarily a chemical process, or whether aerobic organisms develop significant populations at an early stage – or even whether anaerobic microbes become

Table 2.1 Likely permanent changes in sediments resulting from temporary dewatering

Deposit has remained anoxic	After decay episode
Dense texture impedes pore water movement	More open-textured
Abundant buffering by organic matter	May be little buffering
Colloidal fine matrix	Fine particles have flocculated and colloidal structure may not re-establish

active as oxygen combines with toxic substances which formerly inhibited them (so that monitoring may not reveal oxygenation). All of this is eminently testable in the laboratory as well as by field observations. Blocks of sediment can be allowed to decay, and then thin sections compared with sections of the undecayed deposit.

If short decay episodes can reduce the resistance of organics to further decay, then re-wetting is only a limited solution to the problem and the case for detailed excavation of representative 'damaged' deposits before further decay occurs is greatly strengthened.

Conclusion

We will round up by re-stating the obvious: decay is a one-way process and therefore once decay has occurred remains cannot be restored, only ground conditions. We feel that this point is sometimes overlooked, leading to a blasé view of the degrading organic archaeological resource. Secondly, we would argue that monitoring ground conditions, although essential, at best only tells us whether decay may or may not be continuing: but we would also argue that there is barely any understanding of the relationship between ground conditions and the preservation of the full range of biological remains. So two lines of research are needed: firstly into the general issue of how delicate remains decay, and secondly into the relationship between easily measured deposit parameters and the rate of decay of delicate remains. Both are practicable, and without both we will not move much further forward, or be able to argue the case for measures to conserve the resource. Until we understand both of these much better it would be wise to work on the assumption of the worst case, that the resource is actively decaying, with all that is implied. The alternative is inaction, which may lead to the loss of a large proportion of the organic matter in anoxic archaeological deposits, and thus of a unique store of information about the past.

Bibliography

Caple, C, nd [1998] Parameters for monitoring anoxic environments, in *Preserving archaeological remains in situ: proceedings of the conference of 1st–3rd April 1996* (eds M Corfield, P Hinton, T Nixon and M Pollard), 113–23, London

Davis, M, Hall, A, Kenward, and Oxley, J, 2001 Preservation of urban archaeological deposits; monitoring and characterisation of archaeological deposits at Marks and Spencer, 44–45 Parliament Street, York, *Internet Archaeol II*

Hall, A R, Kenward, H K and Williams, D, 1980 *Environmental evidence from Roman deposits in Skeldergate*, The Archaeology of York 14(3), 101–56, London

Kenward, H K, 1978 *The analysis of archaeological insect assemblages: a new approach*, The Archaeology of York 19(1), 1–68 and pl I–IV, London

Kenward, H K and Hall, A R, 1995 *Biological evidence from Anglo-Scandinavian deposits at 16–22 Coppergate*, The Archaeology of York 14(7), 435–797 and xxii and loose figures, York

Kenward, H and Hall, A, 2000 Decay of delicate organic remains in shallow urban deposits: are we at a watershed? *Antiquity* 74, 519–25

Oxley, J, nd [1998] Planning and the conservation of archaeological deposits, in *Preserving archaeological remains in situ: proceedings of the conference of 1st–3rd April 1996* (eds M Corfield, P Hinton, T Nixon and M Pollard), 51–4, London

Pollard, A M, nd [1998] The chemical nature of the burial environment, in *Preserving archaeological remains in situ: proceedings of the conference of 1st–3rd April 1996* (eds M Corfield, P Hinton, T Nixon and M Pollard), 60–5, London

3

Bone as an indicator in the *in situ* degradation of archaeological heritage. Two examples: Apigliano, Italy, and Aartswoud, the Netherlands

HENK KARS,

MATTHEW J COLLINS,

MIRANDA JANS, ANDERS NORD,

PAUL ARTHUR and EVA KARS

Henk Kars
Miranda Jans
Eva Kars
Rijksdienst voor het Oudheidkundig Bodemonderzoek
PO Box 1600, 3800 BP Amersfoort, The Netherlands

Matthew J Collins
Fossil Fuels and Environmental Geochemistry, (Postgraduate Institute)
NRG, Drummond Building, University of Newcastle
Newcastle upon Tyne NE1 7RU, UK

Anders Nord
Riksantikvarieämbetet
PO Box 5405, S-114 84 Stockholm, Sweden

Paul Arthur
Scuola di Specializzazione di Archeologia Classica e Medioevale
University of Lecce,
Via Dalmazio Birago 64, 73100 Lecce, Italy

Abstract

This paper deals with the deterioration of archaeological bone within different climatic zones of Europe. The main objective of the project, subsidised by the European Commission (ENV4-CT98-0712), was to establish methods regarding the long-term preservation of bone in different soil environments, this in light of the verification of current concepts of archaeological heritage management. How do environmental impacts influence the stability of bone? Can an understanding of bone degradation mechanisms be used to predict the preservation potential of a given soil type?

The analyses comprised the macroscopic inspection of bone by the archaeozoologist and a microscopic study by the histologist. This was followed by an inspection of both the mineral and protein phases of bone, together with mercury (Hg)-intrusion porosimetry measurements. In addition to this, more than 300 soil samples were studied.

The investigations have shown that a clear distinction has to be made between the histological and biomolecular information of bone. For instance, assuming taphonomic control, the study has shown that few human bones show good preservation of biomolecules, as compared to proteins in animal bones which have relatively good preservation.

Two case studies are outlined. The first at the medieval site of Apiglianio (southern Italy) where bone preservation is defined by the presence of highly crystalline and altered mineral phases, while the levels of residual collagen in these bones are much lower than is expected for medieval bone material. This shows that not all diagenetic processes in archaeological bone are fully understood.

The dramatic effects of drainage of soils (formerly wetland) on bone preservation are demonstrated by the Neoltihic site at Aartswoud (Netherlands). Differences between the preservation state of bones excavated in 1977 and 2000 are noted, which are confirmed by the micromorphological study of the site. To avoid excavation of the site, it is essential that the water table is raised. This will have the direct effect of putting a stop to bioturbation; however, it cannot be guaranteed that the bone will be preserved.

Introduction

The value of a site or resource of archaeological information may be defined by the sum of its perception value (aesthetic value, historical value), the physical quality (integrity, preservation) and the intrinsic quality (rarity, research potential, group value, representativity), as was described by Deeben et al (1999).

It is all too clear that the physical quality of the resource is the limiting factor in this definition, because it determines to a great extent the values of the two other parameters. This is even more apparent for the invisible part of the heritage, which in north-western Europe makes up more than 90% of our heritage. The perception value of this part is, rationally, close to zero, which means that the value of an invisible site is mainly determined by its intrinsic quality.

This has tremendous consequences in light of the modern concept of archaeological heritage management which requires that our resources shall remain *in situ*, because the information value (V) of the resources will increase with time (Fig 3.1). The assumed reason for this relative increase is that paradigms and research concepts may change and that methods and techniques will improve. At the same time, however, it may be assumed that an absolute loss of information will occur due to the deterioration of one or probably more of the components of a site. The resulting figure demonstrates the dilemma of the decision maker in archaeological heritage management. If it cannot be guaranteed that V2 in Fig 3.1 can be kept at zero, close to zero, or at least is smaller than the completely unknown factor V1, the policy of *in situ* preservation for this particular site has failed.

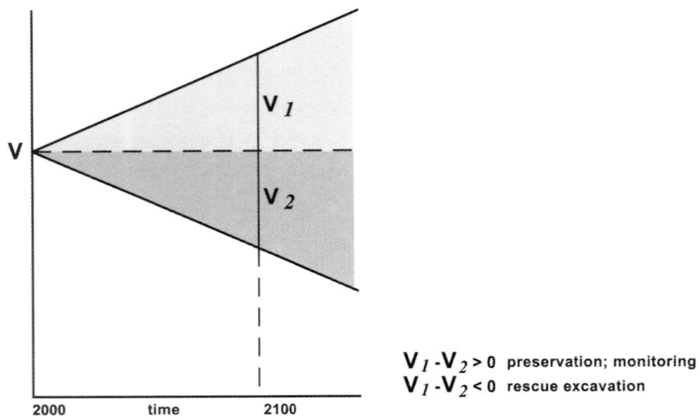

Fig 3.1 *The development of the physical quality of an archaeological resource with time. It is expected that the relative increase of information will have an intermittent character instead of the straight line shown in the figure; the line downwards, representing the decay of the resource, is actually composed of a number of specific curves for different components of the resource*

Since the early nineties this risk of failure has been recognised by several national archaeological heritage institutions in north-west Europe. This has led to a number of national and international research programmes which focused on the degradation of archaeological materials in the soil (cf Mattson et al 1996; Wagner et al 1997; Corfield et al nd; Scharf et al 2000). The common factor in all these studies was that more insight was gained into the complicated physico-chemical system made up of features/artefacts and soil, defined by an almost infinite number of variables. This system, representing a rather unstable equilibrium between archaeomaterials and soil, is very sensitive to changes in the soil environment. Small changes in soil parameters can lead to a dramatic loss of (parts of) the site and therefore of information.

In addition what can be learned from these studies is that for the investigation of the archaeomaterial-soil system the first major task is to reduce the number of variables defining that system. To this end the project presented here (comprehensive final report – Kars and Kars 2001) concentrates on archaeological bone excavated from relevant soil types in Europe.

For bone to survive in the soil, a defined range of conditions must be fulfilled. This range is also assumed to cover at least some of the main conditions responsible for the preservation of other materials which may be present in a site. The main objective of the research was to establish a qualification system for soils which will be used for the assessment of the preservation potential of these soils for bone. Within this objective the following goals were defined: understanding the processes of deterioration of bone under various soil conditions; providing knowledge that may lead to a model that predicts whether archaeological resources are preserved when threatened by actions in the planning process; defining guidelines with regard to mitigation strategies for the *in situ* preservation of sites; providing techniques and methods to monitor sites which have been protected.

Methodology

Bone degradation is a complicated process involving both intrinsic and extrinsic factors. From the moment of death many factors play a role in bone preservation of which bone diagenesis is one, taking place in and influenced by the burial environment. A peculiarity of data derived from human or animal bones is that the record originally present has passed through a complete history by which the archaeological information is reduced and transformed and these processes of taphonomy have also to be taken into account. Moreover, bone is a composite material, consisting of a mineral (bio-apatite) and an organic (mainly collagen) fraction; this means that three alternative pathways of deterioration are possible: chemical deterioration of the organic fraction (primarily collagen), the bone mineral, or biodegradation. It is not clear to what extent any of these three pathways occurs in isolation. The presence of organic molecules will passivate the mineral surface slowing the rate of mineral solution reaction. Conversely dissolution of the mineral phase will expose collagen to biodegradation, thereby increasing the rate of its decomposition. These three pathways represent end-members, useful in terms of understanding the complex issues related to bone diagenesis. In different burial environments, one or other of these three processes may dominate. If the bone lies close to thermodynamic equilibrium with the soil, then rates of mineral dissolution will be slow, and there will be sufficient time for the chemical deterioration of collagen to dominate. Under such circumstances collagen loss leads to an increase in porosity.

From this it becomes clear that the state of preservation of archaeological bone cannot be determined by inspecting bone on the macroscopic level alone. Consequently, there is a need for microscopic and chemical techniques to determine the state of preservation of archaeological bone.

Within this project all bone samples have been investigated in terms of their archaeological information and have been studied in terms of their preservation state from the macroscopic point of view. The macroscopic quality of the bone is judged by a number of criteria, ranging from very bad to very good preservation and is quantified, preferably on site, according to the categories of Gordon and Buikstra (1981). Their categories were developed for human bone in graves, making fragmentation an important evaluation criterion. However, animal bone, which is often deposited as individual bones, is often fragmented before or during deposition. In these cases, quality is determined by inspecting the bone fragment, and fragility and wear of the fragment are considered more important criteria than fragmentation.

The macroscopic inspection was followed by the histological study of samples. In archaeology histology can be used for identification of mineralised tissues, histological age or species determination, and palaeopathology. Furthermore, histology has been useful in identifying diagenetic alteration of bone (Jans et al in prep, herein further references).

Histological analysis produces mainly visual data. To compare histological results with the results of the other analyses, these data were quantified where possible. The amount of microbial attack on the bone seen in thin section is represented in the Histological Index (HI), developed by Hedges et al (1995). The percentage of bone unaffected by microbial attack is represented in a scale from five (virtually perfect bone) to zero (no original microstructure left). Because the HI value does not relate to macroscopic preservation and cracking was an important feature in several sites, quantification of this diagenetic alteration was necessary. A value expressing microcracking for a bone section, comparable to the HI, was devised. Other parameters studied under the microscope are mineral alterations, seen by the presence of calcite or pyrite, the presence or absence of birefringence and infiltration of materials into the bone microstructure.

For the physico-chemical characterisation of bone a large number of organic and inorganic techniques may be used, while the porosity of bone also provides useful information.

The dominant organic matter component in bone is protein (there are some lipids and carbohydrate). Fresh bone contains around 22% by weight of protein, of which nearly 90% is collagen. The amount of protein remaining can be estimated using whole bone powder; the measurement used is the percentage of nitrogen left in the bone (% N). In addition to this the C/N ratio is probably the most commonly used determinant to assess the quality of collagen preservation in archaeological bone.

Fourier transformed infrared spectroscopy (FTIR) has been used to assess the state of bone mineral preservation. The method has a number of particular advantages, requiring small samples and minimal preparation; data is obtained on aspects of both organic and mineralogical alteration. In this study we chose not to quantify the protein contribution by FTIR and instead used the more quantifiable assessment of collagen presence based upon organic nitrogen analysis. FTIR is used here to determine the crystallinity. Crystallinity changes are defined using the so-called infrared splitting factor (SF). Work carried out using FTIR has shown that these measurements can be correlated to both porosity and protein degradation (cf Nielsen-Marsh 1997), strengthening the hypothesis for association between protein and mineral at this structural level.

Archaeological bone often shows an increase in the carbonate content due to the incorporation of diagenetic carbonate from the surrounding soil and groundwater. Carbonate content can be semi-quantitatively determined using FTIR spectra of bone with a technique similar to that used for obtaining SF measurements. The ratio between the main carbonate and phosphate peaks at 14.15m-1 and 10.35m-1 respectively is calculated and calibrated against known carbonate amounts obtained from manometric measurements of released CO^2 from the reaction of bone powder with phosphoric acid.

If diagenetic carbonate is incorporated structurally into the crystal lattice, crystallinity and therefore SF is affected. Adsorption onto the hydroxyapatite surface will not affect the crystal lattice, nor will the presence of calcite in the pore spaces. Porosity will also be affected by diagenetic CO^{32-} because of its influence on the crystallinity (by structural exchange), and through blocking of pore-spaces by crystallisation of secondary, diagenetic minerals, such as calcite.

Water sorption analyses have proved extremely informative, correlating strongly with both mineral and proteinaceous degradation. New research however, has found an alternative to the water sorption technique that allows much more detailed analyses of the pore size distribution of bone in a much shorter time. Although an established technique in the analysis of soils and building materials, mercury intrusion porosimetry is a very new application in bone diagenesis studies. Early experiments, however, have enabled a middle range of pore sizes (defined as mesoporosity) to be observed for the first time in archaeological bone (Nielsen-Marsh 1997; herein further references). Data from Hg-intrusion porosimetry of archaeological and modern bone strongly suggest that certain pore size distributions can be related to particular physiological and diagenetic features.

The pore diameter range obtained during this study starts at 7nm and extends to 70μm and the terms 's', 'm' and 'l' porosity were introduced. These pore ranges are defined as follows. Small 's' porosity, 100nm–7nm diameter, represents sub-collagen fibril porosity, previously reported for deproteinated bone (Nielsen-Marsh and Hedges 1999). This porosity arises from chemical hydrolysis of collagen over archaeological timescales. Medium 'm', 100nm–8.5μm diameter, includes spongiform porosity caused by micro-organisms. Large 'l' porosity represents pores above 8.5μm diameter and reflect both physiological porosity and corrosive bone dissolution features.

The nature of the soil is an important factor influencing the deterioration of unexcavated archaeological bone material.

It is therefore essential to evaluate the chemical and physical properties of the soil samples taken near the bone material and (for reference) some distance from the archaeological feature or construction in question. The classification of the soil included the soil chemistry, involving pH, resistivity, organic matter contents, cations and anions. Even if the physical properties like grain sizes and pore distribution have been almost unchanged for thousands of years, the soil chemistry may have changed since the bone was buried in the ground. Accordingly, the data of the present-day soil sample will therefore only give a picture of the chemical situation at present. It is therefore imperative that the technical examination be complemented by a careful assessment of the archaeological context, and by a characterisation of the environment in general in terms of geology, geography, vegetation, land use, local polluting sources, meteorological conditions, and sensitivity towards acidic deposition.

Whenever possible, a relevant soil sample was taken as near to the bone sample as possible. Sometimes, additional samples were taken from the feature that contained the bone eg grave or refuse pit. Further, a reference sample outside the feature, or a set of 6–10 samples at various depths below the ground surface were taken to gain insight into the local soil variables.

More than 300 bone samples were selected from 50 sites in Sweden, England, the Netherlands, Italy and Turkey representing parts of animal skeletons or isolated bone fragments, and parts of human skeletons from inhumations and cremations. The presence or absence of other materials, like leather, wood, textile, and metals and the preservation state of these materials was also noted. In addition some 300 soil samples of these sites were studied and analysed using the above mentioned methods (see Kars and Kars 2001).

Results

In evaluating the data two approaches have been taken. The first one was the statistical evaluation of data, carried out by different methods of multivariate analysis. It was soon realised, however, that there was serious risk that the conclusions in some cases would be based on more or less 'forced' or biased explanations, because of the fact that the data-set represents a high number of sites with often a restricted numbers of samples per site. Moreover, the complexity of the bone material, combined with the high number of soil science and environmental variables has made it extremely difficult to predict on a statistical evaluation alone which soil environment is 'good' or is 'bad' for bone to survive. Nevertheless the statistics confirmed the positive correlations between soil acidity and bone degradation, and aeration of soil and bone degradation. It was further shown that bone with flesh (human bone and some animal bone) is preserved better than bone without flesh (most animal bone). Moreover the preservation conditions for the macroscopic appearance and quality of bone may differ from those for the histological quality of bone, which again may be different from the chemical quality, ie the collagen content. This means that with regard to archaeological resource management, a quality assessment of bone by the traditional archaeozoologist is insufficient and has to be followed by the study of the microstructure of bone, which in some cases should be accompanied by the determination of some physico-chemical parameters, including mercury porosimetry measurements. This approach is also essential for the development of monitoring programmes for archaeological sites containing bone.

The latter two conclusions were also endorsed by the second approach in which all data per site were combined and interpreted thus giving an insight into the particular preservation conditions of a site. This approach also acted as a key in determining artificial factors in the statistics which could have led to misinterpretations of the data. Two of the case studies are summarised below.

Apigliano, Italy

Archaeological and geographical context

The deserted medieval village of Apigliano lies in the territory of Martano (province of Lecce, southern Puglia). Excavations have been carried out for about a month each year, between 1997 and 2000. The site developed from the 8th century AD as a Byzantine village or 'chorion' and by the 10th century it probably covered about 2 ha. By the later eleventh century it is likely to have fallen to the Normans, who conquered the whole of southern Italy. However, very little has yet been found dating from between the Norman invasion and the 13th century. Excavations have brought to light a later medieval church and cemetery that appear to be of late 13th- and early 14th-century date. Some 53 tombs and 16 charnel pits have so far been excavated, and 82 individuals have been identified (Arthur 1999).

The site lies on a *terra rossa* plough soil in a flat, windswept, area of poor stony land with limestone bedrock breaking the surface in patches. The thickness of the soil above the bedrock is between 0.25–0.50m, and the area is well drained. The land is now used for the cultivation of tobacco and linseed or poor grasses, and occasionally for pasture. The distance to the groundwater is more than 1m.

The sample area has never been ploughed. The surrounding soil is full of stones, much of which is derived from drystone wall rubble. Ten samples (IT-API-hb-001-010) were taken from strong complete bone to extremely fragmented bone. The chalky white bone was in almost all cases quite brittle.

Conclusions and discussion

Histological preservation at this site is very good on the whole with the exception of two samples (006 and 010), the latter one (of a newborn infant) showing extreme microbial attack. This implies that collagenous material was immune to microbial attack. The main degradation type visible in thin section in these bones seems to be cracking (Fig 3.2). The calcite crystals present in some of the cracks and natural pores of the bone could indicate that the fissures are caused by remineralisation processes. Some of the bones are also affected by microbial attack with the infant bone (sample 010) being an extreme example.

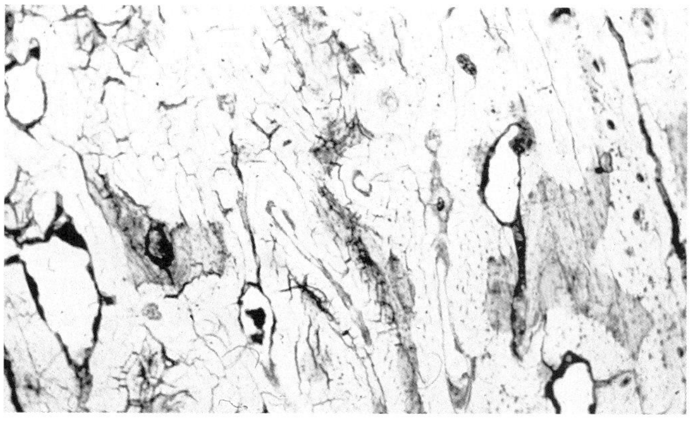

Fig 3.2 Micrograph of sample IT-API-HB-002 showing microfissures following the lamellar anatomy of the bone; Haversian canals are lined with soil material (magnification 100x)

Considering the extreme mineral alteration of the bones, the most likely explanation is that collagen was rapidly lost from the bone, and therefore the bones were not attacked/tunnelled by microbial populations.

This site is characterised by bones with variable organic preservation, and very altered mineral phases, which is the most extreme that we have encountered in human burials. There is evidence for other mineral phases present in the infrared spectra, namely calcite and tricalcium phosphate. Occurrence of the latter is usually associated with high temperature events, which would appear to be contrary to the archaeological evidence (none of the samples are believed to have been cremated/cooked etc).

Five of the samples are children (001, 002, 005, 007 and 008) and one is newborn (010). This does not appear to have any influence over the preservation; the bones from the children do not possess better or worse preservation when compared to those of the adults.

The burial environment does not give any clues as to why these bones should be so extremely altered. The area is flat and the climate very dry; the ground is very stony, and well drained. All these factors accelerate mineral dissolution – however, it is unusual to see such levels of alteration in archaeological material (though for palaeontological material it is common), it is hard to believe that hydrology alone is responsible for the very advanced mineral alteration which has occurred. The area is described as wind swept, susceptible to both erosion and deposition, but this should not have contributed to the poor preservation of these bones, as they are all human, and were buried (presumably the bodies were not left out in the open for any period of time). Possible explanations of this early 'fossilisation' of archaeological bone, displayed by extreme mineralisation and the rapid loss of bone collagen, are considered in more detail in Smith et al (in prep), including a high temperature event, the acceleration of hydrolysis due to liming and extreme wetting and drying cycles.

Aartswoud, the Netherlands

Archaeological and geographical context

The site of Aartswoud in the De Gouw area in the western part of the country represents a late Neolithic settlement of the Single Grave culture dated between 2850 and 2450 cal BC. The inhabitants were farmers, but hunting, fishing, fowling and gathering was important too. The site was situated on a sand ridge bounded by tidal creeks filled with brackish water. The occupation layer comprises the build-up of a number of shell middens alternating with levels of black soil and lenses of ash and clay. This accumulation is some 0.80m thick. The black soil consists of artefacts, ash, charcoal, sand, clay, peat and organic refuse, including the bones studied here.

Due to land consolidation small parts of the site were excavated during seven campaigns from 1972 until 1978. In total an area of $341m^2$ was investigated, this is *c* 1.5% of the occupied area. In 2000 a $1m^2$ test pit was investigated to establish the state of preservation of the site.

The site lies on flat ground and is situated in a former wetland area and is now partly well drained. The soil consists of podsolic silt and there is no underlying bedrock. The present water table is 0.5–1.0m below the surface. The fluctuation of the groundwater between summer and winter is *c* 0.30m. The vegetation indicates moist ground. A hundred years ago the area was much wetter. The soil may contain calcite due to a large concentration of shells. The pedological and hydrological environment suggests that the bones may have lain in predominantly saturated, alkaline soils for much of their burial period. The dark-coloured soils possibly have a high humic content. The land is currently under arable and pasture.

Conclusions and discussion

The bones (NL-AAR-ab-001-003; NL-AAR-hb-004) from Aartswoud are very well preserved with little alteration to any of the measurable parameters, although the preservation is considered average by the field archaeologist. The pedological and hydrological environment suggests that the bones may have lain in predominantly saturated, alkaline soils for much of their burial period – which would account for their good preservation. The anthropogenic soils have a high humic content, which may have helped to preserve the mineral and collagenous component, although collagen may have been contaminated as a result.

However, distinct differences are noted between the preservation state of bones excavated between 1977 and 2000. The histological preservation for samples 001 and 002 (which probably have highest infilling and humic contamination) is excellent (HI score of 5). The more porous samples (in terms of total intrusion) 003 and 004 both have a Histological Index of 3, according to the amount of microbial attack. The type of tunnelling present in 003 and 004 is indicative of the presence of oxygen in the soil (Fig 3.3). Samples 003 and 004 also have more altered mineralogy and less protein content. Although the difference in preservation between samples is rather convincing, more samples from this site need to be examined to validate the apparent dramatic change in preservation potential for bone of this site.

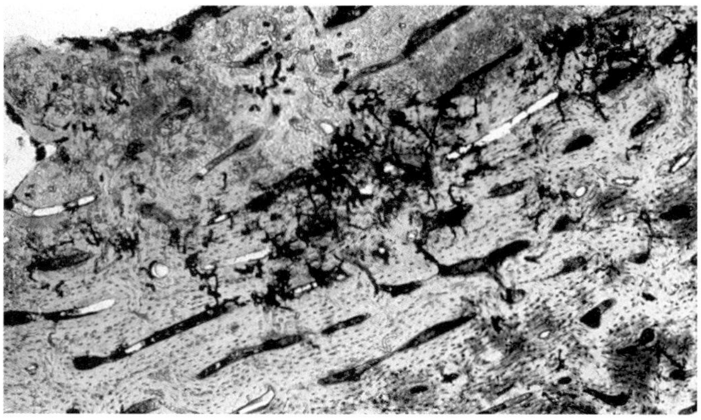

Fig 3.3 Micrograph of sample NL-AAR-AB-003 showing fungal tunnelling invading the bone from the endostal area inward. Fungal attack indicates the presence of oxygen. The whole section shows brown staining due to infiltration of humic acids (magnification 40x)

However, the micromorphological study of the site shows similar results and based on that study the following implications are seen: i) dehydration and ripening by which shrinkage cracks may develop in the anthrosols, ii) oxidation of materials, iii) disintegration of mussel shells by dissolution, and iv) a considerable increase of bioturbation, disturbing the microstratigraphy of features (Kooistra in prep).

It is clear that the unstable equilibrium between archaeomaterial and its environment is disturbed, which will lead to an enhanced loss of archaeological information.

If excavation of the site in the near future is not a realistic option, it is essential that the water table is raised as soon as possible. This will have the direct effect of bringing bioturbation to an end. Furthermore, this measure has to be followed by a monitoring programme to test the level and quality (including continuous low oxygen content) of the groundwater. However, due to the tunnelling seen in the bone, it cannot be guaranteed that bone remains will be preserved.

Future research

More than ever the care of the archaeological heritage is subject to the tensions between modern developments in our society and changing (inter)national legislation. To protect the interests of the archaeological resources ample research is needed into the physical quality of our archaeological property in relation to the soil environment, as a prerequisite for the sustainable development of archaeological heritage.

Future studies have to reinterpret and combine the results of earlier (inter) national studies, while new research programmes have to be developed to contribute to a theoretical basis wherein sophisticated approaches also recognise multiple values of archaeological heritage, as necessary for sustainable development.

With regard to the Netherlands the following initiatives are desirable. Investigations which have been carried out in the early 1990s beg for a follow-up or an updating of the results, for example a national project in which the decay of archaeological features was studied (Meeussen et al 1996–1997) or an international EC-subsidised project on the deterioration of metal artefacts of which the results are not yet fully understood (Wagner et al 1997).

Future new studies will concentrate on a number of carefully selected areas in the Netherlands where the issue of degradation is most apparent. In making this selection the following will need to be taken into account: sustainable development of the area having a high archaeological value as a realistic option; the area contains an assemblage of sites of the same period, and is situated in a rural context; the area also shows a diachronic range of sites, covering a diversity of archaeological phenomena; earlier preservation or monitoring studies have been carried out in the area; historical data regarding water management, change in the use of soil, and preservation conditions of archaeological materials are available; the area is representative of different paleolandscapes/soil environments. It is reasonable to assume that the results of this study will have a direct effect on the practical management of the area (such as the availability of funds for the implementation of proposed measures).

We will concentrate on organic material of a botanical origin, ie remains of (cultivated) food plants (mainly seeds) and weeds, pollen, wood, textiles etc. These ecofacts are present at nearly every archaeological site and are an important resource of (bio) archaeological and ecological information. It is an empirical fact that these materials remain well preserved in a reducing soil environment but deteriorate dramatically with the access of oxygen, where a low pH may have an inhibiting effect (Meeussen et al 1996–7). The oxidation processes may have a different effect on the mainly unaltered macroremains, compared to the humified compounds present in the archaeological soil.

Since very little is known about the degradation mechanisms of organics, this research will focus on the organic chemistry of archaeological remains in relation to the state of preservation.

Acknowledgements

The paper summarises objectives and a selection of results of an Environment and Climate research project (Project no.

ENV4-CT98-0712) which was subsidised within the Fourth Framework Programme of the Directorate General for Research and Development of the European Commission, Brussels. The project was undertaken by a team of experts in archaeology, archaeological science, conservation science, osteology, microbiology, soil science, (in)organic geochemistry and mineralogy of the following organisations: Rijksdienst voor het Oudheidkundig Bodemonderzoek, Amersfoort (the Netherlands); Fossil Fuels and Environmental Geochemistry, NRG, Drummond Building, University of Newcastle, Newcastle upon Tyne (UK), Riksantikvarieämbetet, Stockholm (Sweden); Scuola di Specializzazione di Archeologia Classica e Medioevale, University of Lecce, (Italy) and Land Quality Management, SChEME, University of Nottingham, Nottingham (UK).

The geological classification of the soil samples was carried out by Dr G Ch. Borg of Chalmers University of Technology, Göteborg, Sweden.

Bibliography

Arthur, P (ed), 1999 *Da Apigliano a Martano (LE). Tre anni di archeologia medioevale (1997–99)*, Congedo Ed, Galatina

Corfield, M, Hinton, P, Nixon, T, and Pollard, M (eds), nd [1998] *Preserving archaeological remains* in situ: *proceedings of the conference of 1st–3rd April 1996*, London

Deeben, J, Groenewoudt, B J, Hallewas, D P, and Willems, W J H, 1999 Proposals for a practical system of significance evaluation in archaeological heritage management, *European J Archaeol* 2, 177–99

Gordon, C C, and Buikstra, J E, 1981 Soil pH, Bone Preservation and Sampling Bias at Mortuary Sites, *Amer Antiquity* 46, 566–71

Hedges, R E M, Millard, A R, and Pike, A W G, 1995 Measurements and relationships of diagenetic alteration of bone from three archaeological sites, *J Archaeol Sci* 22, 201–9

Jans, M M E, Kars, H, Nielsen-Marsh, C M, Smith, C I, Nord, A G, Arthur, P, Earl, N, in prep *In situ* preservation of archaeological bone: a histological study within a multidisciplinary approach, *Archaeometry*

Kars, H, and Kars, E A K (eds), 2001 The degradation of bone as an indicator in the deterioration of the European archaeological property, Final report Project ENV4-CT98-0712 (DG Research and Development) European Commission, Brussels

Kooistra, M, in prep Micromorfologisch onderzoek op vindplaats Aartswoud, in *Kwaliteitsbepalend onderzoek ten behoeve van duurzaam behoud van neolitische terreinen in West-Friesland en de Kop van Noord-Holland* (eds R M van Heeringen and E M Theunissen), Intern Rapport ROB, Amersfoort

Mattson, E, Nord, A G, Tronner, K, Fjaerstad, M, Lagerlöf, A, Ullén, I, and Borg, Ch, 1996 *Deterioration of archaeological material in soil: results on bronze artefacts*, Konserveringstekniska studier, Riksantikvarieämbetet och Statens Historiska Museer, Rapport 10, Stockholm

Meeussen, J C L, Boer, G, Exaltus, R, and Kars, H, 1996–7 Effects of soil acidification and declining groundwater tables on the decay of archaeological features in acidic sandy soils, *Berichten van de Rijksdienst voor het Oudheidkundig Bodemonderzoek* 42, 475–90

Nielsen-Marsh, C M, 1997 Studies in archaeological bone diagenesis, unpub DPhil thesis, Univ Oxford

Nielsen-Marsh, C M, and Hedges, R E M, 1999 Bone porosity and the use of mercury intrusion porosimetry in bone diagenesis studies, *Archaeometry* 41, 165–74

Scharf, W, Arnold, Ch, Gerwin, W, Huesmann, I, Menzel, K, Pötzsch, A, Tolksdorf-Lienemann, E, and Tröller-Reimer, A, 2000 Schutz archäologischer Funde aus Metall vor immissionsbedingter Schädigung, *Materialhefte zur Archäologie* 57, Landesdenkmalamt Baden-Württemberg, Stuttgart

Smith, C I, Nielsen-Marsh, C M, Jans, M M E, Arthur, P, Nord, A G, and Collins, M J, in prep The strange case of Apigliano: early 'fossilisation' of medieval bone in southern Italy, *Archaeometry*

Wagner, D, Kropp, M, Abelskamp-Boos, K A N, Dakoronia, F, Earl, N, Ferguson, C C, Fischer, W R, Hills, C C, Kars, H, Leenheer, R, and Meijers, H J M, 1997 *Soil archive: classsification of European excavation sites in terms of impacts of conservability of archaeological heritage*, Final report Project EV5V-CT94-0516 Environment and Climate Programme, European Commission, Brussels

4

The acoustic characteristics of archaeological wood

STEPHANIE ARNOTT, [1,2,3]

JUSTIN K DIX, [1,2]

ANGUS I BEST [2] **and**

DAVID GREGORY [3]

[1] Department of Archaeology
University of Southampton
Avenue Campus, Southampton SO17 1BJ, UK

[2] Southampton Oceanography Centre
Empress Dock, Southampton SO14 3ZH, UK

[3] Centre for Maritime Archaeology
Havnevej 7, Post Box 317
Roskilde DK-4000, Denmark

Abstract

In response to the demand for either the *in situ* or laboratory based preservation and management of our submerged cultural heritage, it is desirable to develop a non-intrusive technique capable of both identifying and assessing the state of preservation of waterlogged wooden materials. Essential to this task is the characterisation of the physical and acoustic properties of archaeological wooden material in varying states of degradation. This task is being achieved through a series of laboratory and *in situ* experiments. These include quantifying the effect of varying states of saturation and more importantly macrofaunal and microfaunal degradation, on the physical properties of modern and ancient wood samples. Preliminary results suggest that acoustic impedance and attenuation variability may be sufficient for high-frequency acoustic sources to detect the change in preservation state over a site or within a single artefact. Such techniques could have a significant impact on the management and conservation of our cultural heritage.

Introduction

Over the last fifteen years the archaeological community has moved away from traditional excavation methods toward a strategy of *in situ* preservation of our cultural resource. This trend has been politically galvanised in Article 6 of the ICOMOS *Charter for the Protection and Management of the Archaeological Heritage* (ICOMOS 1990), which states that 'The overall objective of archaeological heritage management should be the preservation of monuments and sites *in situ*'. Further, this directive also emphasises that the choice of conservation strategies to be deployed should be achieved through non-destructive, non-intrusive survey rather than excavation. However, to date the majority of management decisions have been made on the basis of qualitative interpretation and site specific expertise. What is currently absent from the methodologies available to archaeologists and conservators is the ability to assess rapidly and non-destructively the state of deterioration of a range of archaeological materials *in situ* (Gregory 1998). Wood is especially important to consider as it occurs in a wide range of environments, from wetland to marine, in large quantities and is particularly demanding to conserve.

Geophysical survey techniques such as seismic reflection and sonar have been used for many years to locate underwater sites. In some cases, however, they have not proved successful. For instance, the wreck in St Peter Port harbour, Guernsey is not seen on profiles taken over the area even though timbers are exposed above the seabed (J Dix, pers comm). Clearly more information on the fundamental acoustic properties of wood is required to improve imaging of sites. Preliminary work by Quinn et al (1997) attempted to demonstrate that wooden artefacts could be imaged by a high-resolution seismic system. This was done by the theoretical calculation of reflection coefficients ie contrasts between the acoustic impedance of wood and that of the surrounding substrate. However, they used the properties of dry fresh wood, which vary considerably from those of waterlogged archaeological wood.

The condition of buried wood is controlled by two main factors. These are the degree of saturation and extent of degradation. Knowledge of both parameters is important to the conservator for the choice of conservation strategy. As both factors influence the acoustic properties it is proposed that a laboratory method of determining saturation and degradation state can be developed.

In addition to this, there is currently no method of extracting quantitative information from surveys as to the condition of the material. Quinn et al (1997) also found that the current instrument resolution is too low to obtain the level of detail required by archaeologists and conservators for site management. Adaptations of ultrasonic acoustic techniques for what are generally small-scale archaeological sites provide a potential tool for non-intrusive quantitative analysis of such sites *in situ* through the development of a hand held system.

In order to further develop such techniques, it is necessary to understand both the acoustic properties of archaeological wood and their variability with preservation state. This paper aims to present a method for extending our understanding of the acoustic properties of archaeological wood, thus providing a platform for future attempts to use acoustic sources to determine the preservation state of archaeological material, both in the laboratory and *in situ*.

Although it is anticipated that techniques discussed in this paper will be applicable to any archaeological wood, the initial work has focused on oak in the marine environment. From the Mesolithic to the medieval period in north-west Europe oak, ash, elm, hazel, alder, beech, yew, lime, birch, willow and pine have been used for boat building. However, oak generally seems to have been preferred for the main structural elements whenever it was available, from the mid 2nd millennium BC to the early 19th-century ships. Whilst the survival of boat finds is undoubtedly biased as oak has a greater chance of survival than most other species, nevertheless this apparent preference for oak is confirmed by documentary evidence (McGrail 1998).

Rationale

Acoustic characterisation of unconsolidated sediments

Geologists now use acoustic methods to discover more than the basic layered structure of sediments. The remote determination of the physical properties of the seabed and the immediate sub-surface (<50m) has been the subject of extensive research over the last two decades. This has resulted in the commercial development of a number of sediment classification systems. It has been identified from this literature that P-wave velocity (ie acoustic velocity), impedance (the product of density and velocity) and attenuation are the acoustic properties that have the strongest correlation with physical properties of the seabed (eg bulk density and porosity) (Bachman 1985; Richardson and Briggs 1993). It is proposed that such techniques could be used for the identification and characterisation of archaeological wood as these properties have been shown to be affected by degradation (Okyere and Cousin 1980; Schniewind 1990; Ouis 1999).

Propagation of sound through wood

Wood is an immensely complex medium, a state induced by the specific deposition of anatomical elements during the life of a tree (Hearmon 1948). On a macroscopic level the dominant cell direction is oriented vertically in the tree stem, resulting in a pronounced grain direction and the characteristic growth ring structure when seen in cross section (Fig 4.1a). Softwoods have a simpler structure than hardwoods, with approximately 90% of cells aligned vertically in the tree stem. The other 10% of cells are oriented perpendicular to the stem, forming the flattened ribbons of tissue known as rays, which radiate outward from the pith at the centre of the trunk and cross the growth rings (Fig 4.1a). Hardwoods only contain approximately 70–80% of longitudinal cells (G Daniel, pers comm). The complex structure of wood results in it being elastically anisotropic; that is, elastic wave properties vary according to propagation direction.

To assist the physical description of sound propagation in wood, three axes are defined that relate to the macroscopic structure. They are the longitudinal axis oriented parallel to the tree stem (L), the radial axis perpendicular to the growth rings (R) and the tangential axis parallel to the growth rings (T). These axes are located within the three planes of elastic symmetry, the L-T, L-R and R-T planes (Fig 4.1b). Wood elasticity theory assumes these planes to be mutually perpendicular and that wood is perfectly elastic and homogeneous. In reality, wood only approximates to this ideal structure eg the L-T plane is actually roughly cylindrical. However, if a sample is taken far enough from the centre of the tree compared with the sample size, this surface is more truly planar (Hearmon 1948).

Measurement of the acoustic properties of wood

To date ultrasonic velocity measurements have been made on wood for three main purposes. The first of these was to establish values for the elastic constants of wood, for use in

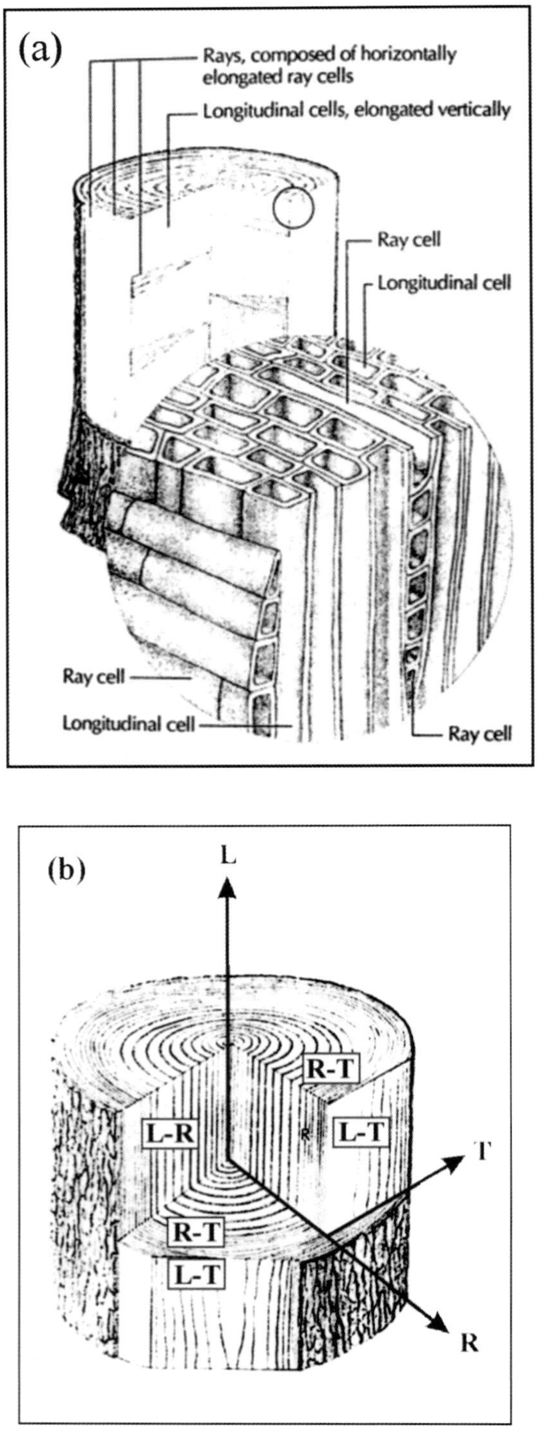

Fig 4.1 a – major structural components of wood; b – the three mutually perpendicular planes of elastic symmetry and their principal axes (modified from Hoadley 1990)

the construction industry and for the manufacture of musical instruments (Bucur 1988; Bucur and Chivers 1991). Secondly, research for the forestry industry has focused on the use of velocity measurements for the detection of decay in standing trees (Breeze and Nilberg 1974; Okyere and Cousin 1980) – focusing on identifying internal decay pockets. Such large internal decay features are not present in degraded cut timbers, although marine timbers can be extensively attacked by marine borers (eg *Teredo navalis*) which can create tunnels up to 1.0m or more in length (Tsoumis 1991). Finally, velocity measurements have been made through drying timber to try and establish a non-destructive technique of measuring moisture contents in order to control the kiln drying of lumber and prevent warping (Simpson 1998).

Owing to the prevalence of softwood timber in the construction and forestry industries most investigation into the acoustic properties of wood has centred on these species. The majority of measurements have been made on dry timbers with some being made on green or live wood. There have been very few measurements made on saturated wood. These experiments on *in situ* or cut timbers typically use a simple direct sound transmission technique, using frequencies in the range of 0.25–1.0MHz.

The hygroscopic nature of wood results in a constant change in the water content of a sample as it attempts to maintain equilibrium with the ambient hygrometric conditions. In the timber industry the water content of wood is expressed in terms of an equilibrium moisture content (EMC), which is equal to the weight of water in a sample, at a known relative humidity, as a percentage of its oven-dry weight (Dinwoodie 1981; Tsoumis 1991). The age and species of the timber affect the EMC, with freshly cut green timbers recording values in the range 35–300% (Ouis 1999). Early work (Kollmann and Krech 1960) on P-wave velocity measurements through timber samples show a 25% reduction in velocity values between 0 and 50% moisture content. This reduction is a result of a decrease in the modulus of elasticity and an increase in density with increasing EMC (Tsoumis 1991). Consequently, the majority of velocity measurements in the literature are quoted with reference to a known EMC (typically 10–12%).

All velocity measurements quoted within the referenced literature are for the three possible end-member cases V_L, V_R and V_T representing incident propagation along the three principal axes. These data identify that the longitudinal velocity (V_L) is much higher than those along the other two axes, probably because the majority of cells are oriented parallel to this direction. The radial velocity (V_R) is slightly higher than the tangential (V_T) because the rays are oriented in this direction and there is no corresponding alignment of cells in the tangential direction (Tsoumis 1991). Typical values for end-member P-wave velocity for air-dry European white oak are given in Table 4.1 (Bucur 1988).

Density of wood

In addition to the literature available on the propagation of sound through modern timbers, there is also a significant amount on density variation within and between wood types. Four different measurements of density are used in different industries. Oven-dry density is derived from both mass and volume of oven-dry wood. Air-dry density is that found under a controlled environment with an EMC of 12%. Conventional or basic density is calculated from oven-dry weight and water-saturated volume. Finally, apparent density is a simple

Table 4.1 The P-wave velocities for European white oak samples (Bucur 1988); measurements acquired using direct transmission with a 1MHz source and at an EMC of 12%

	Longitudinal velocity V_L (ms^{-1})	Radial velocity V_R (ms^{-1})	Tangential velocity V_T (ms^{-1})
European white oak	5071	2148	1538

measure of mass and volume at an unknown moisture content (Dinwoodie 1981; Schniewind 1990; Tsoumis 1991). Those densities requiring oven- or air-dry values would necessitate the permanent destruction of the artefact and as such are not appropriate for non-destructive measurements.

Moisture content (MC) and structure principally affect the density of wood. Wood mass increases with increasing moisture content up to the fully saturated condition. The volume of wood, however, only increases at first (up to the maximum swelling capacity *c* 30% MC) and then remains constant; however much more moisture is retained. Dinwoodie (1981) gives the approximate relationship that for each 1% increase in moisture content, up to 30%, density increases by around 0.5%. As the volume remains constant above 30% MC, above this level the density is predicted to increase rapidly, albeit there are very few directly recorded measurements of density for high moisture contents.

Structural controls on wood density for an individual species result from a number of spatial variants. These include changes in ring width, the development of reaction wood (formed when a hardwood grows under tension), and the relative proportion of latewood (formed later in the growing season) to earlywood. In addition, the density tends to reduce with height and distance from the pith (Tsoumis 1991). Finally, variation between trees of the same species is influenced by heredity and environmental conditions (climate, soil, spacing). Consequently, there is considerable variation in densities of trees of the same species, with European white oak having quoted densities of 600–770kgm^{-3} (Dinwoodie 1981; Bucur 1988), although in many papers it is unclear what type of density value is recorded.

Wood degradation

Biological degradation of wood is caused by two main groups of organisms, the macrofauna and the microfauna. In the marine environment the macrofauna consist primarily of marine boring molluscs and crustaceans. Of these the *Limnoria* and *Teredo* species are the most damaging (Gregory 1998). *Limnoria* ('gribble') are a few millimetres in length and degrade the outer layers of timbers. They bore tunnels approximately 1–3mm in diameter (G Daniel, pers comm). The most destructive degraders of wood in the marine environment are the *Teredinids* ('shipworms'). These organisms begin to colonise wood at the larval stage and spend their lives burrowing a complex network of galleries within a timber. The worms can grow to over 1.0m long (depending on species and conditions) and can remove the majority of the wood substance (Fig 4.2a). A timber may appear almost intact from the outside, with only a few small holes present, even though it has been extensively degraded by shipworm (Tsoumis 1991). Marine borers require a high oxygen level and so are not found below the surface of the sediment (Gregory 1998). The micro-organisms which attack wood can be divided into two groups: fungi and bacteria. Three main types of wood degrading fungi can be found in terrestrial situations: brown, white and soft rot. Of these, only soft rot is found in submerged timbers as the others require higher oxygen levels (G Daniel, pers comm). Soft rot decay proceeds very slowly with only a few millimetres of wood affected after many years. Both fungi and bacteria degrade the walls of the wood cells (Fig 4.2b), with fungi requiring a higher oxygen level than bacteria. Bacterial decay is the sole biological degradation of timbers that occurs in the almost anaerobic environment below the sediment surface (C Björdal, pers comm).

The degradation of wood alters several physical properties,

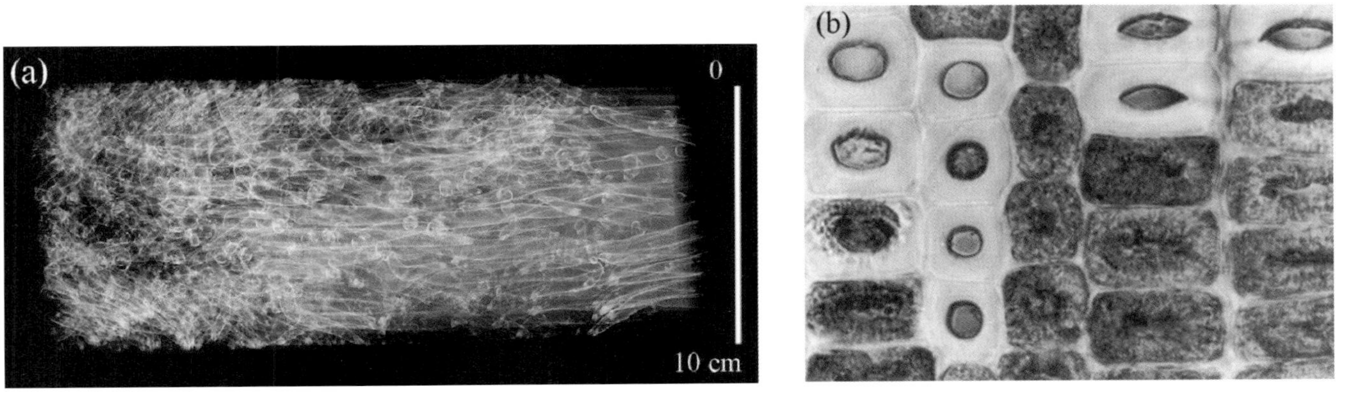

Fig 4.2 a – X-ray of pine pole after 16 weeks exposure to seawater showing galleries of Teredo navalis *(Gregory 1998); b – wood cells degraded by erosion bacteria (black) (Björdal 2000)*

which in turn affect the acoustic properties. The density is reduced and the maximum water content increased by the removal of wood substance. Both the reduction in density and the reduction in the rigidity caused by increased water content will reduce the acoustic velocity. As a consequence of this, the acoustic impedance will also be decreased. In contrast, the attenuation of sound increases with increasing degradation (Ouis 1999).

Experimental method

Archaeological wood has physical characteristics that are distinct from the air-dried material that has been the subject of previous studies. The major difference is the moisture content of wood from aquatic or wetland environments, which is rarely below 90% and values over 800% have been recorded (Rodgers 2000). This is significantly higher than the EMC values of 10–25% that are typical of previous wood measurements (Okyere and Cousin 1980; Bucur 1988; Bucur and Chivers 1991). As outlined above these higher EMC values are predicted to have a major effect on both the velocity and density of the waterlogged material. To assess the impact of such high moisture contents on the physical properties of oak, a series of experiments were conducted. Sapwood oak specimens measuring 100mm x 100mm x 10mm and cut parallel to the L-T, L-R and R-T planes were used, thus enabling the direct measurement of V_R, V_T and V_L respectively. These samples were saturated under pressure to achieve 100% saturation and thus replicate EMC values of marine archaeological wood as recorded by Hoffmann (1981) for a series of marine artefacts dated from the 1st to the 17th century. The maximum water content is that which would occur if the wood were fully saturated compared to the actual water content (or EMC), with the percentage saturation being the actual water content represented as a percentage of the maximum value. From the water contents given in this work, most timbers tested were between 85% and 100% saturated.

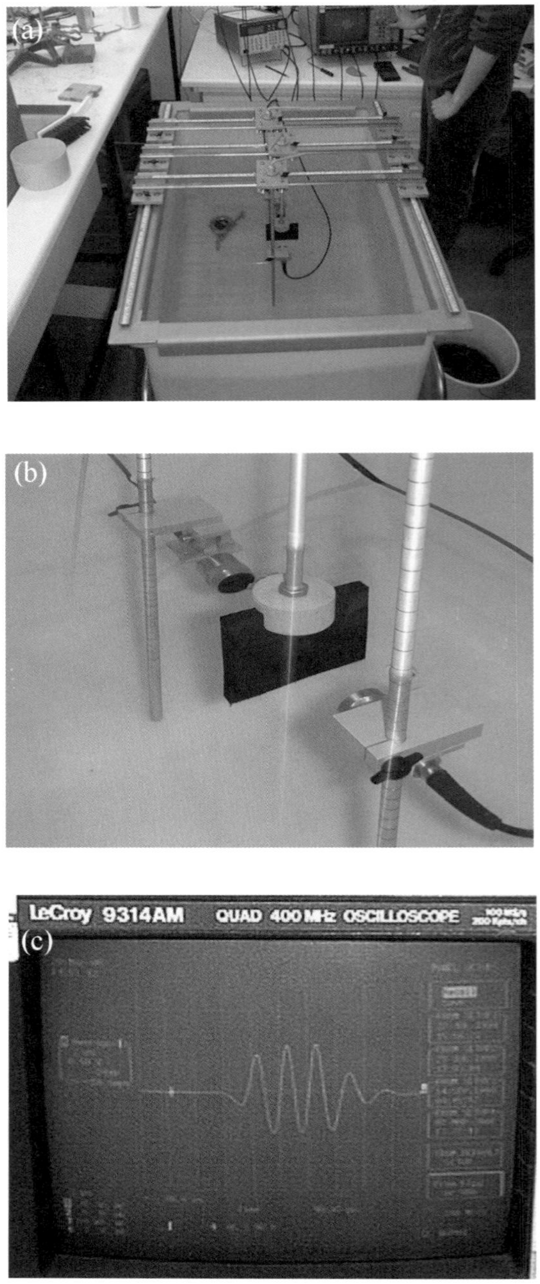

Fig 4.3 a and b – images of the water bath experiments on modern saturated samples and archaeological oak; c – received signal displayed on oscilloscope

P-wave velocity measurements

A direct transmission method, with the samples placed within a water bath, has been used for the velocity measurements. The measurements were conducted using a modular system, consisting of: a TG1010 Programmable 10MHz DDS function generator which produced a sinusoidal pulse three cycles long; a pair of Panametrics V301 transducers (transmitter and receiver) with a resonant frequency of 0.5MHz (Fig 4.3a and b) and a TERA Mk1 transducer and receiver amplification unit to amplify the received signal. Data was logged using a LeCroy 9314 AM Quad 400MHz oscilloscope (Fig 4.3c), with an average of 50 stacks being taken for each measurement. The velocity was calculated by a comparison between the time to the first positive peak value through water and the same peak through a water/wood/water travel path.

The dimensions of the wood samples were chosen to reduce the problems of edge effects and significant attenuation loss following the methodology of Best (1992). The transmission times were recorded to an accuracy of 10ns and the dimensions of the samples were measured with a precision of ± 0.1mm.

Calibration of the system was undertaken on an aluminium standard (Kaye and Laby 1986) with an error of ± 0.1% being recorded. Further, an initial series of experiments were undertaken to determine the frequency dependency of measurements taken on saturated oak samples. Velocity measurements calculated between 200kHz and 800kHz at 100kHz increments showed a variance of only

± 0.6%. All subsequent experiments used 500kHz, the resonant frequency of the transducers.

Multiple readings were taken for the three principal transmission directions (V_L, V_R and V_T) and for comparative purposes three oriented sections were cut from a well preserved 17th-century (1640–4) timber taken from 'Wreck 5' of the BandW Christianshavn excavation, Copenhagen.

Density measurements

As the principal aim of these experiments is to derive a method by which archaeological material can be interrogated without disturbance, density measurements that required oven- or air-drying of samples were considered inappropriate. Consequently, apparent densities were calculated at the known saturation values of 100%. Dimensions were measured with a precision of ± 0.1mm and mass with a precision of ± 0.01g.

Results

Table 4.2 shows the P-wave velocity and density measurements obtained for fully saturated oak samples and the 17th-century artefact for the longitudinal, radial and tangential axes. Also shown are values for the wood velocities and densities of both Bucur (1988) and Quinn et al (1997) for comparison but these were taken at 12% EMC.

The results demonstrate that the anisotropy exhibited by seasoned timbers is maintained when fresh oak samples are fully saturated. The absolute velocity values are significantly reduced for the longitudinal and radial axes measurements of waterlogged wood. However, the tangential axis shows only a small reduction in velocity (within measurement error of the Bucur (1988) value). As would be anticipated, the fully saturated apparent densities have increased significantly. Furthermore, the velocity (particularly V_R and V_T) and density values for the archaeological material correspond well with those recorded for the 100% saturated material, indicating it has undergone little degradation.

These measurements were taken at 500kHz. Typically, the P-wave velocity of a material is dependent on the frequency but measurements made on dry oak samples between 200–800kHz showed a very small variation of ± 0.6%. This suggests that the velocity is independent of frequency over this frequency range.

Table 4.2 Summary table showing the P-wave velocity and density values for 100% saturated oak and a 17th-century timber; values taken from Bucur (1988) and Quinn et al (1997) are shown for comparison

Source	Density (kgm^{-3})	V_L (ms^{-1})	V_R (ms^{-1})	V_T (ms^{-1})
100% saturated	1161 ± 28	2882 ± 110	1668 ± 10	1528 ± 15
17th-century wood	1172 ± 6	3564	1527	1383
Experimental Bucur (1988)	600	5071	2148	1538
Theoretical Quinn et al (1997)	660	3120	1960	1230

Degradation experiments

This study aims to assess the change in acoustic properties with degradation state. To this end, a large number of oak samples have been deployed at Lynæs Sands, Denmark. A total of 240 oak samples were prepared, 80 perpendicular to each of the L, R and T axes to allow sound propagation along these directions. Of these, 30 samples are to act as controls and the remaining samples were attached to plastic mats with cable ties in groups of 15 (5 of each direction) with half deployed at each of two sites approximately 25m apart. Two sites were used to minimise the risk of loss of samples from fishing or other boat activity. Plastic pipes filled with concrete were placed vertically in the seabed to form uprights. The mats of samples were then attached in pairs to other weighted pipes, three of which were attached horizontally to the uprights at each site approximately 1.0m above the seabed (Fig 4.4). The remaining mats were buried below the sediment with a small fisherman's float attached to enable relocation. The samples were put down on 1 June 2001 and one mat from each site will be recovered after 1, 2, 4, 6, 9 and 12 months with the samples under the sediment also being recovered after 12 months. The samples under the sediment will be used as a comparator and very little degradation is expected. The samples exposed in the water column will be progressively more degraded and indeed the first two sets have already been recovered and show colonisation by seaweeds and barnacles amongst other organisms. Degradation is expected only by macrofauna as the timescale is too short for degradation by micro-organisms to occur (D Gregory, pers comm).

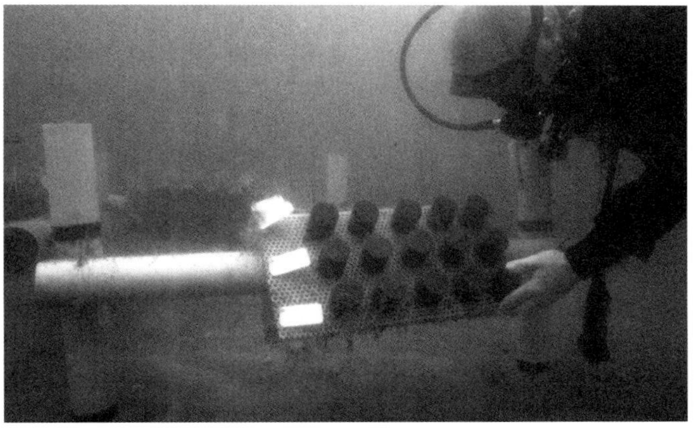

Fig 4.4 Deploying samples at Lynæs Sand

Discussion

The transmission measurements taken along the three principal axes of fully saturated oak samples demonstrate that at these high saturation states (>85%), P-wave velocity is significantly affected in the longitudinal and radial directions when compared to measurements on seasoned wood. At 100% saturation, V_L and V_R values reduced by 43% and 22% respectively, compared to measurements at 12% EMC. By comparison, the tangential velocity exhibits no significant variation between air-dry and fully saturated conditions, thus emphasising the importance of cell structure on sound propagation in wood.

The use of apparent density measurements resulted in significantly greater values than densities commonly recorded in the literature, and more significantly, those used by Quinn et al (1997). The authors consider that apparent density measurements at 100% saturation are more representative of the material *in situ*, and consequently are more indicative of the bulk density values required for the impedance/reflection calculations.

Calculated reflection coefficients, based on the measured P-wave velocities and apparent densities, predict that wood buried in the marine environment should still have characteristically large reflection coefficients, thus supporting the original assertions of Quinn et al (1997). Indeed, the values for oak in sand calculated with these values are closer to *in situ* values recorded from the *Invincible* wreck in the Solent than are those calculated by Quinn et al (1997). However, the longitudinal value is significantly reduced, suggesting artefacts oriented with the longitudinal axis approximately vertical and buried in sands may be difficult to image. Since the dominant timber components of a wreck site are planking, it is far more likely that the radial and tangential orientations will be encountered using normal incidence reflection profiling in a moderately undisturbed wreck.

The saturated measurements, and the comparable set made on waterlogged archaeological material, suggest that the use of a direct acoustic transmission technique in a water bath provides accurate and reproducible measurements of P-wave velocity. It is suggested that the use of these more realistic velocities, in tandem with apparent density measurements, should be used for the calculation of impedance and hence reflection coefficients. This would significantly improve upon the simplistic concept of using values calculated or measured from seasoned timbers.

This first set of recorded measurements of P-wave velocity and density for waterlogged material will provide the basis for extending this experimental work into the analysis of degraded archaeological material. Current research by the authors is attempting to establish if the structural changes to wood caused by macro- and microfaunas are reflected in their acoustic signature. We also intend to study the impact of saturation in more detail.

Measurements in the laboratory are made at frequencies of several hundred kilohertz but sub-bottom profilers typically use <20kHz. In order to investigate acoustic properties at these lower frequencies we will use the Biot model (eg Geertsma and Smit 1961) for sound propagation in porous media to extend our measurements down to seismic frequencies.

Conclusions

This paper presents a method for accurately measuring P-wave velocity on waterlogged archaeological wood. Further developments will enable conservators to rapidly assess the integrity of wooden artefacts in the laboratory. It has been demonstrated that measurements of both P-wave velocity and apparent density of fully saturated oak samples are more representative of waterlogged archaeological wood than the dry wood values previously used. These recorded values have enabled the calculation of realistic reflection coefficients for oak buried in sand and these values have proved close to those recorded *in situ* (Quinn et al 1997). This work provides a technological platform for future research into the variation of acoustic properties with changing degradation state. The development of a system for the *in situ* determination of degradation state of timbers in a wide range of environments is planned. The ability not only to identify buried archaeological material, but also to non-destructively assess its state of preservation, will represent a major advance in our ability to manage our cultural heritage.

Acknowledgements

The authors would like to thank the financial support of the National Museum of Denmark, the University of Southampton, NERC and the Royal Commission on the Historical Monuments of England. Thanks are also due to Claus Pedersen and others at the National Museum of Denmark for their help with preparing wood samples and deploying those at Lynæs.

Bibliography

Bachman, R T, 1985 Acoustic and physical property relationships in marine sediment, *J Acoust Soc Amer* 78, 616–21

Best, A I, 1992 The prediction of reservoir properties of sedimentary rocks from seismic measurements, unpub PhD thesis, Univ Reading

Björdal, C G, 2000 Waterlogged archaeological wood: biodegradation and its implications for conservation, unpub PhD thesis, Swedish Univ Agric Sci, Uppsala

Breeze, J E, and Nilberg, R H, 1974 Predicting by sonic measurements the strength of logs and poles having internal decay, *Forest Prod J* 21, 39–43

Bucur, V, 1988 Wood structural anisotropy estimated by acoustic invariants, *Int Assoc Wood Anatomists Bull* ns 9, 67–74

Bucur, V, and Chivers, R C, 1991 Acoustic properties and anisotropy of some Australian wood species, *Acustica* 75, 69–74

Dinwoodie, J M, 1981 *Timber, its nature and behaviour*, Wokingham

Geertsma, J, and Smit, D C, 1961 Some aspects of elastic wave propagation in fluid-saturated porous solids, *Geophysics* 26, 169–81

Gregory, D, 1998 Re-burial of timbers in the marine environment as a means of their long-term storage: experimental studies in Lynaes Sands, Denmark, *Int J Nautical Archaeol* 27, 343–58

Hearmon, R F S, 1948 *The elasticity of wood and plywood*, London

Hoadley, R B, 1990 *Identifying wood: accurate results with simple tools*, Newton, USA

Hoffmann, P, 1981 Chemical wood analysis as a means of characterising archaeological wood, in *Proceedings of the ICOM waterlogged wood working group conference* (eds D W Grattan and J C McCawley), 73–83, Ottawa

ICOMOS 1990 International Conference of Monuments and Sites, Charter for the protection and management of the archaeological heritage, http://www.international.icomos.org/e_archae.htm

Kaye, G W C, and Laby, T H, 1986 *Tables of physical and chemical constants*, 15 edn, New York

Kollmann, F, and Krech, H, 1960 Dynamische messung der elastischen Holzeigenschaften und der Dämpfung, *Holz als Roh- und Werkstoff* 18, 41–51

McGrail, S, 1998 *Ancient boats in north-west Europe: the archaeology of water transport to AD 1500*, London

Okyere, J G, and Cousin, A J, 1980 On flaw detection in live wood, *Materials Evaluation* 38, 43–6

Ouis, D, 1999 Vibrational and acoustical experiments on logs of spruce, *Wood Sci Tech* 33, 151–84

Quinn, R, Bull, J M, and Dix, J K, 1997 Imaging wooden artefacts using Chirp sources, *Archaeol Prosp* 4, 25–35

Rodgers, B A, 2000 The ECU conservator's cookbook, Chapter 2: waterlogged wood http://www.ecu.edu/maritime/publications/cookbook/the-conservation-cookbook.htm

Richardson, M D, and Briggs, K B, 1993 On the use of acoustic impedance values to determine sediment properties, *Proc Inst Acoust* 15, 15–24

Schniewind, A P, 1990 Physical and mechanical properties of archaeological wood, in *Archaeological wood: properties, chemistry and preservation* (ed R J Barbour), Advances in Chemistry Series 225, 87–109

Simpson, W T, 1998 Relationship between speed of sound and moisture content of red oak and hard maple during drying, *Wood Fibre Sci* 30, 405–13

Tsoumis, G T, 1991 *Science and technology of wood: structure, properties, utilization*, New York

5

Can the process of investigative conservation elucidate information relating to preservation *in situ*: Owmby-by-Spital (Lincolnshire), a case study

DYLAN COX and KARLA GRAHAM

Dylan Cox
Research Fellow in Archaeological Conservation
Department of Archaeology
University of Southampton
Avenue Campus, Southampton SO17 1BJ, UK

Karla Graham
Archaeological Conservator
English Heritage Centre for Archaeology
Fort Cumberland, Fort Cumberland Road
Eastney, Portsmouth PO4 9LD, UK

Abstract

This paper will cover the Late Iron Age and Roman settlement of Owmby-by-Spital, Lincolnshire (a Scheduled Ancient Monument). English Heritage has been measuring and monitoring the effects to the archaeological resource of agricultural activity (ploughing) and illegal metal detecting on the site. It briefly outlines the project as a whole to date, examines the material assessed by conservation and explores possible relationships between state of preservation and site activity.

Introduction

The main aim of this paper is to present an attempt to elucidate information (damage through ploughing, metal detecting, archaeology) from two groups of material (nails and coins) recovered from Owmby-by-Spital. It briefly outlines the project as a whole to date, examines the material assessed by conservation and explores possible relationships between state of preservation and site activity.

Site background

Owmby is a Late Iron Age and Roman settlement located in the parish of Owmby-by-Spital, Linconlnshire (SK 97108648). The site, located on arable land, covers 41.3ha and lies in the dip slope of the limestone plateau. The soils in this area are generally well-drained calcareous loams but change to clays on the northern slope of the valley (McAvoy 1996).

This Scheduled Ancient Monument has a class consent for cultivation. The arable farming regime includes cereals and root crops (McAvoy 1996). The 1906 Ordnance Survey map shows the land was primarily woodland. However by 1945 aerial photographs showed that this had been removed and ridge and furrow are visible.

The site has been known as a rich source of metalwork since at least the 18th century. It continues to attract metal detectorists, usually coinciding with or following agricultural activity at the site. This may be seen as evidence that new material is being brought up by ploughing, but the relationship is not proven (McAvoy 1996).

The immediate impetus for the project came with the uncovering of a stone sarcophagus during ploughing in 1994 on land within the eastern scheduled area. The discovery of the sarcophagus highlighted evidence for incremental erosion of the resource (morphology and sequence of structures, deposits and artefacts) into the ploughsoil by continuous agricultural activity. This provided an opportunity for English Heritage to measure and monitor both the effects of this activity and illegal metal detecting to the archaeological resource.

Project Director Fachtna McAvoy is currently assessing the level of site preservation and the options for long-term protection that can be practically instigated at this and similar sites. The project's main aims were to: determine the character and significance of the archaeological resource; identify and understand the threats to this resource; determine the means by which the resource can best be protected. To achieve these the following were carried out: a topographic model of the field was created; an evaluation survey was conducted to establish the condition and nature of the structural resource; trial trenches were excavated to determine the depth of the plough soil so that material would not be recovered from stratified deposits; plough depth was monitored to determine the depth of the current ploughing regime and monitor the movement of artefacts in the plough zone; detailed transcription of aerial photographs; deterrence of unauthorised metal-detector activity; systematic metal detecting survey to characterise the nature and distribution of the resource within the ploughsoil which involved a 100% retrieval strategy.

The collection of artefacts lead to the involvement of conservation in this project. It is important to note however that conservation's input was only requested after the fieldwork phase had concluded.

Conservation

Before addressing the specific issues of conservation in this project, it is worth summarising the contributions conservation can make throughout the archaeological process.

Stage 1: project planning

Conservators meet with other members of the project team to discuss the project as a whole and provide specialist advice such as: the likely nature and effect of the burial conditions on the artefacts; their analytical potential; the resources required to undertake the project.

Stage 2: lifting

Fragile or complex objects are retrieved from the site either by a conservator or under advice from a conservator.

Stage 3: X-radiography

Objects are X-rayed to determine form, condition and technology such as non-ferrous inlays and decoration.

Stage 4: investigation

Selected objects are analysed to provide information relating to burial conditions, preservation, technology and use. For example non-ferrous inlays are characterised using X-ray fluorescence; mineral preserved organics are identified though the use of the scanning electron microscope.

Stage 5: dissemination

The conservation findings are published in the monograph and can also published as separate papers on specific observations and conservation research.

Case studies: methodology

A number of pilot studies were carried out which aimed to find links between the levels of preservation of the assemblage and site activity in terms of agriculture and metal detecting. This was conducted through two pilot studies of coins and nails. This paper will concentrate primarily on the findings of the nail pilot study.

Case study 1: coin pilot study

A pilot study was devised and conducted by a number of placement students (based at Ancient Monuments Laboratory). 1992 copper-alloy and silver coins were X-rayed and then categorised (Fig 5.1). The five categories devised were based upon the extent of obscuring soil and corrosion products (Fig 5.2). Where the detail was clearly preserved in an unbroken patina with virtually no soil or corrosion upon the

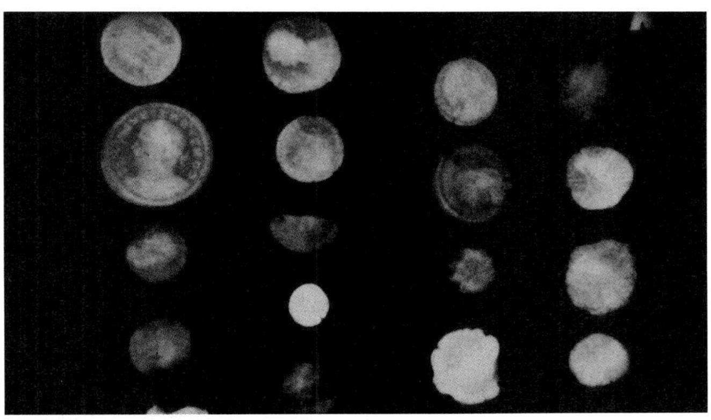

Fig 5.1 X-ray of coins

Fig 5.2 Coin categories; Category 1(left) and Category 2(right)

surface, these were classed as Category I. Category 5 coins were mineralised with friable patina, heavily encrusted with corrosion and with no visible surface detail.

Case study 2: nail pilot study

A 10% sample of the iron nails (719 items) recovered from the ploughsoil through controlled metal detecting of transects 301–30 and 491–520 on the site formed the basis of the study (Graham and Cox 2001). Most appeared to be in similar condition with little or no apparent obscuring or active corrosion products.

This pilot study aimed to deduce the analytical potential of the nail assemblage in terms of typology (after Manning 1985) and condition (completeness and degree of flaking). All the items were X-rayed (Fig 5.3). The next stage was to devise a method for recording typology and condition. For this, a table was created which would serve as a guide for incorporating codified data into a spreadsheet (Table 5.1).

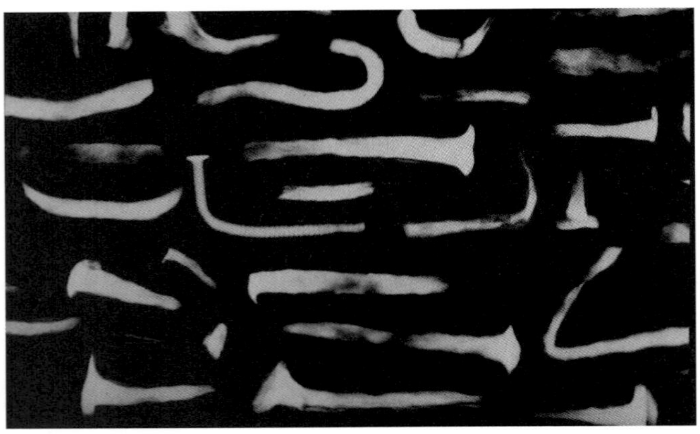

Fig 5.3 X-ray of nails

Typology

The Manning typology for nails (1985) was selected by Nicola Hembrey and Fachtna McAvoy (Centre for Archaeology) since it is considered the most comprehensive nail typology available (Fig 5.4).

Further typological categories were devised by the authors to describe objects which did not fall within the Manning typologies (Graham and Cox 2001).

INDETERMINATE

Nails having a lack of sufficient features to assign them to a specific Manning typology, for example items with the remains of possible nail heads or tapering rod like sections.

NOT APPLICABLE

Objects lacking diagnostic nail features.

MODERN

The recovery of the material through metal detecting had resulted in modern nails being incorporated into the assemblage.

Table 5.1 Table of values for recording typology and condition of nails

Typology	Completeness	Degree of flaking
Manning (1985) typology	1 complete	A no flaking
10 types	2 incomplete	B partial flaking on <50% surface area
		C partial flaking on >50% surface area
		D complete flaking
Indeterminate (Ind)	1 complete	categories A to D
Unassignable to Manning typology (1985)	2 incomplete	
N/A	N/A	categories A to D
cannot be confidently identified as a nail or part of a nail		
Modern (M)	1 complete	categories A to D
Modern	2 incomplete	

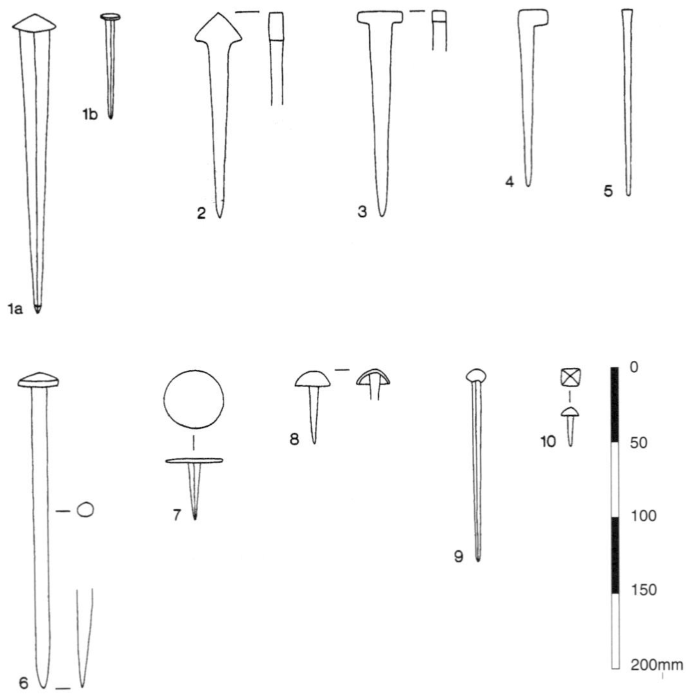

Fig 5.4 Manning nail typology (1985, 133)

Condition

The condition of the items was subdivided into completeness and degree of flaking. For items positively identified as nails (comprising Manning categories 1–10, indeterminate and modern), two classes were created: complete (1) and incomplete (2). Complete refers to a nail that appears to have sustained no morphological losses. Incomplete refers to a nail where morphological losses are apparent. The completeness was determined by observing the nail in conjunction with the X-rays.

The degree of flaking was subdivided into four levels:

Level A

An item that had suffered practically no flaking.

Level D

The opposite of A in that the item had suffered near complete flaking.

Levels B and C

Deemed to have undergone partial flaking, B referring to less than 50% of the surface area loss and C greater than 50%.

The numbers of levels for degree of flaking were kept to a minimum in order to decrease the margin of error in terms of statistics due to the increased subjectivity.

Case studies: results and discussion

Case study 1: coin pilot study

Limited analysis was carried out on the corrosion products and composition to attempt to show a relationship between burial environment and preservation. In terms of preservation, X-ray fluorescence (XRF) analysis showed a possible relationship between the tin to lead ratio and preservation: condition deteriorates as the ratio of tin to lead decreases (Edwards 1997). However, the study emphasised that care must be taken when interpreting XRF results as it must be realised that they are only representative of the stable corrosion compounds that are formed upon the surface not a true depiction of the composition.

Discussion

The coin category system provided information concerning how well the surface detail could be interpreted by a numismatist as well as degrees of corrosion and preservation. The lack of distinction between the levels of visible surface detail and coin condition however entailed that for example, coins with a high level of visible surface detail but poor level of preservation could potentially be classed as the highest category 1. A comparative example would be a copper alloy ring with incised decoration. If this decoration was obscured by stable corrosion products, it would not necessarily be classed as poorly preserved because preservation applies to the material and not the decoration.

The coin categorisation would have been more specific to the research objectives of this project if a distinction had been made between surface detail (numismatic information) and the actual condition of the substrate.

Case study 2: nail pilot study

All the nail data was analysed statistically in terms of typology and condition.

Typology

Of the 719 selected items just under 50% were classed as N/A; 28% were indeterminate and *c* 20% could be fitted into the Manning typology (Fig 5.5).

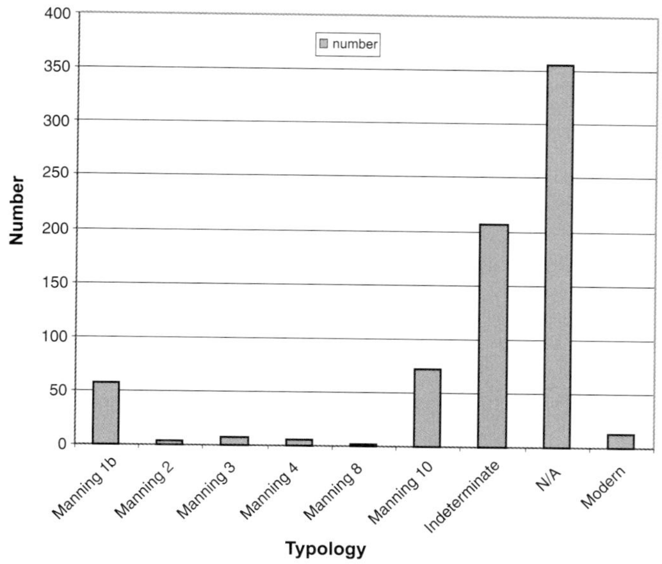

Fig 5.5 Nail results: typology

Completeness

The majority of the nails and indeterminate examples are incomplete, with fewer of the nails being incomplete (79%) compared to the indeterminate ones (96%) (Fig 5.6).

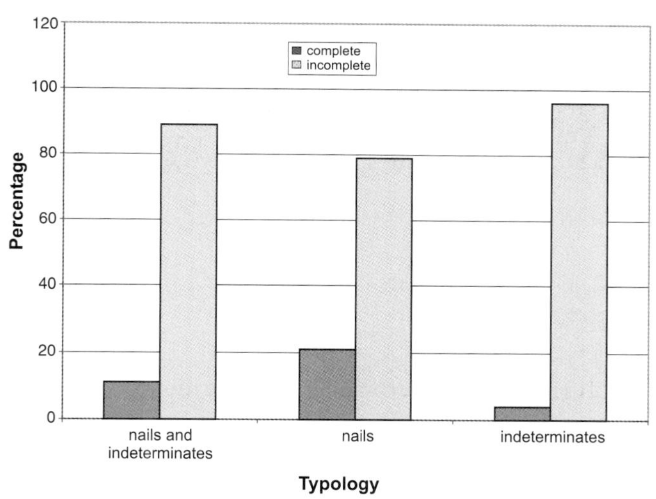

Fig 5.6 Nail results: degree of completeness

Degree of flaking

For both the nails and indeterminate examples, the majority display partial flaking (Fig 5.7). However, the majority of the nails suffered less than 50% partial flaking whereas the indeterminate group are evenly spread between the two partial flaking categories. The degree of flaking and therefore the amount of recognisable material left would appear to have affected whether a typology could be determined.

Discussion

Nails were chosen as opposed to other iron artefacts

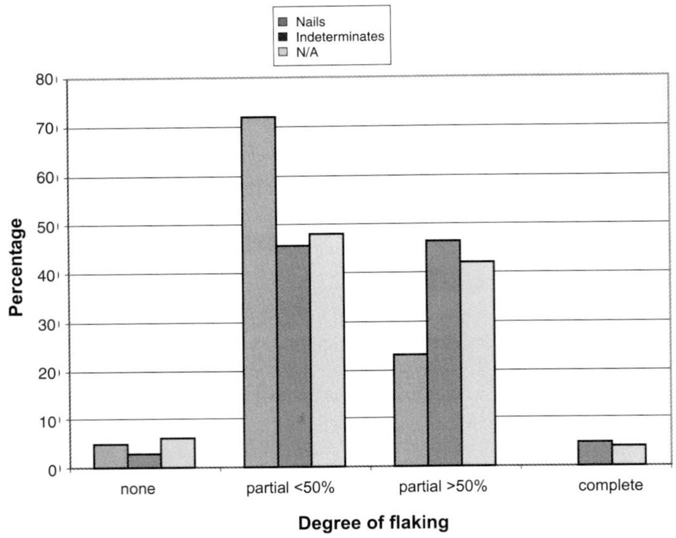

Fig 5.7 Nail results: degree of flaking in all items

because they are plentiful and also have distinct morphology comprising a shaft and a head. This should have made the degree of completeness easier to ascertain. However, whereas for the degree of flaking of approximate surface loss could be recorded, this was not the case with completeness where the original size of the nail was unknown.

Visual examination of items was able to confirm the typology in the majority of cases because of the low levels of obscuring or active corrosion products. X-rays were essential for the identification of a certain number of items for example, sections of barbed wire that were originally classed as nails. X-rays did not however provide information other than condition in cases where the typology had already been identified visually. In many cases the X-rays did not show the most informative plane, as the majority of items were X-rayed lengthways (as is common practice with X-rays of large assemblages).

Conclusions

Normally, archaeological conservators work on material which has been retrieved from secure stratified deposits. Information is provided which relates objects to the burial environment in terms of soil chemistry, associated finds and features. This helps to determine the possible factors affecting preservation.

In the case of Owmby, the context has been disrupted and no attempt to collect soil samples for chemical analysis was undertaken. It is therefore difficult to understand which factors in the original context have contributed to the preservation or deterioration of the objects.

The following factors in the agricultural process can result in both physical and chemical damage (Davies 1998).

Ploughing

Ploughing inverts the topsoil to bury the weeds and crop residue remains. The fresh soil and all the artefacts contained within are therefore open to weathering.

The extent of physical damage to objects can be quite varied: on the one hand it occurs when objects are hit or flung by the plough; on the other, there is evidence that objects have been protected by surrounding clods of soil. For example, numerous copper alloy brooches from Owmby have been found with the delicate pins preserved and still attached.

Subsoiling

Subsoiling 'shakes' the soil and oxygen and water permeate the soil, affecting metal artefacts previously contained in stable conditions.

Fertilisers

Chemicals are added to the land to achieve specific pH levels with most crops requiring a more acidic level.

Agricultural drainage

The purpose of drainage is to increase the flow of water through the soil by making the subsoil free draining. As the water table falls, the soil above the water table becomes exposed to capillary forces that pull the soil particles into a closer packing leading to possible mechanical damage. It also increases the flow of added materials through the soil which affects the nature of the corrosion process. It is obvious from these factors, that the ploughzone is clearly a very unstable environment both physically and chemically.

Beyond this there are other factors which can affect the levels of preservation and include: the composition of the object; the function of the object during its lifetime; damage incurred when the object was discarded; the burial environment and natural fluctuations prior to the agricultural land use; acidic rainfall which can affect objects located in the ploughsoil near the surface. In addition damage can be caused by: illegal metal detectorists excavating and discarding unwanted objects; collecting, processing and storing objects and even, during the process of X-radiography!

What this study has shown is that it is difficult to show at this site (or similar sites) what specific factors have affected the level of preservation; there are simply too many variables in this particular equation. For example, are the flakes in a bag containing an object the result of poor packaging or do other factors need to be considered? Is a rod-like section a nail without a head and hence evidence of plough damage or is it simply a section from another type of artefact?

There is clearly then a lack of information associated with the artefact assemblages at this site and hence suppositions about state of preservation and the likely causative factors are almost impossible to deduce.

As a last note, we would like to make reference to an interesting observation which emerged during the nail assessment. An unusual pattern of damage was displayed by a number of nails showing the loss of two opposing faces on the shaft resulting in a thin, sheet-like shaft (Fig 5.8).

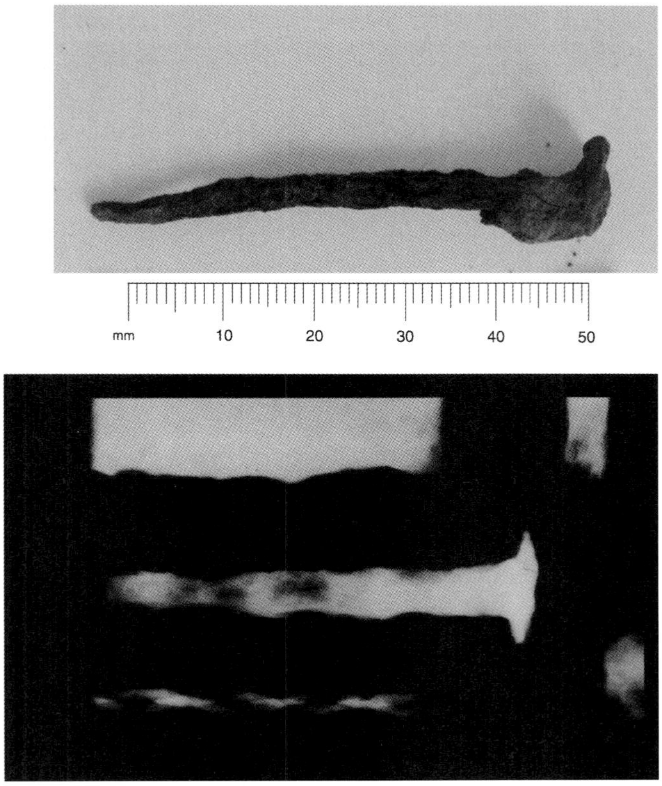

Fig 5.8 Sheet-like shaft nail; photograph (top) and X-ray (bottom)

This could possibly be either a physical indication of plough damage or freeze-thaw action. It will be interesting to compare this type of damage with other sites in the future, as further investigation would be required to confirm this observation.

Acknowledgements

Many thanks to Fachtna McAvoy and Nicola Hembrey for asking us to undertake the project.

Bibliography

Davies, J, 1998 From the other end of the microscope: numismatic considerations, in *Looking after the pennies: numismatics and conservation in the 1990s* (eds D Goodburn-Brown and J Jones), 73–80, London

Edwards, N, 1997 Owmby coin analysis, unpub MA portfolio, Univ Durham.

Graham, K and Cox, D, 2001 Owmby nails pilot study: Owmby-by-Spital, Lincolnshire, unpub Engl Heritage Centre for Archaeology rep 26/2001

Manning, W, 1985 *Catalogue of the Romano-British iron tools, fittings and weapons in the British Museum*, London

McAvoy, F, 1996 Owmby-by-Spital project review, unpub Engl Heritage Central Archaeol Service rep

6

Damage to inclusions in sand subjected to one-dimensional compression

ADRIAN HYDE

Abstract

The probability of damage occurring to archaeological remains is needed in order to make critical decisions for the future protection of our archaeological heritage. An artefact can be regarded as an inclusion in a sedimentary matrix. The degree of damage which will occur will depend on the nature and duration of the applied stresses, the composition of the sedimentary matrix, the boundary conditions of the stratum and surrounding soils, and the strength, stiffness, relative size, angularity and variability of the inclusion. One-dimensional compression tests on silica sand samples seeded with marked particles were carried out in order to examine the statistics of individual particle damage taking into account particle size and overall grading. Five levels of damage for marked particles were obtained from microscopic observations of the particles before and after testing. A statistical analysis was carried out on the data for each damage level to investigate the variation, with increasing applied stress, of the cumulative frequency for each level. In the case of uniformly graded sand, there was a sudden catastrophic onset of major splitting between the yield stress p_y and the point at which the compression index C_c reached a maximum. It was observed that for this sand, after yielding, 50% of the particles were damaged. In the case of well-graded sand, the crushing during yielding was characterised by a slow increase in the breakage of asperities for relatively large particles and major splitting of small particles. Single particle strengths were compared with a characteristic tensile stress for a particle embedded in the soil matrix. The logarithm of the characteristic tensile stress at maximum compression index for a particle embedded in a granular matrix could be related to the logarithm of the single particle strength multiplied by a proportional factor.

Introduction

Archaeologists and engineers working together require an indication of the probability of damage occurring to archaeological remains in order to make critical decisions for the future protection of our archaeological heritage. An artefact can be regarded as an inclusion in a sedimentary matrix. Geotechnical engineers have been studying the breakage of individual particles in soil for many years. Most of this work has however been high pressure one-dimensional compression testing on sand carried out by researchers such as Terzaghi and Peck (1948), De Souza (1958), Hendron (1963), Hagerty et al (1993) and Yamamuro and Lade (1996). Terzaghi and Peck (1948) testing sands first described the phenomenon of crushing at high stresses while De Souza (1958) also testing sands at stresses up to 138MPa first defined a yield point beyond which changes in grain size distribution showed substantial particle breakage. Many researchers have gone on to examine this yield point. De Souza (1958) and Hendron

Adrian Hyde
Senior Lecturer
Department of Civil and Structural Engineering
University of Sheffield
Sheffield S1 3JD, UK

(1963) both suggested that the yield stress increased with increasing density. Later researchers such as Hagerty et al (1993) have examined particle crushing in terms of initial void ratio, particle size, angularity and material composition. More recently McDowell et al (1996) developed a numerical model for crushable aggregates using a statistical function and work equation to predict the evolution of a fractal grain size distribution and linear e-logσ' compression line.

Particle crushing can be defined for increasing levels of damage such as grinding of the grain surface, breakage of an asperity and splitting of a particle. A careful investigation of individual particle crushing would reward us with a greater understanding of the susceptibility of artefacts to damage when loaded in a soil matrix.

Recent work by Nakata et al (1999a) has related individual particle crushing in triaxial sand samples to the soil particle strength variability using a Weibull function. Using data from one-dimensional compression tests on samples seeded with marked particles, this paper describes a continuation of this work by examining further the statistics of individual particle crushing taking into account particle size and overall grading.

It has been established in recent years that soil compressibility can be linked to particle crushing. But the relationship between these two factors is neither straightforward nor simple. One-dimensional compression tests have therefore been conducted to examine the effects of initial void ratio and grain size distribution on soil crushability and consequently the soil's compressibility. Although this data was obtained from high-pressure one-dimensional compression tests on silica sand, whose quartz particles have a very high strength; the findings have relevance for lower stress levels applied to sediments containing weaker inclusions such as archaeological artefacts.

Particle crushing under one-dimensional compression

Materials and experimental methods

One-dimensional compression tests have been carried out on air-dried silica sand containing only quartz particles. The oedometer specimens were 50mm in diameter and 10mm in height and the tests were carried out at a constant displacement rate of 0.01mm/min. In order to examine the relation between microscopic particle crushing and the compression curve, twelve dyed particles were seeded in each sample. The digital images of each particle were recorded prior to placing in the oedometer. Following the compression test, the particles were removed and a new digital image recorded. In addition all of the specimens were sieved after testing to determine any changes in particle size grading. The same techniques were used by Nakata et al (1999a) when investigating the crushing of individual particles in triaxial specimens. Two different particle size distributions, shown in Fig 6.1, were tested. One was a single size uniformly graded sand with particle sizes between 1.4mm and 1.7mm. The other was a well-graded material containing particle sizes from 0.25mm to 2.0mm. This was in order to examine the influence of particle size distribution on the crushing characteristics. The compression curves for the one-dimensional compression tests can be seen in Fig 6.2. Samples were prepared in a dense state to similar void ratios e = 0.6±0.03. Having established the compression curves, a subsequent series of tests was carried out also seeded with dyed particles but these tests were each terminated at the different stress levels indicated by the data points on Fig 6.2. In these tests, point P corresponded to the yield stress and point A_1 the maximum compression index C_c (defined as the slope of the e-logσ_v line). Observations of the nature of the crushing of dyed particles were carried out for all these tests apart from the test terminated at point P for the uniform sample where only a sieving test was carried out.

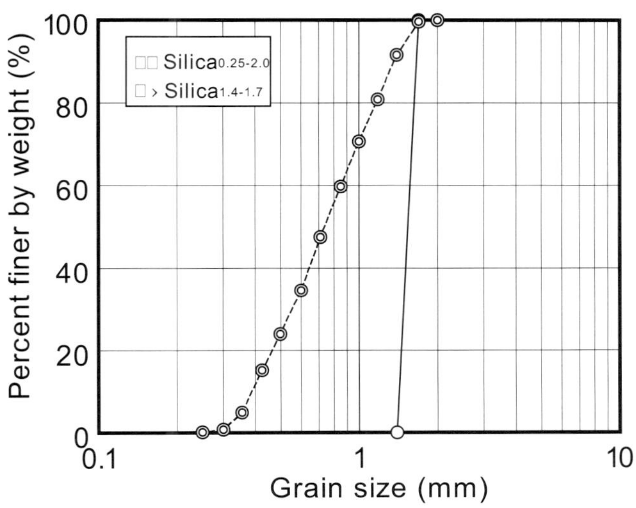

Fig 6.1 Grain size distribution curves for well-graded and uniform silica sand

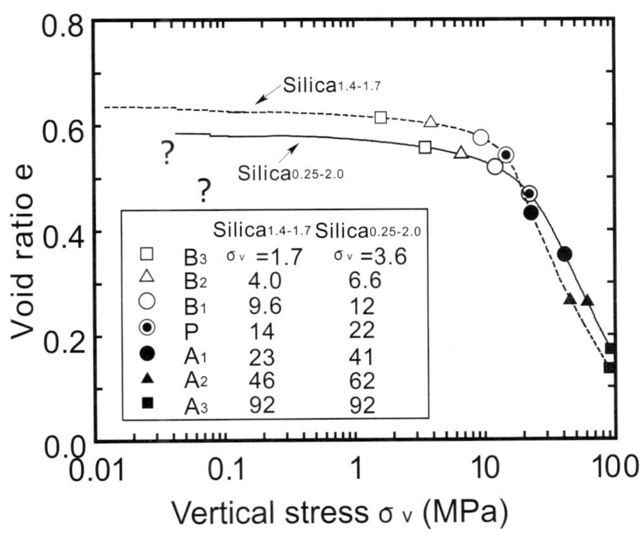

Fig 6.2 One-dimensional compression curves for well-graded and uniform sand

One-dimensional compression behaviour

Fig 6.2 shows the relationship between void ratio and logarithm of vertical stress for both materials. An examination of Fig 6.2 shows that the yield stress given by the point of maximum curvature (point P) is larger, at 22 MPa, for the well graded material compared with 14 MPa for the uniform sand and that the maximum curvature was greater for the uniformly graded material. Researchers such as Hagerty et al (1993) have already indicated that the steeper the curve after yield, the greater is the degree of crushing that is occurring. It is also well known that under an isotropic pressure the compressibility of sands increases with particle crushing (Miura and Yamanouchi 1977; Miura et al 1984; Yasufuku et al 1991).

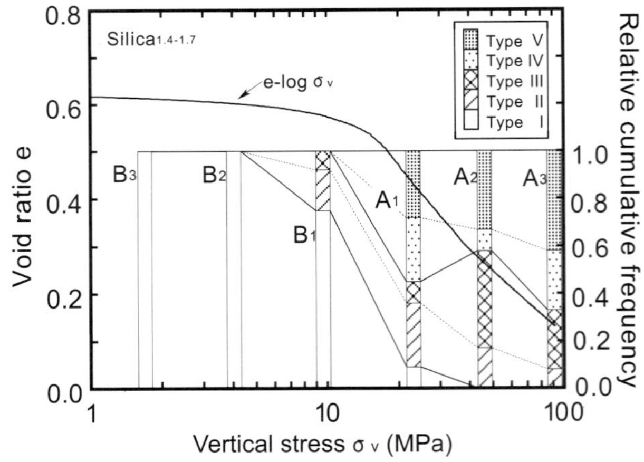

Fig 6.3 *Relative cumulative frequency of damage for uniform sand*

Particle crushing for uniformly graded samples

In the case of the uniform sand the dyed grains were the same size as the grains making up the matrix (1.4–1.7mm). Particles were examined for damage before and after testing. No visible damage was defined as Type I damage. Type II damage was defined as a single abrasion or breakage of an asperity. The next level of damage was Type III, which was defined as more than one abrasion or asperity fracture. Type IV damage consisted of the major splitting of a grain into two or more particles. Finally Type V was where it was not possible to completely recover all the sub-particles as any attempt at their removal resulted in fragments crumbling to powder.

In the case of damaged particles (Type II to Type V), the cross sectional areas were determined from digital images taken before and after testing using computer scanning software. The change in area was defined by the ratio:

$$R_A = \frac{\textit{Change in Area}}{\textit{Original Area}} \times 100$$

The resolution of the technique for determining R_A was dependent on the digital image being taken at exactly the same orientation before and after testing. This was clearly not always possible and even for Type I particles R_A values varying from -2% to 6% were obtained. The resolution decreased as the particle damage increased. The definition of the different types of particle damage relied largely on visual observation and required a measure of judgement. The boundary between Type III and Type IV was sometimes difficult to judge and in that case an additional measure of R_A = 30% was used to delineate the two types of damage.

A statistical analysis has been carried out of the particle crushing data. Each category of particle damage was treated as a class and the frequency was defined as the number of particles in each class. The relative cumulative frequency of each class has been plotted against the vertical stress in Fig 6.3. The e-log p' consolidation curve has been imposed over the frequency data in this figure. Particle crushing can be seen to initiate at Point B_2 (4.0MPa). At Point B_1 (9.6MPa) approximately 80% of the particles are still undamaged (Type I). However when the yield point was passed at Point A_1 (23MPa) this reduced to less than 10%. Thus more than 90% of the particles had some kind of damage and about 50% of the particles had undergone splitting. After yielding occurred the relative cumulative frequency curves began to level out indicating that the rate of increase of particle crushing slowed down considerably.

Fig 6.4 shows the evolution of the particle size distribution curves with increasing vertical stress. The sudden increase in particle crushing as the stress passes the crushing threshold (point B_1) and moves towards 23MPa (point A_1) can be seen clearly. An examination of Fig 6.4 shows a similarity between the curves as they deviate from the original distribution. Thus the curves appear to unzip as the small particles are broken first.

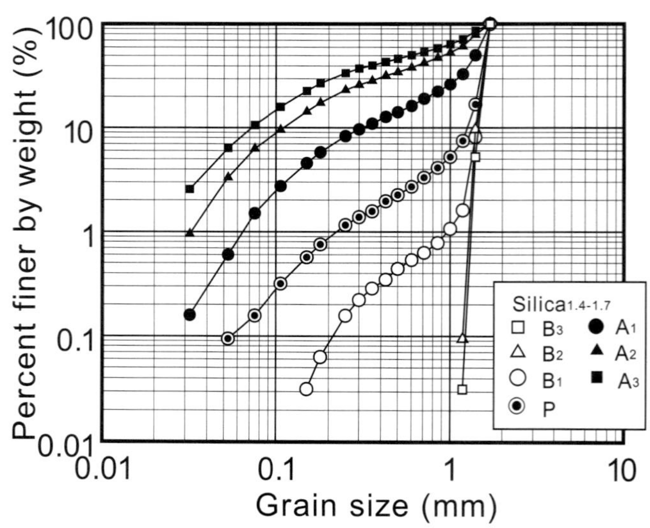

Fig 6.4 *Evolution of particle size grading curves for uniform sand (log-log scale)*

Particle crushing for well-graded samples

Observation of the crushing of dyed particles was carried out in the same way for well-graded samples as for the uniformly graded sand. The sizes of the dyed particles which were inserted were 1.4–1.7mm, 0.85–1.0mm, 0.71–0.85mm, 0.5–0.6mm and 0.3–0.355mm and were defined by the sieve sizes used. So for example 1.4–1.7mm passed a 1.7mm sieve but was retained on a 1.4mm sieve. The particles of size 0.3–0.355mm were at the limit of optical observation and particles smaller than this were difficult to find after testing. Fig 6.2 shows the full compression curve and the stresses at which individual tests were terminated to examine particle crushing.

The damage accumulation data for the largest, 1.4mm-1.7mm, intermediate, 0.71mm–0.85mm, and the smallest 0.3mm–0.355mm particles has been plotted as a relative cumulative frequency plot in Fig 6.5a–c respectively. Superimposed on each of these figures is the compression line for the sand as shown in Fig 6.2.

Considering Fig 6.5a for the largest observed particles, a small degree of damage was observed at the yield point but overall it can be seen that the worst damage that occurred was of Type II or Type III, with no splitting of grains occurring. Even at the highest stress levels 50%–60% of the particles remained undamaged. For the intermediate sized particles it can be seen in Fig 6.5b that some Type II damage commenced before the yield stress was reached. As the stress increased from the yield point P at 22MPa to point A_1 at maximum gradient C_c the degree of Type III and V damage increased. After this point the relative cumulative frequency for each degree of damage began to level out and remained almost constant, with approximately 70%–80% of particles undergoing some form of damage and 20% of the particles suffering Type V major splitting. The observed data for the smallest particles is shown in Fig 6.5c. As for the intermediate particles, damage commenced before the yield point P was reached and major splitting of grains (Type IV and V) rapidly accelerated between the yield point P and the maximum compression index C_c at A_1. Damage continued to worsen after this point and finally more than 80% of the grains underwent major splitting and 100% of the particles underwent some form of damage.

It is interesting to compare these data for well-graded soils with that for uniform samples shown previously in Fig 6.3. If we define the coordination number as the number of particle contacts experienced by a single grain, then in the case of the well graded sand the coordination number of large particles surrounded by large numbers of smaller particles will have been very high and the opposite will have been true for the smallest particles. In this case the tensile splitting stress for the large particles will have been relatively low while on the other hand that of the small particles with a low coordination number will have been much higher. So that even though larger particles are statistically weaker (Braddick 1963; Billam 1972; Lee 1992; Nakata et al 1999a; Nakata et al 1999b) their probability of survival in the matrix is much higher because of their higher coordination number. So it can be seen that the large particles, 1.4–1.7mm, in the well-graded matrix were protected by the smaller particles which underwent splitting as a result of their low coordination numbers. Thus

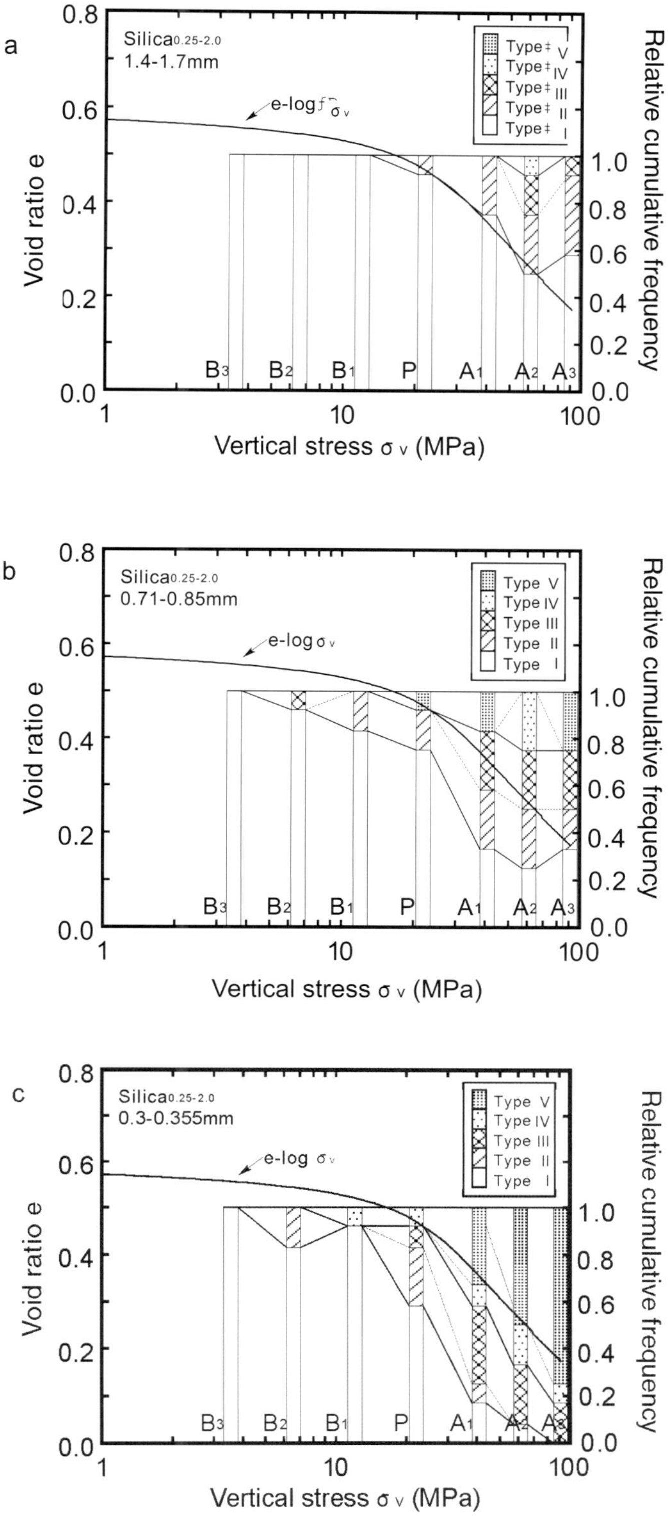

Fig 6.5 Relative cumulative frequency of damage for well-graded sand; a – 1.4–1.7mm particles; b – 0.71–0.85mm particles; c – 0.3–0.355mm particles

in Fig 6.5a there was little damage to the 1.4–1.7mm particles compared to Fig 6.3 and the high degree of damage was reserved for the smaller particles as seen in Fig 6.5c. Thus for archaeological artefacts it can be concluded that damage will depend on grading of the sediment and relative size of the artefact with relatively small artefacts being more liable to

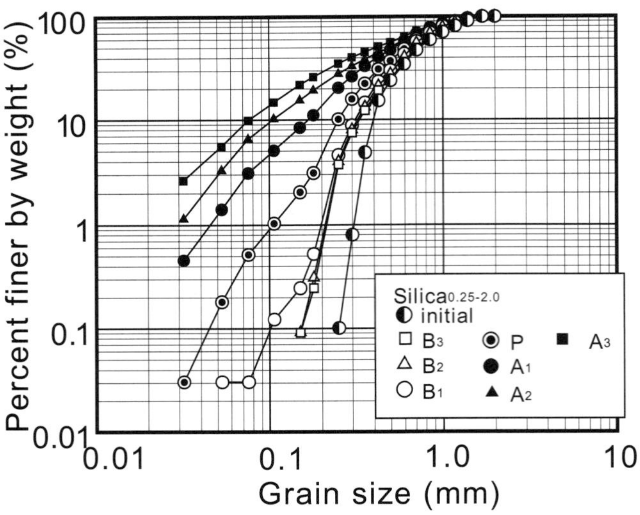

Fig 6.6 Evolution of particle size grading curves for well-graded sand (log-log scale)

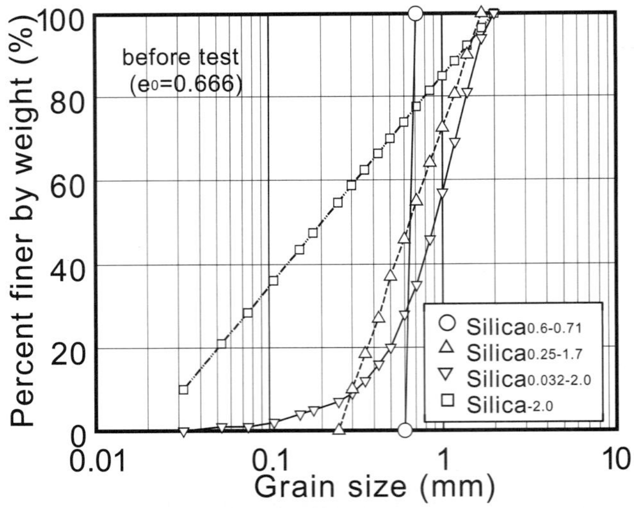

Fig 6.7 Particle size grading curves for modified sands before testing

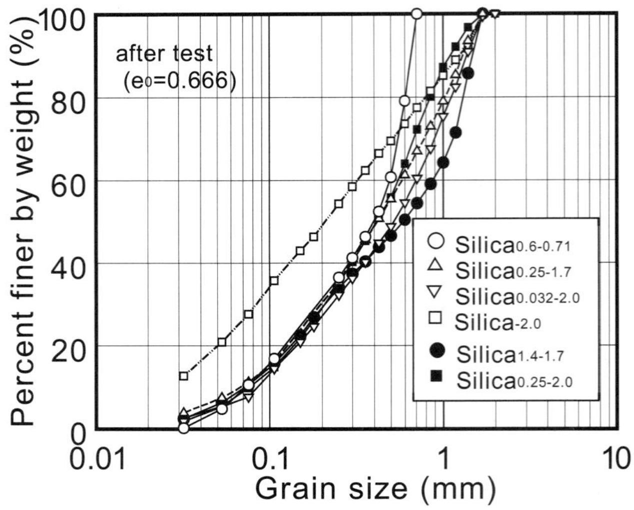

Fig 6.8 Grain size distribution curves for silica sands after testing

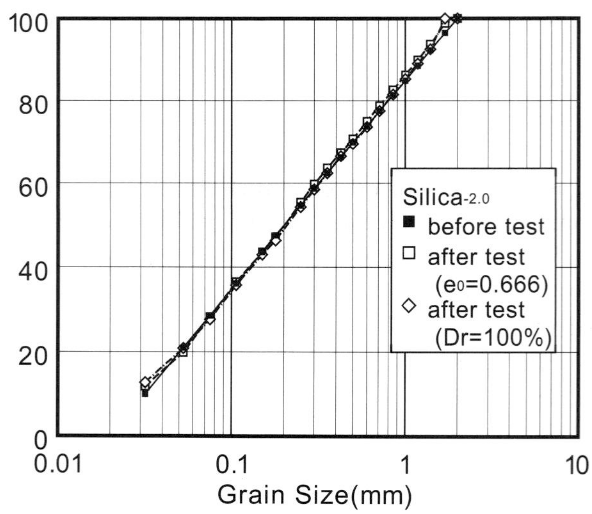

Fig 6.9 Grain size distribution curves for $Silica_{-2.0}$ before and after test

suffer damage.

Fig 6.6 shows the evolution of the particle size distribution for the well-graded sand. Compared with the uniform sand the curves develop more gradually although the final curve for the stress level A_3 is similar to that for the uniform sand at the same stress level.

Particle crushing occurring between the stresses at yield and maximum compression index for $Silica_{0.25\text{-}2.0}$ could be characterised in the case of relatively large particles, by a marked increase in the breaking of asperities and grinding of the surface (Type II and III). For smaller particles on the other hand there was a clear increase in the major splitting of particles (Type IV or higher) rather than the breaking of asperities. The increment during yielding of the frequency of splitting of small particles is higher than that for the breakage of asperities on large particles but is slightly lower than that for the splitting of the particles inserted in the uniform material. As the particle size decreases, so the levels of damage and frequencies for each type of crushing increase.

Effect of initial particle size distribution on particle breakage

Four different particle size distributions of the same silica sand were prepared as shown in Fig 6.7. $Silica_{0.32\text{-}2.0}$ was a well-graded sand; $Silica_{0.6\text{-}0.71}$ was a single size material with particle size 0.6mm-0.71mm; $Silica_{0.25\text{-}1.7}$ had a particle size distribution from 0.25mm to 1.7mm which plotted as a straight line on the semi-logarithmic scale of Fig 6.7; $Silica_{-2.0}$ had an artificially well-graded straight line distribution curve from <0.03mm to 2mm; $Silica_{0.6\text{-}0.71}$ and $Silica_{0.25\text{-}1.7}$ had the same D_{50} value. Each sand was prepared to the same initial void ratio $e_0 = 0.666$ prior to compression testing. Fig 6.8 shows particle size distribution curves after testing for these samples with an initial void ratio $e_0 = 0.666$. The figures also show the particle size distribution curves for $Silica_{1.4\text{-}1.7}$ and $Silica_{0.25\text{-}2.0}$ which were discussed above in the context of the characteristics of individual particle crushing in the matrix. It can be seen that the particle size distribution curves for sizes above 0.4mm are dependent on the original curves but below 0.4mm the curves merge into one single distribution as small particles are produced and dominate the crushing behaviour.

The curve for $Silica_{-2.0}$ stands apart from the others as shown in Fig 6.9 where the grading curves for the $Silica_{-2.0}$ before and after testing seemed to change little for samples prepared either with an initial void ratio of 0.666 or an initial relative density of 100%. It is thought that some particle crushing occurred which would have consisted of the grinding of the surface and the breaking of small asperities. It would appear reasonable to hypothecate a critical particle size distribution where no particle splitting occurs in one-dimensional compression. Below this line the grading curves evolve and merge as they move towards the critical line. The composition and content of small particles would be key determinants of the critical line. McDowell et al (1996) suggested that for simple triangular shaped particles the theoretical evolution of particle size distributions for crushing under high pressures is self-similar or fractal.

Characteristic tensile stress acting on a single particle during a test

In order to investigate the particle crushing behaviour in more detail, it is necessary to examine the average force or stress acting on a single grain. The average force acting on a particle embedded in a sample has been calculated using the following simplified approach (Nakata et al 1999a). The volume of solids V_S in a unit volume of soil is given by:

$$V_s = \frac{1}{1+e} \qquad (1)$$

where e is the void ratio. The number of sand particles per unit volume is then given below by making the simplifying assumption of a soil with single size particles:

$$N = \frac{V_s}{V_{sp}} \qquad (2)$$

where V_{sp} is the volume of a single particle.

In a unit cubic volume containing N particles, any straight line of unit length will cut through $N^{1/3}$ particles. Thus the number of particles per unit cross sectional area can then be calculated as $(N^{1/3})^2$. The force F_{sp} acting on a single particle in the specimen is then given by dividing the normal stress σ by the number of particles across a plane of unit cross section.

$$F_{sp} = \frac{\sigma}{N^{2/3}} \qquad (3)$$

In order to calculate the particle tensile stress, the average volume of a single particle is taken as:

$$V_{sp} = \frac{1}{6}\pi\bar{d}^3 \qquad (4)$$

Where d is the mean particle diameter.

The tensile stress in a sphere subject to point loads F_{sp} across the diameter can be shown to be:

$$\sigma_t = \frac{\chi F_{sp}}{d^2} \qquad (5)$$

where it was suggested by Billam (1967) that $\chi = 2.0/\pi$ and by Hiramatsu and Oka (1966) that $\chi = 0.9$ and d is the diameter of the particle.

The author has therefore defined the tensile stress in terms of a characteristic stress, σ_{sp}, following Jaeger (1967), Lee (1992), Shipway and Hutchings (1993), McDowell et al (1996) and McDowell and Bolton (1998).

$$\sigma_{sp} = \frac{F_{sp}}{d^2} \qquad (6)$$

Substituting equations (1), (2) and (4) into (3) gives the force on a single particle embedded in a soil matrix:

$$F_{sp} = \sigma\left(\sqrt[3]{\frac{(1+e)\pi}{6}}\right)^2 \bar{d}^2 \qquad (7)$$

and the corresponding characteristic tensile stress

$$\sigma_{sp} = \frac{F_{sp}}{\bar{d}^2} = \sigma\left(\sqrt[3]{\frac{(1+e)\pi}{6}}\right)^2 \qquad (8)$$

It can be seen in equation (8) that for a given set of applied stresses the average characteristic stress is not dependent on the particle size but instead is a function of the void ratio.

Single particle crushing strength

Fig 6.10 shows a schematic diagram of the apparatus used for particle crushing (Nakata et al 1999a). The test was carried out by placing a particle between hardened platens and then

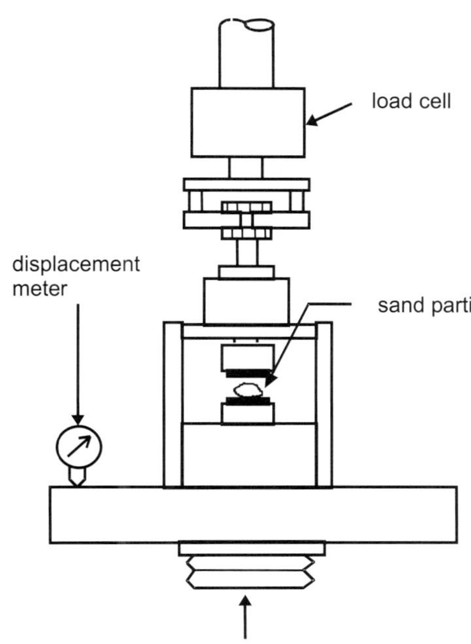

Fig 6.10 Schematic diagram of particle crushing apparatus

moving the lower platen at a constant rate of displacement to crush the particle. The diameter of the removable hardened platens was 30mm. The load measuring capacity was 500N with a resolution of 0.01N. A displacement rate of 0.1mm/minute was used. During the tests force and displacement were measured.

Fig 11 shows a typical sawtooth load displacement relationship for a silica particle. A consistent definition of the force required to crush a particle is not straightforward. Two levels are marked on the figure. F_c is the first peak and represents the breakage of an asperity while F_f is the maximum peak loading where major splitting of the particle has occurred. A characteristic particle tensile splitting strength σ_f was defined similar to equation (8):

$$\sigma_f = \frac{F_f}{d_0^2} \qquad (9)$$

Where d_0 was the distance between the platens at the start of the particle crushing test and F_f was the load at failure as defined in Fig 6.11.

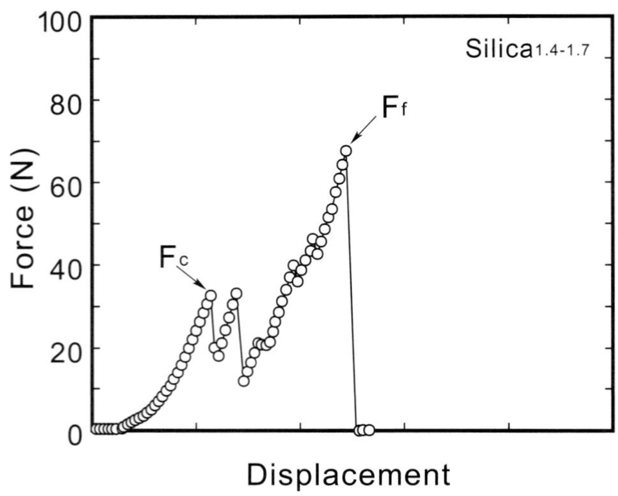

Fig 6.11 Typical load; displacement relationship (Silica 1.4–1.7mm)

Single particle crushing stress related to matrix characteristic tensile stress

Fig 6.12 shows the relationship between the mean single particle strength σ_{fmean} and the characteristic tensile stress $\sigma_{spCcmax}$ of a particle embedded in the granular matrix defined as:

$$\sigma_{spCc\,max} = \frac{F_{sp}}{\bar{d}_0^2} = \sigma_{vCc\,max}\left(\sqrt[3]{\frac{(1+e)\pi}{6}}\right)^2 \qquad (10)$$

The data in the figure is obtained from tests on a number of different materials. These included Silica, Toyoura and Aio quartzitic sands, Masado a decomposed granite, glass beads (G.B.) and angular glass (A,G.) The dashed line in the figure is where the mean single particle strength would be equal to the characteristic tensile stress of a particle embedded in the soil matrix. It can be seen in this figure that the mean particle strength is greater than the characteristic tensile stress calculated for a particle embedded in the granular matrix. However the logarithm of the characteristic tensile stress at maximum compression index, of a particle embedded in a granular matrix, can be evaluated from the logarithm of the mean single particle strength multiplied by a proportional factor.

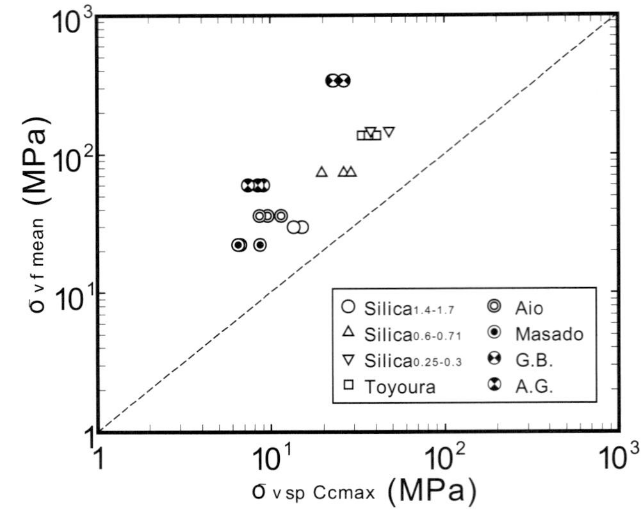

Fig 6.12 Relationship between mean single particle characteristic strength and characteristic tensile stress at maximum compression index

Conclusions

High pressure one dimensional compression tests on silica sand samples seeded with marked particles were carried out in order to examine the statistics of individual particle crushing taking into account particle size and overall grading. Five levels of particle damage were defined and obtained from microscopic observations of the particles before and after testing. A statistical analysis was carried out on the data for the observed levels of damage to investigate their frequency variation with increasing applied stress. Additional one dimensional compression tests were conducted on samples with a range of particle size grading curves. The changes in these curves were monitored in order to examine the effects of initial void ratio and grain size distribution on particle crushing.

From observations of the coloured particles inserted in an almost single size uniformly graded sand it was found that major splitting of particles occurred mainly between the yield stress p_y and the point at which the compression index C_c reached a maximum. It was observed that for this sand, after yielding, 50% of the particles were damaged. As the sand was further compressed, C_c became constant and levels of particle damage became steady. 100% of the seeded particles underwent some kind of crushing and about 60% of the particles suffered major grain splitting. On the other hand for the well-graded silica sand, there was a distinct difference in the crushing behaviour depending on the particle size. As the particle size decreased so the frequency of occurrence of higher damage levels increased. Thus for archaeological artefacts it can be concluded that damage will depend on grading of the sediment and relative size of the artefact with relatively small artefacts being more liable to suffer damage.

Based on the fact that the particle size distribution curves evolved towards a common line and that an initially well graded particle size distribution lying above this line suffered little or no particle crushing, it would appear reasonable to hypothesise a critical particle size distribution for which no particle crushing occurs in one dimensional compression. Below this line the grading curves evolve and merge as they move towards the critical line. The single particle strengths were compared with the characteristic tensile stress for a particle embedded in the soil matrix. The logarithm of the characteristic tensile stress at maximum compression index of a particle embedded in a granular matrix could be evaluated from the logarithm of the single particle strength multiplied by a proportional factor.

Bibliography

Billam, J, 1972 Some aspects of the behaviour of granular materials at high pressures, *Proc Roscoe Memorial Symposium*, 69–80, Cambridge

Braddick, H J J, 1963 *The physics of experimental method*, London

De Souza, J M, 1958 Compressibility of sand at high pressure, unpub MS thesis Massachusetts Institute of Technology, 63–4, Cambridge, Massachusetts,

Hagerty, M M, Hite, D R, Ullrich, C R, and Hagerty, D J, 1993 One-dimensional high pressure compression of granular media, *J Geotechnical Engineering* ASCE 119(1), 1–18

Hendron, A J, 1963 The behavior of sand in one-dimensional compression, unpub PhD thesis, 50–89, Univ Illinois, Illinois,

Hiramatsu, Y and Oka, Y, 1966 Determination of the tensile strength of rock by a compression test of an irregular test piece, *Int J Rock Mechanics Mining Sci* 3, 89–99

Jaeger, J C, 1967 Failure of rocks under tensile conditions, *Int J Rock Mechanics Mining Sci* 4, 219–27

Lee, D M, 1992 The angle of friction of granular fills, unpub PhD dissertation, Univ Cambridge

McDowell, G R, Bolton, M D, and Robertson, D, 1996 The fractal crushing of granular materials, *J Mechanics Physics Solids* 44(12), 2079–102

McDowell, G R, and Bolton, M D, 1998 On the micromechanics of crushable aggregates, *Geotechnique* 48(5), 667–79

Miura, N, and Yamanouchi, T, 1977 The effect of the particle-crushing on the shear characteristics of a sand, *Proc Japan Soc Civil Engineering* 260, 109–18

Miura, N, Murata, H, and Yasufuku, N, 1984 Stress-strain characteristics of sand in particle-crushing region, *Soils Foundations* 24(1), 77–89

Nakata, Y, Hyde, A F L, Hyodo, M, and Murata, H, 1999a A probabilistic approach to sand particle crushing in the triaxial test, *Geotechnique* 49(5), 567–83

Nakata, Y, Hyde, A. F L, Kato, Hyodo, M, and Murata, H, 1999b Single particle crushing and the mechanical behaviour of sand, IS Torino, *Second international seminar on pre-failure deformation characteristics of geomaterials* 1(1), 221–8

Pestana, J M, and Whittle, A J, 1995 Compression model of cohesionless soils, *Geotechnique* 45(4), 611–31

Shipway, P H, and Hutchings, I M, 1993 Fracture of brittle spheres under compression and impact loading, I. Elastic stress distributions, *Phil Mag* 67(6), 1389–1404

Terzaghi, K, and Peck, R B, 1948 *Soil mechanics in engineering practice*, New York

Yamamuro, J A, and Lade, P V, 1996 One-dimensional compression of sands at high presuures, *J Geotechnical Engineering* ASCE 122(2), 147–54

Yasufuku, N, Murata, H, and Hyodo, M, 1991 Yield characteristics of anisotropically consolidated sand under low and high stresses, *Soils Foundations* 31(1), 95–109

7

Making sense of diagenesis? The development of integrated laboratory and geochemical modelling strategies in studies of diagenetic change

LYN WILSON and MARK POLLARD

Lyn Wilson
Research Technician

Mark Pollard
Professor of Archaeological Science

Department of Archaeological Sciences
University of Bradford
Bradford BD7 1DP, UK

Introduction

The survival or otherwise of archaeological material in the burial environment must be understood in order to allow systematic interpretation of the archaeological record. Studies of post-depositional change, or diagenesis, are therefore essential in determining optimum mitigation strategies for archaeology at risk. The dynamic nature of terrestrial burial environments and the complexity of aqueous fluid-mediated transformation mechanisms have hindered diagenetic research to date. The novel use of combined experimental and theoretical geochemical modelling strategies has the potential to make progress towards a quantified understanding of the diagenetic fate of inorganic and bioinorganic archaeological materials. By envisaging a series of simplifications, aspects of behaviour in the natural environment can be simulated. Only geochemical reactions are considered here.

Experimentation and geochemical modelling: making sense of diagenesis

Our approach stems from the belief that a holistic understanding of the geochemical aspects of diagenesis is achievable by integrating laboratory-derived data and geochemical modelling using *The Geochemist's Workbench* 3.2™ (Bethke 1994). Geochemical models can simulate many reactions, but from an archaeological perspective, reaction-path modelling of water–rock (ie minerals) interactions has most value. Geochemical modelling is therefore suitable for visualising diagenesis of archaeological materials, with a mineral composition, in aqueous burial environments, at an assumed local thermodynamic equilibrium (van der Lee and De Windt 2001). Fig 7.1 outlines the philosophy of our approach.

Geochemical modelling: example

Replication of published experimental studies using *The Geochemist's Workbench*™ (React programme), allows a comparison of predicted and observed diagenesis.

Lead uptake by hydroxyapatite

Hydroxyapatite (HAP) was exposed to a lead nitrate solution at pH 6 and at a constant temperature of 22°C. This resulted in simultaneous dissolution of HAP and precipitation of Hydroxypyromorphite ($Pb_5(PO_4)_3OH$) (HPY). Geochemical modelling predictions confirm experimental results (Fig 7.2). This example indicates that the digenetic uptake of lead from groundwater into bone can be quantitatively modelled (Lower et al 1998).

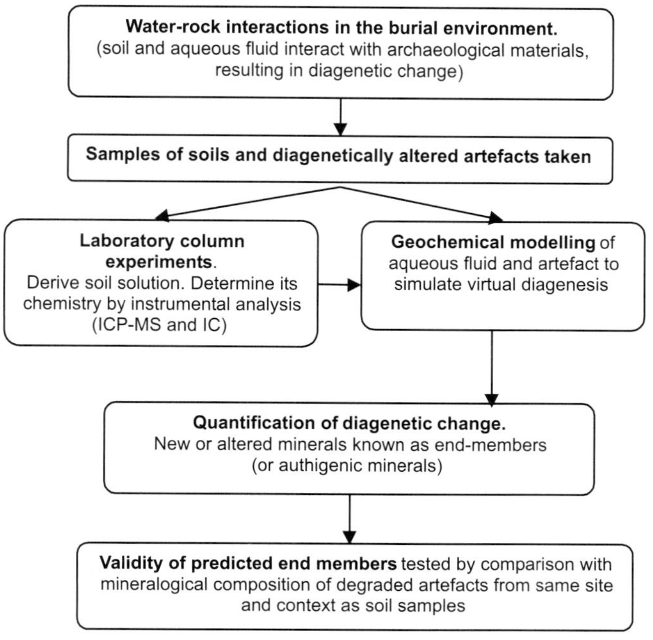

Fig 7.1 Philosophy of terrestrial burial environment diagenetic investigations; the laboratory methodology involves the extraction of soil solution from field-moist, unsterilised soils retained in small column microcosms, following established column displacement procedures (Wolt 1994)

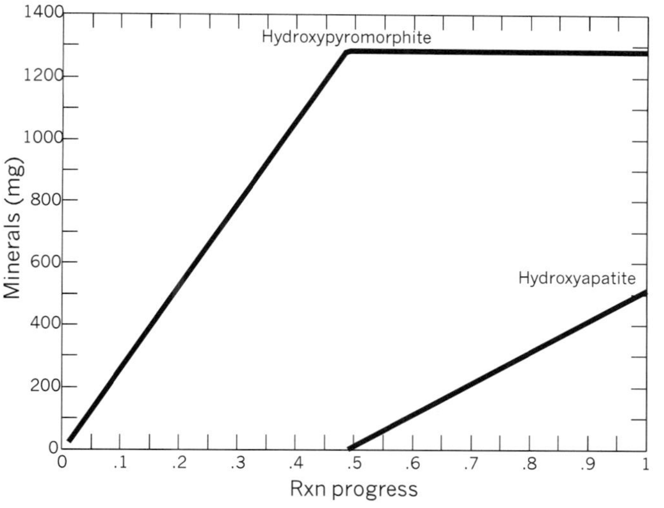

Fig 7.2 Lead uptake by hydroxyapatite, alteration in bone mineralogy; experimental and modelling results indicate that aqueous lead can be sequestered via uptake by HAP and its subsequent transformation to HPY (the x-axis follows the reaction progress to theoretical thermodynamic completion)

Concluding points

Geochemical modelling is a powerful instrument and compatible to studies of archaeological diagenesis in terrestrial and marine burial environments. The integrated experimental and theoretical modelling approach advocated here is feasible and necessary for a quantitative analysis of diagenesis. This research is still at a preliminary stage however, and there are many steps necessary to enable construction of valid and useful models.

Bibliography

Bethke, C, 1994 *The Geochemist's Workbench™*, v 3.2 released 2001, Univ Illinois, Urbana-Champaign

Lee, J van der, and Windt, L de, 2001 Present state and future directions of modelling of geochemistry in hydrogeological systems, *J Contaminant Hydrology* 47, 265–82

Lower, S K, Maurice, P A, and Traina, S J, 1998 Simultaneous dissolution of hydroxyapatite and precipitation of hydroxypyromorphite: direct evidence of homogeneous nucleation, *Geochimica et Cosmochimica Acta* 62, 1773–80

Wolt, J, 1994 *Soil solution chemistry applications to environmental science and agriculture*, New York

8

The response of archaeological sediments and artefacts to imposed stress regimes as a consequence of past, present and future anthropogenic activity

E J SIDELL, T HIGUCHI, R J ALLISON and A J LONG

E J Sidell
Senior Research Fellow
Institute of Archaeology
University College London/English Heritage
31–34 Gordon Square, London WC1H 0PY, UK

T Higuchi
Senior Research Associate in Geography

R J Allison
Professor of Geography

A J Long
Reader in Geography

Department of Geography
University of Durham
Science Site, South Road
Durham DH1 3LE, UK

Abstract

This paper presents the aims, objectives and initial results of a current National Environment Research Council (NERC) funded project (Urban Regeneration and Environment (URGENT) theme) directly geared towards establishing suitable geotechnical parameters for construction work over archaeological sites designated for preservation *in situ*.

This project is a response to a perceived serious deficiency in the archaeological community's understanding of how archaeological sediments and the artefacts contained therein have and will respond to a range of loading and unloading scenarios. Curatorial archaeologists are currently in the invidious position of attempting to select sites for preservation without this knowledge. The additional complexities of fluctuating hydrology and soil chemistry within archaeological deposits makes predicting accurately which sites will survive in perpetuity an almost impossible task (see Banwart nd; Pollard nd). This project aims to generate comprehensive data-sets on the fundamental geotechnical properties of archaeological sediments. For instance, the conditions at which each sediment type and artefact class combination will deform, fail or obliterate the stratigraphy and artefacts. This should assist the curatorial task of marrying development schemes with the preservation of archaeological sites.

The paper begins with a description of how the project is being undertaken using a combination of field investigation, laboratory geotechnical analyses and modelling developed through critical state soil mechanics (see Atkinson and Bransby 1978). The field investigation to date has concentrated on archaeological sites in London, selecting those in a range of sedimentary environments including the dry gravel terraces and the floodplain. Laboratory testing has thus far concentrated on examining the sampled sediment using conventional methods and then using stress-path testing to examine the geotechnical characteristics of key materials. Vibration patterns have been collected from a range of piling solutions and plant movement employed on archaeological and non-archaeological sites. Results from the monotonic compression tests and the initial interpretations to date will be presented. As yet, no results are available for vibration tests.

Introduction

The project was conceived in 1999 during the call for applications to the National Environment Research Council (NERC) Urban Regeneration and Environment (URGENT) theme, specifically targeted at aspects of urban regeneration. The research proposal directly addresses the aim of the URGENT programme by using science-based archaeology to focus on an important urban regeneration issue at the heart of the country's largest conurbation. The research encompasses a range of approaches highlighted by URGENT including: diagnosis and characterisation of a specific problem in relation to the natural background; evaluation of risk in order to understand the underlying patterns and processes; development of strategies for remediation and management within the urban environment.

A number of the scientific aspects grouped under the umbrella term preservation *in situ* were considered to be poorly understood, particularly with respect to sediment diagenesis over the archaeological record. This was notwithstanding the data presented at the 1996 Preserving archaeological remains *in situ* conference, particularly the papers by Shilston and Fletcher (nd) and Nixon (nd). Insufficient information was available regarding exactly how sediments react to the varying loading scenarios imposed upon them in the various schemes designed to ensure preservation of sites selected for burial. Yet, this information is fundamental to establishing whether sites will or will not be preserved – a factor which is particularly important within a culture which currently favours preservation of at least nationally important sites (*sensu* PPG16, DoE 1990). Furthermore, those making the decisions whether to preserve or excavate tend to be individuals lacking the technical expertise to predict how a site will respond under the scenarios presented for a site's future, yet they have the responsibility of deciding when and how to preserve. Often these decisions have to be made rapidly and with incomplete data-sets.

The current project was conceived in order to gather basic data on how a series of archaeological sediments respond to a range of loading scenarios. In addition, it was decided to include a wide range of artefact types to measure their responses to loading within a range of sediment types. The overarching objective is to provide a guide to how common sediment types, ie silt, clay, sand, gravel and mixtures of these, would react on an increasing scale of imposed stress and more importantly, at which point they would deform, and how a range of artefacts/biological material such as glass, metal, wood, bone and ceramics respond to the same levels of imposed stress bedded within the various sediment types.

In addition to the examination of loading, it was decided to include an examination of vibration properties, by establishing how different sediments respond to a range of piling methods and varying frequencies of vibration patterns. Ground vibration is potentially destructive to buried archaeological deposits. This can occur in a number of ways including plant traversing sites, but piling is the key factor. The effects of piling have been documented previously (Nixon nd; Arup 1992; Fearon O'Neill Rooney 1996; see paper 30).

London was selected as the subject area in view of its status as a city where intense amounts of costly development take place each year in an area of complex and extensive archaeology. London also has a good track record of undertaking preservation *in situ*, ie the Rose Theatre (English Heritage 1994; Fig 8.1, site 1), the Globe Theatre (Hughes

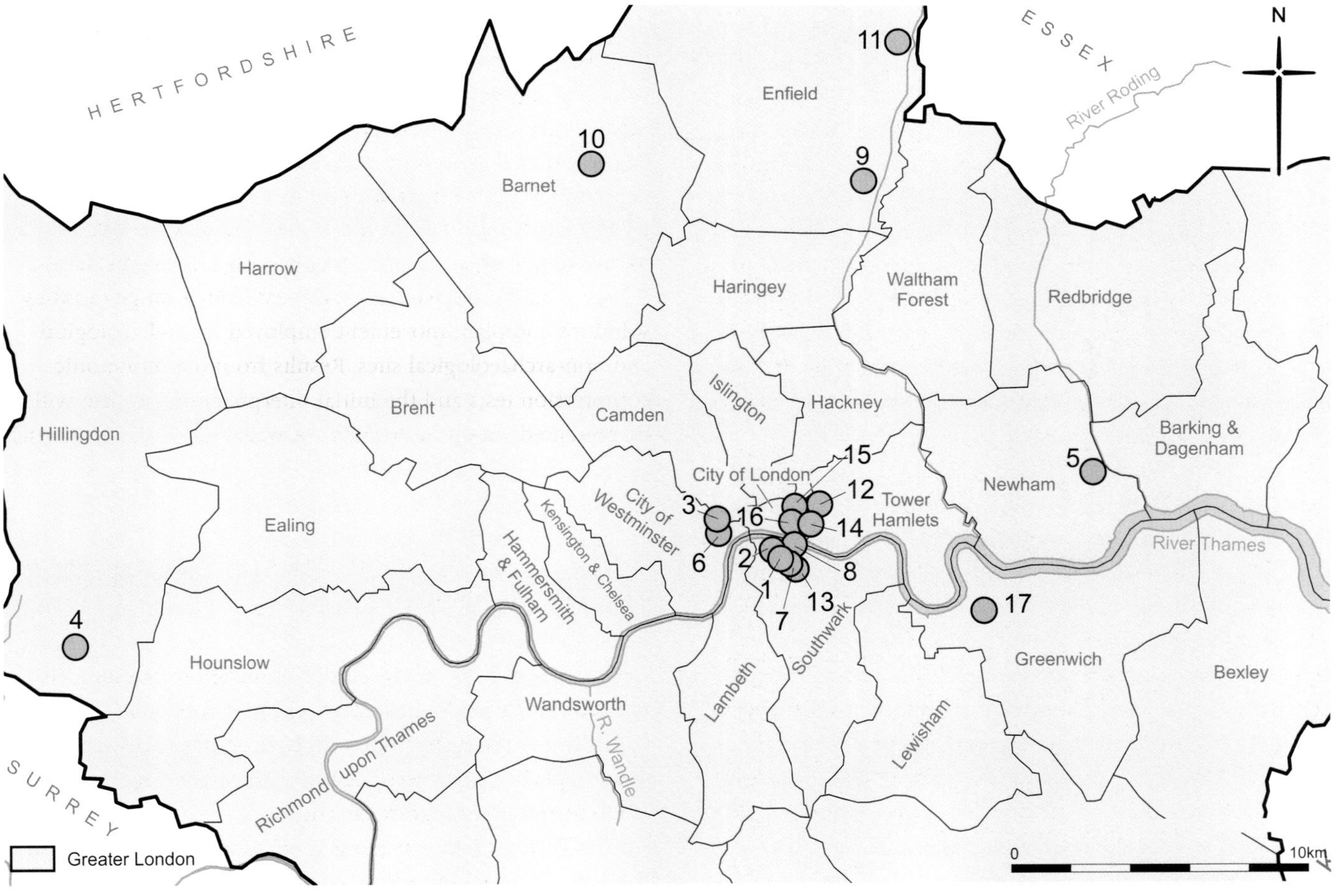

Fig 8.1 Location of sites mentioned (© Crown copyright. All rights reserved. Corporation of London 087254-2004)

1990; Hunting Land and Environment 1996; Fig 8.1, site 2) and a variety of other sites (Goodburn-Brown 1996).

Fieldwork

A series of sites have been visited in order to view the range of sediment types within London; collect bulk samples; collect undisturbed core samples and to collect artefacts. Samples have been collected that are representative of a range of geological types within London, but the sites have also been selected for their relevance outside London in terms of equivalent and typical sedimentary environments. The sediment types being targeted are brickearth, both alluvial and estuarine silt-clay, sand, gravel and the ubiquitous homogenous deposits found on sites and described, variously, as plough soil, garden soil, dark earth etc. These are being collected across London, not just in the historic core of the city.

Brickearth (silt/sand) has been collected from Floral Street, Covent Garden (Fig 8.1 site 3), where it is described by the British Geological Survey as Langley Silt (Gibbard 1994 for a discussion of the Langley Silts). Various types of archaeology are associated with the brickearths in London, including Palaeolithic deposits in the Heathrow area (MoLAS 2000, 38 and 48; Fig 8.1, site 4), Neolithic deposits in East London (Fig 8.1, site 5) and the Middle Saxon settlement of Lundenwic in Covent Garden (Malcolm and Bowsher 2003; Fig 8.1, site 6). Brickearths are mainly found in southern Britain, (Catt 1977; Catt 1978) and tend to be associated with prehistoric archaeology (Wymer 1999).

Estuarine silt-clay has been collected from Thrale Street, Southwark (Fig 8.1 site 7), from the relict channels associated with the River Thames and between the Southwark sand eyots (Fig 8.1, site 8). This area has produced significant amounts of Roman and later material within the channels, and also Roman revetments around the edges of the islands. Estuarine and riverine systems around Britain and indeed the world have produced significant quantities of archaeology within these types of silt-clay, examples including the Trent valley (Noort and Ellis 1998), the Severn estuary (Bell and Neumann 1997; Bell et al 2000), and the Gironde estuary, south-west France (Mellalieu et al 2000).

Alluvial silt-clay was sampled at Deephams Sewage Farm in the Lea Valley (Fig 8.1, site 9), thought to be in-channel or channel marginal sedimentation from the relict Lea River. The Lea is one of the Thames tributaries and its freshwater deposits have yielded much significant archaeology, including Mesolithic material (MoLAS 2000, 55). Alluvial situations are commonly found and often are rich in archaeology (Needham and Macklin 1992; Howard et al 2003).

Gravel has been collected from Totteridge Lane, Barnet (Fig 8.1, site 10), but as yet this has not been examined due to the problems in examining loosely consolidated sediment, primarily maintaining the integrity of the sample between the on-site sedimentary stack to the triaxial cell without the structure collapsing. It is conceivable that it will not be possible to measure gravel without losing the structure, in which case, either gravel is abandoned as a test material, or it is examined from reconstituted samples, with the appropriate caveats in place. Gravel is a commonly encountered facies type upon which archaeology is found, in many river valleys for example (Howard and Macklin 1999). Such sites are often at risk through mineral extraction and are often much more badly protected through mineral planning legislation (although this has improved of late), though the advent of the Aggregates Levy Sustainability Fund will at least help towards research into sites in these areas (http://www.english-heritage.org.uk/archaeology/aggregateslevy/).

Peat has yet to be collected, although several wetland sites in Southwark and Enfield (Fig 8.1, site 11) have been visited to assess their viability for sampling. Peat will be extremely difficult to test owing to its high moisture content and generally acidic nature which corrodes parts of the laboratory machinery. There are also large problems with overconsolidation of peat recovered from archaeological sites. Some peat can be reduced by up to 90% of the original volume (Allen 2000; Heyworth and Kidson 1982), leading to massive changes in volume and original structure. It is likely that peat will need to be collected from an area where there is limited overburden and it is a relatively recent formation in order to minimise the problems caused by consolidation. Nevertheless, wetland sites are potentially the most susceptible to problems of imposed stress and have previously been considered the most vulnerable sites in view of the problems of dewatering destroying artefacts and leading to deformation through shrinkage (Allen 2000). However, it is this very wetness which makes the archaeology found within peat some of the most outstanding in this country and abroad (Coles and Olivier 2001), for instance, the timber circle at Holme-next-the-sea (paper 20), the range of structures in the Somerset Levels (Coles and Coles 1986) and at Nydam Mose (paper 15). Therefore, peat is considered a priority for data collection within the remit of the project, but will be analysed last owing to its complexity.

Artefacts have been collected from several sites, generally from unstratified or residual contexts so that the damage to the archaeological contexts is minimised. Wood, stone and bone was collected from St Mary Spital, Tower Hamlets (Fig 8.1, site 12); whilst glass, brick, tile, pottery, metal, shell and building stone were collected from Totteridge Lane, Barnet; pottery from Borough High Street, Southwark (Fig 8.1, site 13) and animal bone and pottery from Plantation House, City of London (Fig 8.1, site 14). Also it is hoped that glass from the Guildhall cullett dump (Bateman 2000, 89; Fig 8.1, site 15) can be used in additional to prehistoric timber from east London.

Piling has been monitored in several places to date. The equipment has been tested during construction on the Stockton Campus, University of Durham, examining the vibration patterns caused during vibro-compaction and drop-hammer piling on clays and sands. Piling has since been monitored at Gresham Street, City of London (Fig 8.1, site 16) to monitor large bore continuous flight auger (CFA) piling. This was drilling through Roman deposits (mainly sand and silt) and Pleistocene gravel. This also drilled through a particularly fine mosaic (Fig 8.2), but only one tessera was observed during this monitoring. A third site on gravel in Greenwich (Fig 8.1, site 17) was visited to consider small bore CFA piling, but was not finally measured as the vibration noted during the large bore CFA piling appears to be reasonably negligible. The vibration from these forms of piling now needs to be measured on varying sedimentary types. Vibro-compaction is particularly important, as this is the piling type that has frequently been considered as potentially the most damaging to archaeological deposits, particularly those considered to be fragile, such as wetland sites. Moreover it tends to be favoured by contractors as it is relatively cheap.

Methodology

The samples that have been collected are being subjected to a comprehensive analytical programme to establish the physical properties of each of the materials (British Standards Institution 1990). Standard test procedures are being used to determine particle size distribution, plasticity, bulk and dry density, porosity, permeability and field moisture content. This baseline data is vital to understanding the mechanical characteristics of the range of sediments collected. A triaxial stress-path cell with sample diameter of 38mm has been purpose-built and employed to determine the consolidation and monotonic compression characteristics of each sediment type (except gravels, see above). The first stage of the test programme has been investigating isotropic conditions, using stress boundary conditions relevant to the site as determined by the field archaeology and laboratory physical property tests. The second part of the laboratory geotechnical test programme has been developed to take full account of site characteristics relevant to this particular study and it is for this reason that the computer-controlled triaxial cyclic loading facility has been constructed for the project. By varying vertical and horizontal stresses, a broad range of natural (ie compression through a sediment's own weight or compression through subsequent overburden) and engineering-induced site conditions can be reproduced in the laboratory. Both the triaxial stress-path cell and cyclic loading facility were coupled with the ability to measure the pore water pressure generated within soil samples. This means that a range of soil pore water conditions can be recreated, ie the degree of saturation. This methodological approach has been designed to provide a comprehensive picture of the archaeological stratigraphy of the selected field site; a detailed knowledge of the physical properties of the different sediments found at the site; and an understanding of how the sediments behave, relative to their specific point in the overall stratigraphy, under applied stresses that have been determined by site conditions.

Fig 8.2 *Mosaic with secant pile wall at Gresham Street, City of London*
© MoLAS

These data will enable rigorously constrained predictions of how the archaeological sediments will behave under loading scenarios based on individual site construction plans and engineering solutions.

Vibration patterns caused by piling have been measured using a Magus Vibroanalyser (Figs 8.3–8.4), which picks up ground movement in three directions: vertically, horizontally and transversely, as particle velocity output. The recorded vibration patterns are downloaded to the computer, which drives the computer-controlled cyclic loading facility. The combination of cell pressure and axial load acting on the sample to simulate the ground shaking can thus replicate the stress conditions imposed during the various forms of piling measured on site.

As yet, no artefacts have been tested owing to need to design and construct attachments (on-artefact strain gauges and mid height pore water pressure transducer) for the cyclic loading cell to facilitate this aspect of the analysis. The proposed method is to place an artefact within a sediment matrix and consolidate this in the triaxial cell and then subject the sample to a range of stresses imposed from above. Two strain gauges will be attached to the artefact in order to measure the impact of the stress in both vertical and horizontal planes (Fig 8.5).

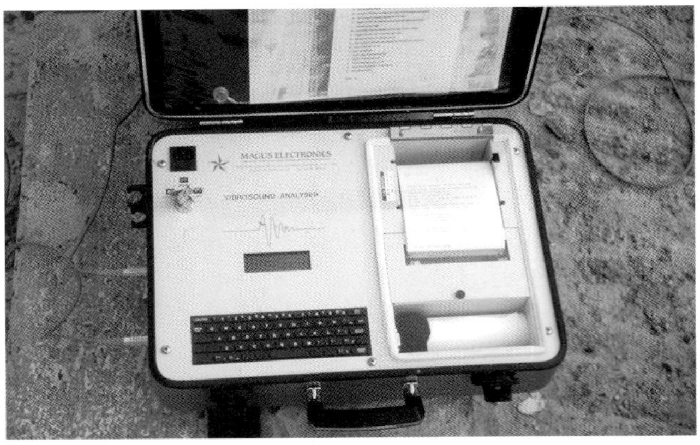

Fig 8.3 *Vibroanalyser on site at the Stockton Campus*

Fig 8.4 *Vibroanalyser in use on site at the Stockton Campus*

Results to date

Vibration patterns

Fig 8.6 shows three vibration patterns obtained from the auger piling (a), vibrodriver (b) and drop hammer (c). All of the patterns were measured 1.0m away (1.0m stand-off) from the piles. The vibration pattern from the vibrodriver recorded the greatest particle velocity and it was felt through field observation that this piling method shakes the ground most. It can be clearly seen that ground shaking due to the auger piling was much less than the other two and seems the ideal method to minimise site disturbance. However of the three methods recorded to date it is the most expensive. Using the results, it was decided to investigate the response of soils and artefacts to stresses imposed by the vibrodriver and drop hammer using the computer-controlled cyclic loading facility.

Soil laboratory tests

Limited isotropic consolidation tests have been carried out using the stress-path cell in order to investigate the compressibility of representative sediment from sites in London: a silt from Thrale Street and silt-sand from Covent Garden. The silt sample was prepared by trimming using a soil lathe, while the silt-sand sample was remoulded keeping

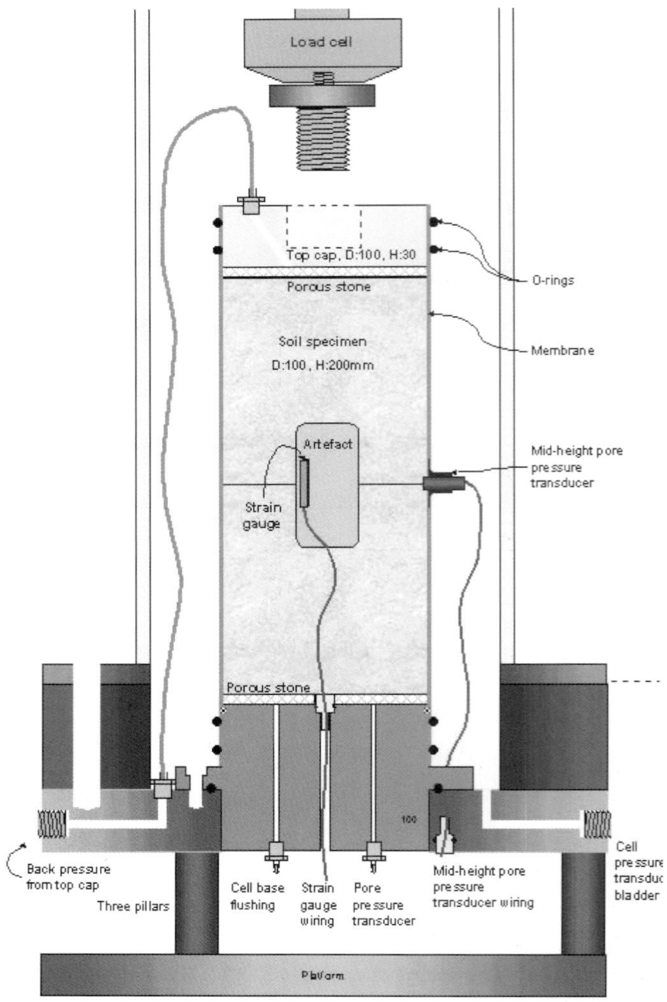

Fig 8.5 *Schematic diagram of modified soil triaxial testing with artefact inclusion*

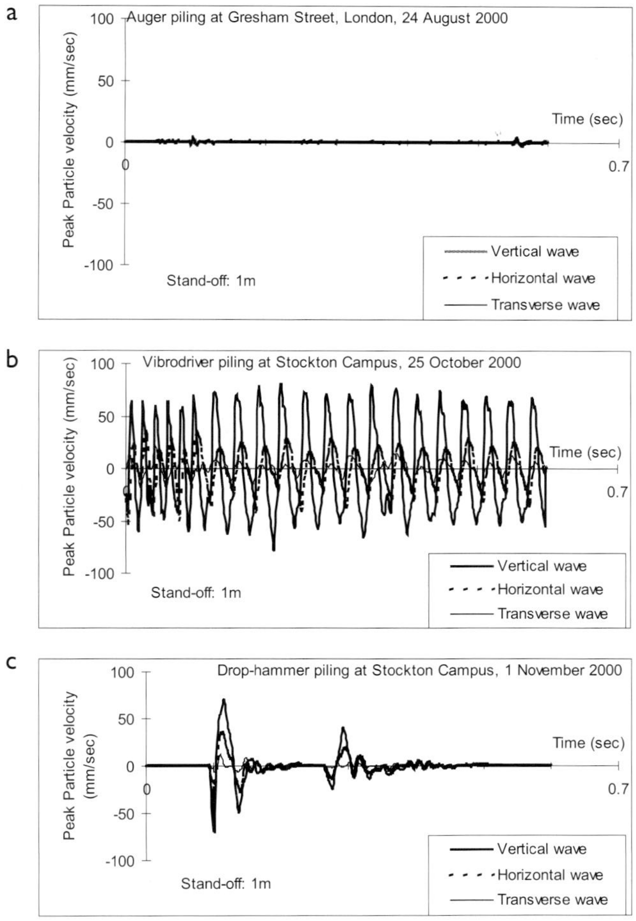

Fig 8.6 Vibration patterns: a – auger piling; b – vibrodriver; c – drop-hammer piling

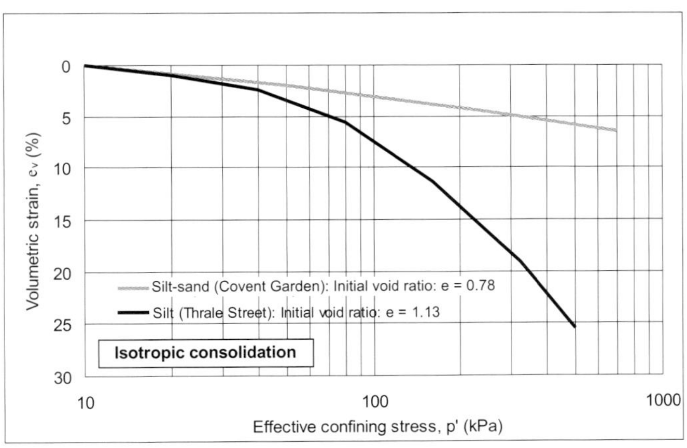

Fig 8.7 Isotropic consolidation

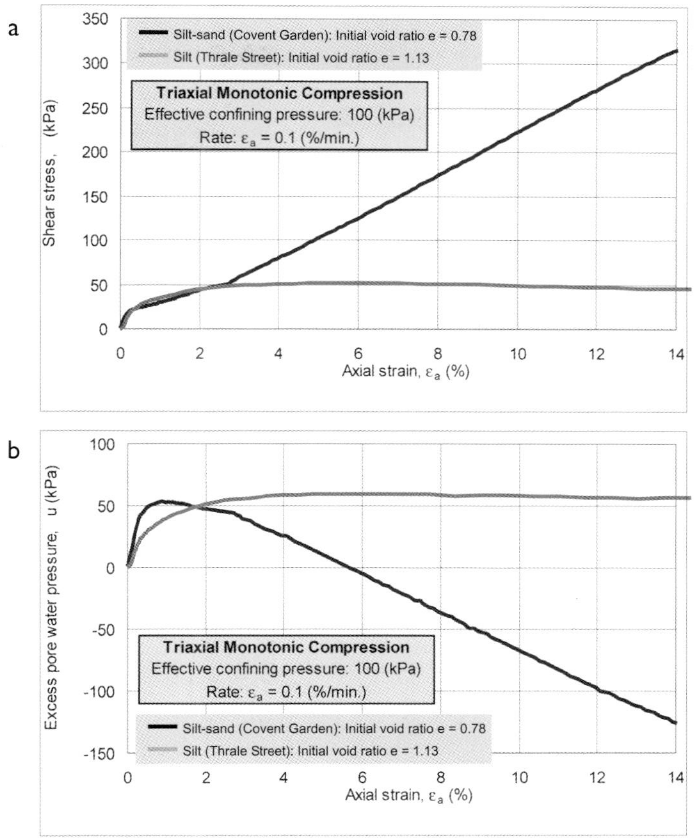

Fig 8.8 Monotonic compression

the same dry density as when collected. Fig 8.7 shows the results from the isotropic consolidation tests and it can be seen that the sand sample was found to be much more incompressible.

Monotonic compression tests were conducted for the same soils using the stress-path cell. The samples were prepared in the same way as described above and subjected to undrained compression at a constant rate of axial strain $\varepsilon_a = 0.1\%/\text{min}$. The test results presented in Fig 8.8a show the stress-strain curves for the two materials. The silt-sand sample exhibited a continuous strain hardening behaviour and did not fail within a strain level of $\varepsilon_a < 10\%$, on the other hand, the silt sample failed at $\varepsilon_a = 2\%$. The pore water pressure response shown in Fig 8.8b gives a different view for what happens in soil under imposed stress. In short, the sample of weaker silt in shear strength had a much greater development in its excess pore water pressure Δu and thus a greater reduction in the effective stress. In terms of the basic mechanical properties of the soils, the granular soil was considered to be a more preferable material to be used for reburial by engineers since it was less compressible and more resistant to applied stresses. However, it was thought that the stresses on the granular materials will be transferred to artefacts in soil more directly and therefore the potential to damage artefacts is higher. These quantitatively clarified soil data will be used as a framework for the later test programme.

Summary

This is an interim progress report from a relatively early stage in the project. Nevertheless, the results indicate there is indeed potential to produce robust and high quality data that will provide vital guidance on how the range of sediment types, structures and artefact assemblages which make up archaeological sites will respond to the types of stress imposed by preserving them under new developments.

The end product is envisaged as a data-set with details of how sites that include archaeological remains respond to a range of loading and vibration scenarios. Which will be of direct relevance to curators, planning authorities, consultancy companies, archaeological contractors, civil engineers and developers. Ideally, it will provide the basis of establishing good practice in the context of preserving sites and following Government regulatory guidelines, to strike the right balance between preservation which can be very costly and destruction by a development for which there may be a well established and justified need. The work will, for the first time, provide a rigorous, quantifiable framework for future excavations. It will complement and support PPG16 (DoE 1990) which has the weakness of being slightly ambiguous with reference to the criteria for preservation *in situ*.

Although much of the data will be site specific as archaeological deposits vary considerably in their composition and character, the overall effects of loading on archaeological remains will be established, and a framework set for undertaking similar work at other sites where important remains are discovered as a consequence of development and construction. The logical conclusion will be to develop this framework into a simple brief or specification, whereby a level of integrated geoarchaeological and geotechnical detail suited to a specific site and development is routinely sought by the planning authority. Much of the subjectivity presently surrounding archaeological resource management decisions would thus be eliminated, giving a more structured and consistent framework for preserving at least a representative sample of what is an increasingly fragile and finite resource.

Acknowledgements

This research is funded by NERC for awarding a grant to this project (GST022623). We thank the many archaeological organisations, field archaeologists and curators who have contributed by providing sites, sediments and artefacts for sampling as well as providing valuable suggestions for our sampling programme. We are also grateful to Dr Alan Selby of the School of Engineering, University of Durham, for assisting with equipment for measuring vibration and the technicians in the Department of Geography, University of Durham, for their assistance with the project.

Bibliography

Allen, J R L, 2000 Holocene coastal lowlands in northwest Europe: autocompaction and the uncertain ground, in *Coastal and estuarine environments: sedimentology, geomorphology and geoarchaeology* (eds K Pye and J R L Allen), The Geological Soc Spec Pap 175, 239–52, London

Arup, 1992 *The York development and archaeology study*, Ove Arup and Partners and York Univ in association with Bernard Thorpe, York

Atkinson, J H, and Bransby, P L, 1978 *The mechanics of soils: an introduction to critical state soil mechanics*, Maidenhead

Banwart, S A, nd [1998] Groundwater geochemistry in the burial environment, in *Preserving archaeological remains in situ: proceedings of the conference of 1st–3rd April 1996* (eds M Corfield, P Hinton, T Nixon and M Pollard), 66–72, London

Bateman, N 2000 *Gladiators at the Guildhall: the story of London's Roman amphitheatre and medieval guildhall*, London

Bell, M, and Neumann, H, 1997 Prehistoric intertidal archaeology and environments in the Severn estuary, Wales, *World Archaeol* 29(1), 95–113

Bell, M G, Caseldine, A, and Neuman, H, 2000 *Prehistoric intertidal archaeology in the Welsh Severn estuary*, CBA Res Rep 120, York

British Standards Institution, 1990 *Method of test for soils for civil engineering purposes*, London

Catt, J A, 1977 Loess and coversands, in *British Quaternary studies: recent advances* (ed F W Shotton), 222–9, Oxford

Catt, J A, 1978 The contribution of loess to soils in lowland Britain, in *The effect of man on the landscape: the lowland zone* (eds S Limbrey and J G Evans), CBA Res Rep 21, 12–21, London

Coles, J, and Coles, B, 1986 *Sweet track to Glastonbury*, London

Coles, B, and Olivier, A, 2001 *The Heritage management of wetlands in Europe*, EAC Occas Pap 1, Belgium

DoE, 1990 Department of the Environment, *Planning and policy guidance note 16: archaeology and planning*, London

English Heritage, 1994 An environmental evaluation at the Rose Theatre, Southwark, London, unpub English Heritage CAS site rep 441

Fearon O'Neill Rooney, 1996 Feasibility study of archaeological /engineering interfaces in a city environment for Cork Corporation, unpub consultancy rep

Gibbard, P L, 1994 *Pleistocene history of the Lower Thames Valley*, Cambridge

Goodburn-Brown, D, 1996 A review of some conservation procedures for the reburial of archaeological sites in London, in *Archaeological conservation and its consequences* (eds A Roy and P Smith), preprints of the IIC 16th international congress: Copenhagen 26–30 August 1996, London

Heyworth, A, and Kidson, C, 1982 Sea-level changes in southwest England and in Wales, *Proc Geol Assoc* 93(1), 91–112

Howard, A J, and Macklin, M G, 1999 A generic geomorphological approach to archaeological interpretation and prospection in British river valleys: a guide for archaeologists investigating Holocene landscapes, *Antiquity* 73, 527–41

Howard, A J, Macklin, M G, and Passmore, D G, 2003 *The alluvial archaeology of north-west Europe and the Mediterranean*, Balkema, Rotterdam

Hughes, R, 1990 A note on the potential conservation techniques for the Globe Theatre structural remains and associated *in situ* soils, unpub Ove Arup rep

Hunting Land and Environment, 1996 Study of sands for use in the preservation backfill at Anchor Terrace Car Park, 1–15 Southwark Bridge Road, London SE1, unpub, Hunting Technical Services rep no. R1112

Malcolm, G, and Bowsher, D, with Cowie, R, 2003 *Middle Saxon London: excavations at the Royal Opera House 1989–99*, MoLAS Monogr Ser 15, London

Mellalieu, S J, Masse, L, Coquillas, D, Alfonso, S, and Taset, J-P, 2000 Holocene development of the east bank of the Gironde estuary: geoarchaeological investigation of the Saint Ciers-sur-Gironde marsh, in *Coastal and estuarine environments: sedimentology, geomorphology and geoarchaeology* (eds K Pye and J R L Allen), Geol Soc Spec Pap 175, 317–42, London

MoLAS, 2000 *The archaeology of Greater London: an assessment of archaeological evidence for human presence in the area now covered by Greater London*, MoLAS, London

Needham, S, and Macklin, M G, 1992 *Alluvial archaeology in Britain*, Oxbow Monogr 27, Oxford

Nixon, T, nd [1998] Practically preserved: observations on the impact of construction on urban archaeological deposits, in *Preserving archaeological remains* in situ: *proceedings of the conference of 1st–3rd April 1996* (eds M Corfield, P Hinton, T Nixon and M Pollard), 39–46, London

Pollard, M, nd [1998] The chemical nature of the burial environment, in *Preserving archaeological remains* in situ: *proceedings of the conference of 1st–3rd April 1996* (eds M Corfield, P Hinton, T Nixon and M Pollard), 60–5, London

Shilston, D T, and Fletcher, S L, nd [1998] Geotechnical engineering for the *in situ* preservation of archaeological remains, in *Preserving archaeological remains* in situ: *proceedings of the conference of 1st–3rd April 1996* (eds M Corfield, P Hinton, T Nixon and M Pollard), 8–15, London

Wymer, J J, 1999 *The Lower Palaeolithic occupation of Britain*, Salisbury

9

Deterioration of prehistoric bronzes as an indicator of the state of preservation of metal antiquities in the Danish agrarian landscape: preliminary results

HELGE BRINCH MADSEN, JAN HOLME ANDERSEN and LONE BRORSON ANDERSEN

Helge Brinch Madsen
Head of the Agrarian Landscape project

Jan Holme Andersen
Assistant on the Agrarian Landscape project

Lone Brorson Andersen
Assistant on the Agrarian Landscape project

The Royal Danish Academy of Fine Arts
School of Conservation, Esplanaden 34
DK 1263 Copenhagen, Denmark

Abstract

In 1998 the Danish Research Council initiated a project with a view to examining the state of preservation of antiquities in soil. The aim of the project is to formulate a classification system, which can be used by conservation authorities, natural environmentalists and archaeological museums. This paper discusses parameters such as archaeological age, find contexts, soil types and time of excavation as influential in the rate of deterioration of prehistoric bronzes.

Background

The Danish cultural landscape changed dramatically over the last century as Denmark was transformed into a primarily urban society. At the same time we have seen expansion and intensification of agriculture and forestry, which have been restructured for rationalised large-scale production. This has had the result that ever-larger areas of the traditional cultural landscape have been annexed, so that central identity-related and natural values are disappearing. At the same time these cultural environments are attractive to the general population as recreational destinations.

Against this background the Danish Research Council launched a four-year basic research project 'The Agrarian Landscape' in 1998. The aim of this project is to illuminate aspects of the development of the cultural landscape past and present through five main topics. The objective is the drawing up of guidelines for an action plan. This plan should contribute to a balancing of the relationship between areas used for agriculture, forestry, construction and recreational purposes, among other ways by registering whole cultural environments with a view to the protection and preservation of selected parts as nature reserves or through fallowing. The cultural values also include archaeological antiquities above and below the soil, and one of the five main projects of the research programme is an 'Investigation of the state of preservation of antiquities and barrows on agricultural areas'. This project consists of three sub-projects, one of which will be presented here.

State of preservation of antiquities in soil

The state of preservation of antiquities in soil is investigating the deterioration of iron, bronze and amber artefacts in various soil types. The object is to develop methods and systems for assessing the state of deterioration of antiquities and describing the relationships between the degree and rate of deterioration and the environment of their contexts. This is being done in order to draw up a classification system for the use of cultural heritage and environmental bodies and archaeological museums. It is to be achieved through two investigations.

One of the investigations is a systematic review of antiquities excavated in a selected study area over the past 170 years and an assessment of their state of preservation. The information gathered will be used to assess: whether some soil types are more aggressive than others; whether there are find categories that are more threatened than others; and, whether there has been a change in the state of preservation over the last 170 years.

The second investigation involves the development of methods for measuring the corrosive effects of soil types on iron and bronze artefacts, as well as establishing field stations on different soil types for local measurements of corrosivity over a period of years with a view to identifying aggressive soil conditions.

As a study area, the county of South Jutland (North Slesvig) has been chosen, because it is one of the places in the country where almost all the soil types prevalent in Denmark are represented. In addition the region is relatively rich in archaeological finds from the whole of Denmark's prehistory. The antiquities are in the National Museum of Denmark and the main prehistoric museum in North Slesvig County, the Haderslev Museum, which has collections of antiquities that once belonged to the museums in Ribe, Tønder, Aabenraa and Sønderborg.

The data collection phase is now complete, and this article presents an overall picture of the methodology and results of the systematic review of the state of preservation of 2225 prehistoric bronzes from these museums.

Methodology

The state of preservation of the 2225 bronzes has been assessed visually according to a grading scale and their condition has been described. They have all been documented photographically with a digital camera and the pictures have been transferred to CD-ROMs. A comprehensive selection of bronzes exhibiting various corrosion phenomena has been photographed through an optical microscope. In addition a small selection of 15 tin-bronzes has been investigated with the aid of scanning electron microscopy (SEM).

At the same time find information was obtained about the objects from the museum's report files and from conversations with the local archaeologists specialising in the study areas. We have documented the following where possible: object number, object type, provenance, find date, acquisition date, archaeological age, find context, find circumstances and causes, landscape and soil type, places of preservation, description of condition, any conservation and analyses, as well as references to publications in which the finds are mentioned. The information has been registered in a relational database with a view to statistical processing. Similarly, the photo numbers have been entered with a view to localisation and display of the images along with the find information. Wherever the find information is accurate and sufficiently precise the museums have given coordinates for the find-spots (excavation sites). With the aid of these coordinates it is possible to plot the find-spots on a map of the country.

In Denmark chronological and concise descriptions of our ancient monuments and the excavation reports of the museums are gathered in the Cultural-Historical Central Register (DKC) under the auspices of the State Antiquary in the so-called 'parish description'. For each parish the individual artefact is assigned an SB number (*sognebeskrivelse*, parish description and number), by which it can be uniquely identified in conjunction with its site number.

The object numbers and this site SB number form the primary keys in the project database. By correlating the site SB numbers of the find spots and the DKC database it has been possible to retrieve the coordinates of the find spots and incorporate them in the project database. By then correlating these coordinates with the electronic soil type map of the Geological Survey of Denmark and Greenland (GEUS) in a geographical information system (GIS) program, it has further been possible to incorporate the soil types of the find-spots.

The landscape of the study area

North Slesvig forms the southernmost part of the Jutland Peninsula, which lies between the North Sea in the west and the Baltic to the east, and it borders with Germany on the south (Fig 9.1a). The region is some 60 x 60 km, an area of 3938 km^2, making up about 10% of the total area of Denmark. The geological development of the country has been very important to its utilisation and settlement over the centuries. North Slesvig is one of the areas in Denmark that has the highest level of rainfall. For example certain areas in West Jutland receive up to 800mm a year, while the rainfall on Zealand is about 500–600mm. The chloride precipitation follows the same pattern with 80 kg per ha a year in West Jutland, 44 kg per ha in East Jutland and as little as 23 kg per ha on Zealand. Finally, North Slesvig today is one of the most intensively cultivated areas in Europe.

The landscape was mainly created during the last two ice ages and their warmer interglacial as well as the postglacial period. The penultimate ice age, the Saalian Ice Age (195,000 to 127,000 years ago) covered the whole area with huge ice masses, while the Weischelian Ice Age (c 127,000/75,000 to c 15,000 years ago) did extend far enough west for the whole of Jutland to be covered by glaciers. For an extended period the ice margin lay along a north–south line down through

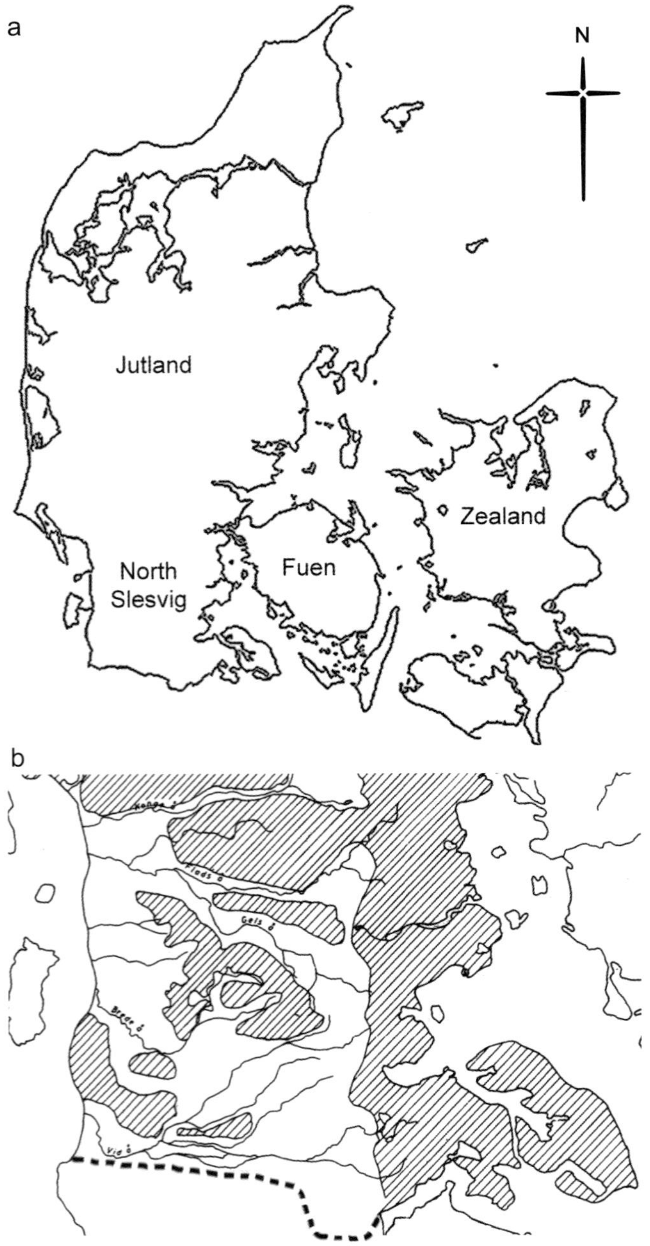

Fig 9.1 *Location of the study area: a – outline map of Denmark; b – map of the study area, North Slesvig, with the moraine landscape shown by the hatched areas (drawing by Haderslev Museum)*

Jutland, and there it has left a striking dividing line where the main watershed also runs (Fig 9.1b).

North Slesvig is characterised by four different landscape types. Towards the east lies the high, hilly landscape that was deposited during the last ice age. The soil consists mainly of boulder clay (till) and sand, some of which has been disturbed or deposited by ice-dammed lakes. The east coast is intersected by subglacial stream trenches through which the meltwater rivers welled up; today several of them form the fjords of the east coast.

Towards the west spreads a landscape formed by an interplay of the material left by the Saalian ice age and meltwater rivers during and after the last ice age. Thus the moraine landscape of the next-to-last ice age has been eroded away by these meltwater rivers, which have left a strip of hill islands running from the midland area towards the west. The soil of these hill islands consists of till and sand, but of a different, more leached-out composition than the young moraine landscape to the east. In the north these hill islands tower up between the meltwater valleys and in the south-west above the relatively low and flat meltwater river plains (Fig 9.1b). From the Saalian ice age, too, come swathes of diluvial sand along the ice front line and around and on the hill islands.

From their sources in the eastern moraine the meltwater rivers first deposited stones and gravel, then sand and, farthest to the west and in the sea, clay particles. In our own time the torrential rivers of earlier ages have dwindled into several widespread rivulet systems spreading from east to west with related wetlands and bogs. The bogs are mainly concentrated in the southern part of North Slesvig, where one also finds larger concentrations of pyrites and bog iron ore in the large river plain. In the western and southern part of the area one also finds scattered deposits of drift sand. Farthest to the west comes the low, flat marsh landscape built up by marine tidal deposits (Meier 2000).

Deterioration

Many of the bronzes have pitting and selective corrosion, and it is the latter type of corrosion we have looked at in this context. The selective corrosion has been documented by SEM analyses of 15 tin-bronzes carried out by Arne Jouttijävi, Heimdal Archaeometry, an engineering graduate. The ratio of copper to tin was investigated on cross sections through each object from the surface into the core. These analyses all show decuprification.

Fig 9.2 shows (left column) the composition of the

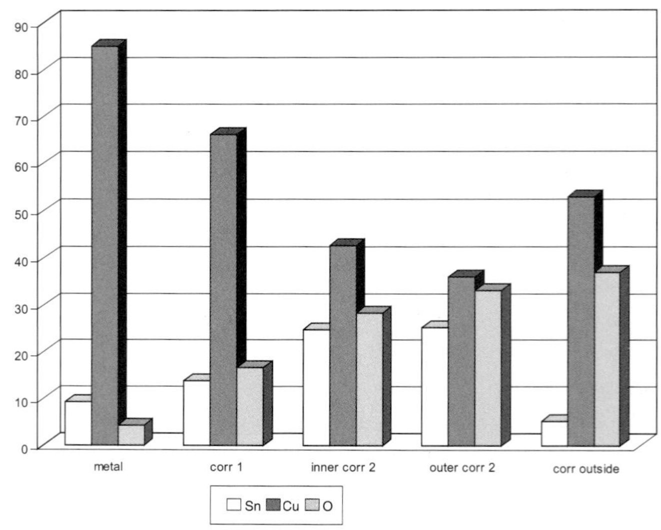

Fig 9.2 *Proportions of tin, copper and oxygen content in a cross section through a prehistoric bronze (HAM 800)*

Fig 9.3 *Cross section through the prehistoric bronze (HAM 800)*

original metal of one of the objects. The graph shows that the composition of the corrosion layer changes from the metallic core out towards the surface. The percentage of copper falls while the relative tin and oxygen content rises. Outside the original surface there is a layer of corrosion products that consists almost exclusively of cupric oxide.

Fig 9.3 shows an SEM photomicrograph of the same cross section of the bronze from Fig 9.2. At the bottom left we see the original metal as light grey. Corrosion Layer 1 can be seen as grey and Corrosion Layer 2 as dark-grey. At the top right we see the outer layer of corrosion products as blackish-grey.

For an earlier investigation of the state of preservation of the antiquities in the Ribe area between 1981 and 1984 the following corrosion scheme was used (Madsen 1991; Madsen 1994, 23–45; Madsen 1998): tarnished surface; original surface covered by a corrosion layer, coloured green; original surface uncovered, corrosion layer washed away, mother-of-peal patina, coloured olive-green, brownish-yellow or grey.

This corrosion scheme could not be used in the present investigation because many of the bronzes had been treated by conservators who had generally removed the corrosion down to the original surface. Instead we chose the following corrosion chart, where the 2225 bronzes were visually evaluated on a scale from 1 to 5. The evaluation is subjective, since no invasive techniques were involved. The gradings are as follows: not tarnished or little tarnished, iridescent red, brown, black or olive-coloured; green layer, thin with mother-of-pearl lustre and original surface; green original surface with underlying corrosion layer; thick corrosion, but with remaining metal in the core; no metal or cuprite left.

It is rare to see a prehistoric bronze with the same degree of corrosion over the whole surface. Often, several grades will occur on the same object. In such cases grades have been assigned according to the 'worst case'. All objects are graded by the same person for consistency. However, a check was made by having a selection of 100 bronzes also graded by a different person. The deviation was less than 5% between the two assessments. In addition the grades of the 15 SEM-analysed bronzes correspond to the visual assessment of the same bronzes.

The assessment scale and procedure used is broadly similar to the one used by a group of Swedish researchers in the large investigation of prehistoric bronzes they conducted in the course of the 1990s (Borg et al 1995). The advantage of using the same scale is that the results can be used to shed light on the preservation conditions in a larger geographic area. Similarities and differences can help us to achieve a better understanding of the many issues that have arisen during the two studies.

Results

We present the preliminary results of the assessment of the 2225 bronzes in the investigation based on the chosen classification system and the find information in relation to four factors: archaeological age (available for 2052 artefacts); find context (available for 1985 artefacts); soil type (available for 1850 artefacts); find date (available for 2067 artefacts).

In most cases some of this information is missing because the bronzes come from private collections of antiquities. For the sake of coherence in the subsequent investigation, the following graphs present only those bronzes where all four factors could be determined for each artefact. This applied to 1785 bronzes. The overall state of preservation grading is evident from Fig 9.4. As can be seen only a mere 3.9% has a little tarnish with Grade 1, 13.7% has a green layer, while Grades 3 and 4 dominate with corrosion layers of respectively 29% and 34.6%. Lastly 18.8% of the bronzes with Grade 5 have only the metallic core remaining.

Although there is correspondence among the subsequent graphs (Figs 9.5–9.8), inasmuch as the data basis is the same, the same is not true of the individual subgroups. This

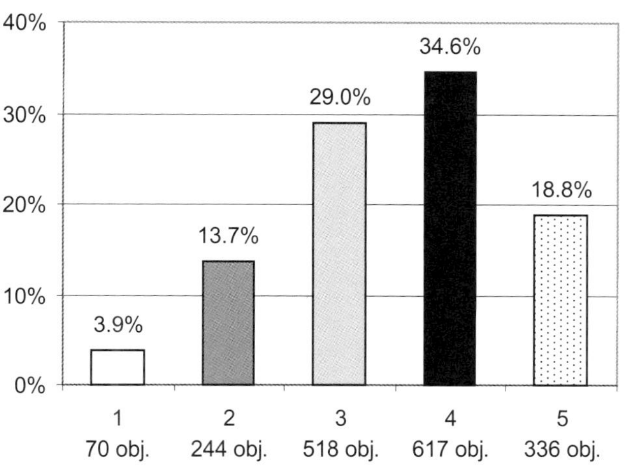

Fig 9.4 *Percentage distribution of the 1785 prehistoric bronzes according to their degree of corrosion, Grades 1 to 5*

is because the number of objects and the combination of the other three factors in each of the subgroups of the individual factors differ. Thus the graphs do not show an absolutely accurate picture of the real influence of the four factors. This can to some extent be compensated for by mathematical calculations – work that is being done at present in the project before the final results can be presented.

Archaeological age

In order to obtain as large as datatset as possible for each of the main parameters we grouped the artefacts into the following main periods arranged in chronological order.

Early Bronze Age	1800–1000 BC
Late Bronze Age	1000–500 BC
pre-Roman Iron Age	500–0 BC
Roman Iron Age	AD 0–400
Germanic Iron Age	AD 400–800
Viking Age	AD 800–1050

Fig 9.5 shows the results on the basis of the classification grading scale. Bronzes from the Late Bronze Age are the best preserved, while finds from the Germanic Iron Age are worst off, with no bronzes assigned Grades 1, 2 and 3, but 71.4% with Grade 4. Second worst off are finds from the Early Bronze Age with the largest number of Grade 5 at 31%. Particularly striking too is the great difference between bronzes from the Early and Late Bronze Age, as well as the fact that finds from the pre-Roman Iron Age, although close to the Late Bronze Age in time, are not as well preserved. It should also be noted that the small number of bronzes from the Germanic Iron Age and the Viking Age is due to the fact that there are in general very few finds of antiquities from these periods in North Slesvig.

Find context

Fig 9.6 shows the grading according to context arranged in order from the worst to best as regards the sum of Grades 1 and 2.

'Settlement finds' are consistently poorly preserved, with 83% with Grade 4 and none with Grades 1 and 2. The category has the smallest number of artefacts of all the find contexts and they are all from the Iron Age.

In the Early Bronze Age one finds inhumation graves in barrows. This form of burial was not used in the Late Bronze Age and the pre-Roman Iron Age, but is then found again in the remainder of the Iron Age, when it takes the form of flat graves. Bronzes from this find category are the most poorly preserved with a total 47.2% with Grade 5 and only 1.6% with Grade 1 and 2.6% with Grade 2.

'Grave finds' include finds from graves where it has not been possible to determine whether they were inhumation or cremation graves. But the appearance of the graph suggests that in most cases these were not inhumation graves.

Cremation graves are found in all the archaeological ages except the Germanic Iron Age in this investigation. Finds from this type of burial can be seen to be in rather better condition than those from inhumation graves, with 19.7% with Grade 5. This might indicate that the decomposition processes that take place in the inhumation graves when the organic material decomposes may have an aggressive effect on the bronzes; conversely the cremated bones, ashes and charcoal in the cremation graves may have a certain protective effect on the bronzes.

In even better condition are the bronzes from urn graves, with 4.1% with Grade 1 and 20.8% with Grade 2, and just 8.1% with Grade 5. This might indicate that the urn has a further protective effect beyond that of the burnt bones, ash

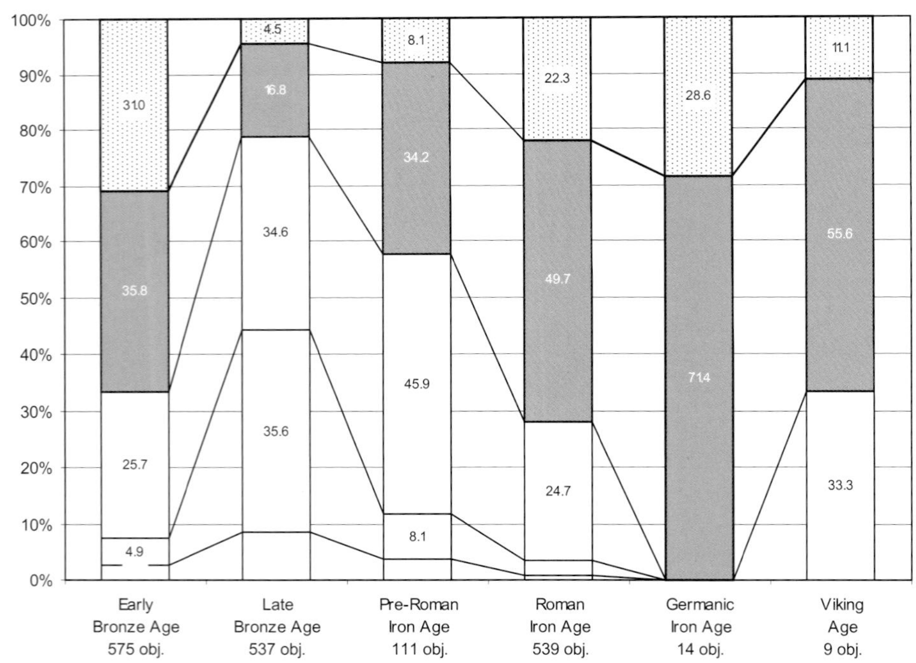

Fig 9.5 Distribution of the 1785 prehistoric bronzes by archaeological age according to their degree of corrosion, Grades 1 to 5 (for key to Grade shading see Fig 9.4)

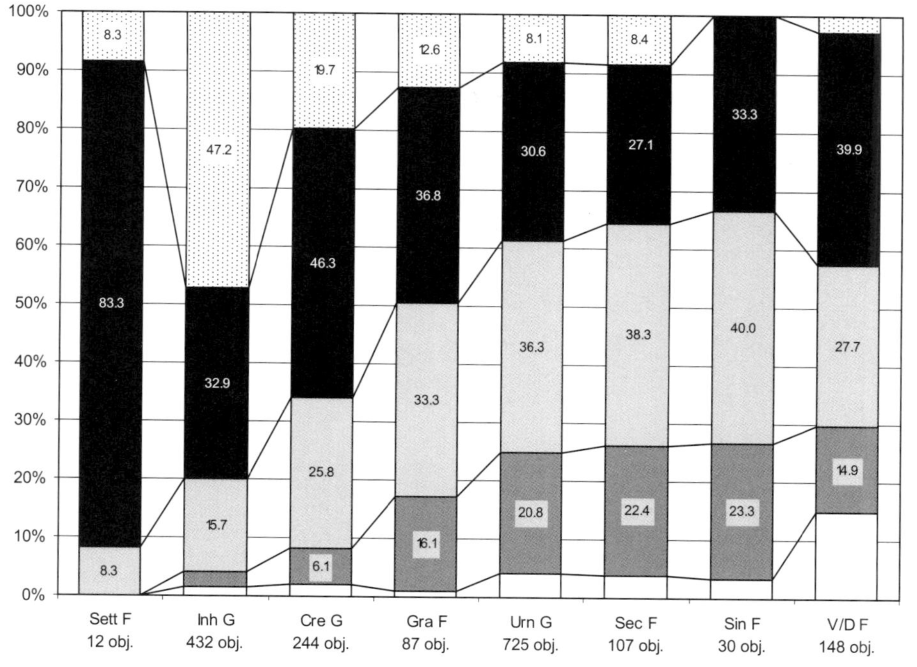

Fig 9.6 Distribution of the 1785 prehistoric bronzes on find context age according to their degree of corrosion, Grades 1 to 5
Key: Sett F settlement finds; Inh G inhumation graves; Cre G cremation graves; Gra F grave finds; Urn G urn graves; Sec F secondary finds; Sin F single finds; V/D F votive/deposit finds (for key to Grade shading see Fig 9.4)

and charcoal. This find context constitutes the one with the largest number of artefacts with 725 bronzes. Finds from urn graves range from the Late Bronze Age up to and including the Germanic Iron Age. If we now compare the finds from the Late Bronze Age and the pre-Roman Iron Age from Fig 9.5 in this respect, we see a difference. The bronzes from the Late Bronze Age were found in barrows – although ploughed-down ones in many cases – while finds from the pre-Roman Iron Age come from hillock graves on flat ground, which are more severely ploughed down. The striking difference in the state of preservation between inhumation graves and urn graves is more understandable given the above mentioned great difference between Early and Late Bronze Age from Fig 9.5. Furthermore Fig 9.5 shows that the state of preservation deteriorates from the Late Bronze Age onwards. This may be because the deposition of urn graves in existing barrows is the prevalent burial custom in the Late Bronze Age, while inhumation graves were again used alongside urn graves on flat ground from the Roman Iron Age onwards.

'Secondary finds' covers finds from secondary levels in barrows or burial sites that cannot be associated with a particular grave – for example a sword ploughed-up from a barrow.

'Single finds' are finds with no known context, for example ploughed-up from a field, and in many cases with no information about the find spot. Most of these finds can however be dated, and for many of them the find date is known. Several of them are prestige objects.

'Votive and deposit finds' are quite clearly the find context with the best-preserved bronzes, with 14.9% with Grade 1 and only 2.7% with Grade 5. And this comes as no surprise to archaeologists, since in most cases these bronzes are prestige objects from the Bronze Age, and most votive finds are recovered from wetland areas such as waterways and bogs. In addition there are some large deposits or hoards from the pre-Roman Iron Age on dry land which are less well preserved. This is reflected in the graphs. The fact that 'single finds' as mentioned above are the next best preserved category may not seem surprising either in this context, since in some cases they may in fact be from votive deposits.

Soil type

1571 bronzes distributed over 431 find spots have an SB number from which the soil type can be identified quite precisely. Where the find spots do not have an SB number, but the find information is so precise that the find spot can be plotted on an official topographical map and a soil-type map, this information is included in the investigation. This applies to a further 93 find spots.

In Denmark the soil types are defined by GEUS on the basis of the FAO standards (FAO 1966). Fig 9.7 shows the distribution of the bronzes for soil types arranged in according to their height above sea level.

It is evident (Fig 9.7) that the state of preservation improves the more we move from the highest-lying moraine soils (MS and ML) over the marginal ridges (DS) to the sandy river plains (TS) and on to the lowest-lying wetland areas (FS). Particularly striking is the difference between the clayey and sandy soils.

At the same time it must be noted that there are proportionally more finds from the Iron Age on the moraine deposits than from the Bronze Age; and that the best preserved bronzes in the FS areas are almost all votive deposits, often the best prestige artefacts of the time. It can further be noted that bronzes found totally immersed in water and bog are in an anaerobic environment which has a different influence on the state of preservation from aerobic environments.

Find date

In most cases, on the basis of the archive material, we have information about the year of the find or of the excavation. In some cases there is only information that it was before the accession year. This applies for example to bronzes from private collections or from events such as periods of war. By dividing the overall period into 20-year intervals we can compensate to some extent for this factor. But at the same time we get a division of the data into suitable groups that are not too small or non-uniform for statistical treatment. The year 1879 was chosen as the first cut-off point, although the first finds were excavated in 1827. This 20-year periodisation also reflects the historical development of museums and the events of the history of Denmark very well.

The resulting graph (Fig 9.8) shows a very striking development where the state of preservation deteriorates with later find dates. However, some of the development may be due to excavation policy factors. For example most finds before the year 1900 are from the Bronze Age, whereas finds from the Germanic Iron Age, Viking Age and Iron Age settlements were first excavated within the last thirty years. It should also be noted that the rise in Grades 1 and 2 in the period 1940–59 is partly due to certain extremely well-preserved hoards.

If we compare the grades for the periods 'pre 1879' and 'post 1980' we see that in the period 'pre 1879' nearly 50% of the bronzes have Grades 1 or 2 and there are none with Grade 5, while in the period 'post 1980' there are none with Grade 1, but 33.5% with Grade 5. In other words this shows an increase of more than 33.5% in the degree of deterioration. This has happened in the course of 120 years compared with what had happened to the state of preservation over the preceding 1000–3600 years. This development seems disturbing, but is confirmed by the local archaeologists, who think that the prehistoric material excavated today is more deteriorated than in the past.

Conclusion

In the investigation 2225 prehistoric bronzes were reviewed from a selected study area, North Slesvig County, which makes up 10% of Denmark's total area. SEM investigation of cross sections of 15 selected bronzes show corrosion by decuprification (Figs 9.2–9.3). The state of preservation of the bronzes was visually assessed in accordance with a deterioration classification grading system from 1 to 5. Their grades were then correlated with their find circumstances in terms of archaeological age, find context, soil type and find date. The results show that the state of preservation of the bronzes is influenced by four main parameters. It can be seen from their characteristic profiles that archaeological age and find context each have a special influence. It can further be seen that the soil type also has an influence, although not a striking one. On the other hand we see a more striking and conspicuous development in the state of preservation of the

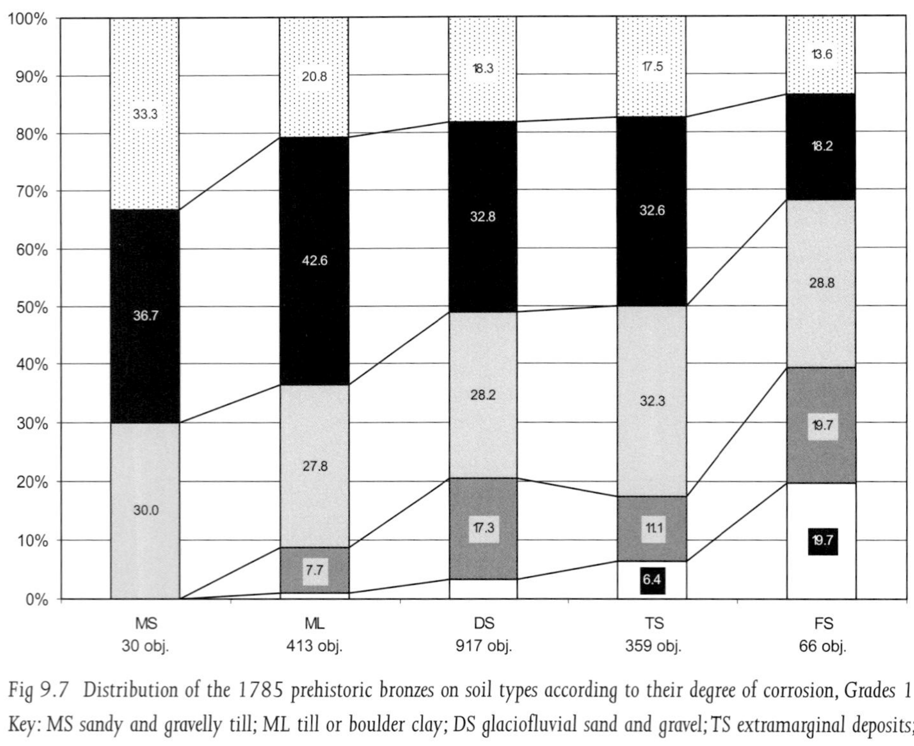

Fig 9.7 Distribution of the 1785 prehistoric bronzes on soil types according to their degree of corrosion, Grades 1 to 5
Key: MS sandy and gravelly till; ML till or boulder clay; DS glaciofluvial sand and gravel; TS extramarginal deposits; FS freshwater deposits – all as defined by GEUS (for key to Grade shading see Fig 9.4)

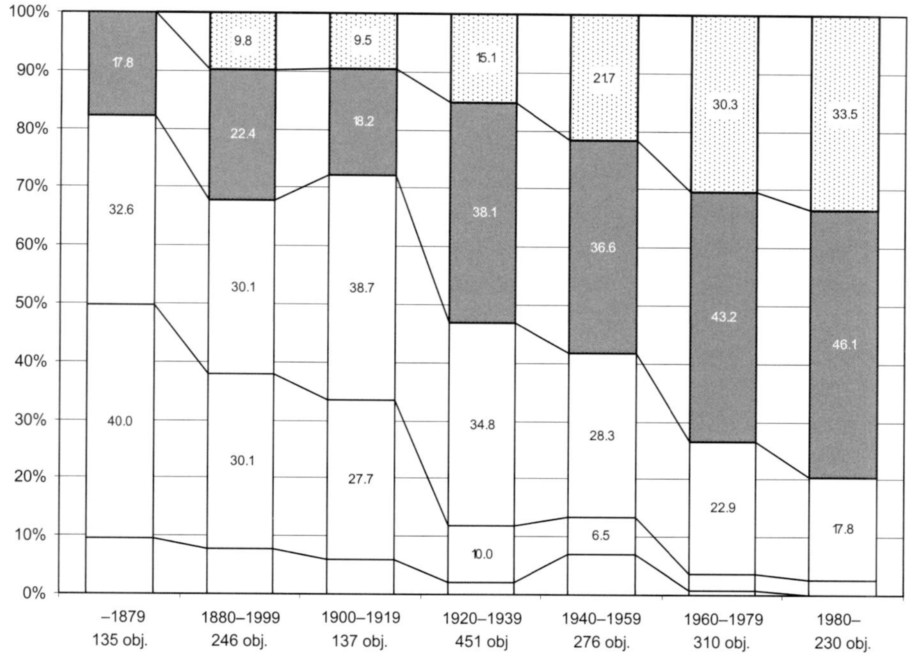

Fig 9.8 Distribution of the 1785 prehistoric bronzes on find date according to their degree of corrosion, Grades 1 to 5 (for key to Grade shading see Fig 9.4)

bronzes when we view it in relation to their find date within the last 170 years.

(Translated by James Manley)

Bibliography

Borg, G CH, Johnsson, L, Lagerlöf, A, Mattsson, E, Ullén, I, and Werner, G, 1995 *Nedbrytning av arkeologiskt material i jord*, Konserveringstekniska Studier RIK 9, Stockholm

FAO, 1966 Food and Agriculture Organization of the United Nations, *Soil map of Europe*, 1:2,500,000, prepared by R Dudal, R Travernier and D Osmond, Rome

Madsen, H Brinch, 1991 Soil as a medium of preservation for artefacts, in *Science, technology and European cutural heritage: proceedings of the European symposium, Bologna, Italy, 13–16 June 1989* (eds N S Baer, C Sabbioni and A I Sors), 284–91, Oxford

Madsen, H Brinch, 1994 *Handbook of field conservation*, Copenhagen

Madsen, H Brinch, 1998 Oldsagernes bevaringstilstand, in *Marsk, land og bebyggelse – Ribeegnen gennem 10.000 år: Vol 1* (ed S Jensen), Jysk Arkæologisk Selskabs Skrifter 35, 43–52, Højbjerg

Meier, D, 2000 Sønderjylland fra natur- til kulturlandskab, in *Det Sønderjyske Landbrugs Historie*, (P Ethelberg, E Jørgensen, D Meier and D Robinson; eds L S Madsen and O Madsen), Skrifter udgivet af Historisk Samfund for Sønderjylland 81, 11–62, Haderslev

II

DECISIONS, CASE STUDIES and MONITORING REGIMES

10

Even the ancient was once young: lessons about biodegradation from the Wareham (Dorset) experimental earthwork

DAVID HOPKINS

David Hopkins
Professor of Environmental Biogeochemistry and
Royal Society of Edinburgh Research Fellow
Department of Environmental Science
University of Stirling
Stirling FK9 4LA, Scotland, UK

Abstract

In this paper understanding of how organic materials decompose in soils and the factors which affect this process are outlined. Although much of the relevant understanding arises from non-archaeological investigations, its relevance to archaeological materials and their preservation *in situ* is illustrated with data from the Wareham experimental earthwork.

Introduction

The emphasis of this paper is a consideration of the factors affecting the decay or survival of organic archaeological materials and, in particular, how investigations into biodegradation carried out in the context of the Wareham experimental earthwork (Bell et al 1996) can provide information relevant to preservation *in situ*. Since the history of the Wareham experimental earthwork is relatively short, it having only been established in 1963, the information presented will of necessity concentrate on recent and relatively rapid, microbiological processes. Such information may not be immediately relevant to the preservation of all archaeological materials *in situ*, but the general principles that influence biodegradation discussed are applicable to those cases where preservation *in situ* involves reburial. There are also the obvious pieces of wider justification for presenting information from short-term investigations: ancient things did not get ancient without surviving the first few hours, days, weeks and years, and early changes in the burial influence subsequent survival or preservation.

Decomposition of organic materials in soils

The decay of organic residues in terrestrial ecosystems has been widely investigated as a result of its relevance to nutrient cycling in ecological, agricultural, remediation and forestry context residues (Fig 10.1). It is my contention that the understanding developed for decay in these contexts is equally relevant to the biodegradation of organic materials of archaeological interest. The process of biodegradation of any organic material involves organisms utilising the material (the residue) in order to fuel their own metabolism, which produces metabolic wastes (most commonly carbon dioxide), and for growth, which results in an increase in the biomass of decomposer organisms (Fig 10.1). The synthesis of new decomposer biomass involves biochemical transformation of compounds in the residue from the original form to new compounds, with a possible change in their recalcitrance, and

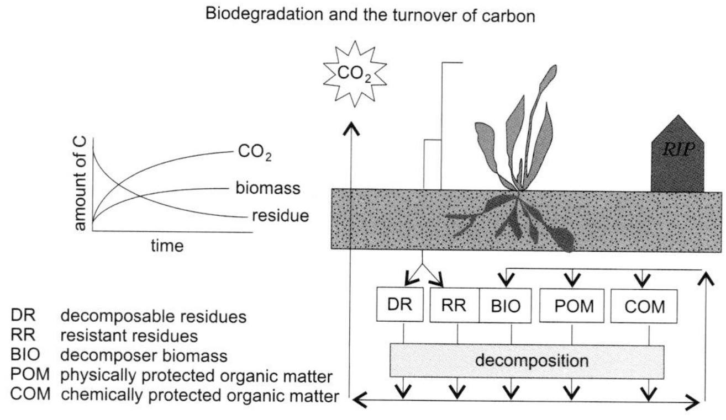

Fig 10.1 Conceptual model of decomposition of organic materials in soil based on the Rothamsted carbon turnover model (see Jenkinson and Rayner 1977), and (inset) a summary of the dynamics of a simple residue, decomposer biomass and carbon dioxide production following a single pulse addition of residue to soil

the production of metabolic wastes. During biodegradation some components may become stabilised by physical or chemical phenomena, which reduce but rarely completely arrest their subsequent decay. In soils and near surface deposits, addition of new organic residues, usually in the form of plant residues, occurs continually added so that organic matter in soils is a complex mixture of materials of different ages and in different states of decay or preservation.

The principle that the decay of material only slows down with time but does not stop is an important one as it means that even so-called 'preserved' materials may be decaying slowly provided environmental conditions permit (see below). This is essentially the same contention as that of Kenward and Hall (paper 2), who question whether organic residues were actually preserved or whether they were always decaying to some extent. Rather than focusing on preservation, the issue is then whether the decay of particular components of the archaeological record is sufficiently slow to lead to no appreciable loss of archaeological information. Over the century to millennium timescale of archaeological 'preservation' answering this question poses significant practical difficulties as it cannot be addressed in full empirically. Although the experimental earthworks project offers one way forward, it seems to me that a combined experimental and modelling approach is required – this is the approach adopted by agricultural and environmental scientists when attempting to quantify the turnover of organic residues in soils. The rate at which materials decay is determined by the interplay of the three sets of factors: the characteristics of the material itself (for example, its solubility in water hydrophobicity, nitrogen and other nutrient content, recalcitrance, water content, physical protection, toughness, toxicity); the presence of organisms capable of decomposing the particular material; the environmental conditions (the water content, acidity or alkalinity, oxygen content, temperature, physical protection).

In the proceedings to the first conference I presented an introduction to the biology of soil organisms, and especially the soil micro-organisms (Hopkins nd). Two important environmental microbiological considerations were outlined in that paper. Firstly, that virtually any known and many unknown species of micro-organism can be isolated from the burial environment provided conditions are manipulated, perhaps unwittingly, to favour them – all things are everywhere. Secondly that micro-organisms are collectively infallible in terms of the ability to degrade all known and many unknown organic compounds and composite organic structures provided that the environmental conditions permit biodegradation – everything is biodegradable.

In reality, it is probable that neither of these bold assertions is absolutely true, but both are sufficiently close to being true to act as guiding principles for preservation *in situ*, and for this reason, 'soil microbiologists would not normally regard soils [or other similar substrates] to be suitable environments in which willingly to [attempt to] preserve materials' (Hopkins nd). Nevertheless, there are many cases outlined in the proceedings to the first conference (Corfield et al nd), where this was attempted with a greater or lesser degree of success. It follows from the 'all things are everywhere' and the 'everything is biodegradable' principles that of the three sets of factors that affect biodegradation, only the environmental factors can be influenced by those attempting to manage organic archaeological materials *in situ*. It is only under extreme environmental conditions, such sub-zero temperatures or anoxic conditions, that organic residues will be effectively preserved.

Anoxia is widely regarded as the most favourable environmental condition for preservation of archaeological remains, but for at least three reasons, this cannot be accepted unquestioningly as a good preservation condition. First, the physiological diversity of micro-organisms means that there are many organisms capable of exploiting a wide range of oxidizable organic resources in the absence of detectable oxygen provided there is an alternative electron acceptor present. Anoxic environments nevertheless favour 'preservation' as decay in them is much slower and is usually incomplete (ie the carbon is often not oxidized completely to carbon dioxide, especially in the case of lignified material). Second, anoxic environments usually exist as a result of being capped off and thus protected from the atmosphere by a medium through which oxygen diffuses only slowly. No detectable oxygen does not necessarily mean that there is no supply of oxygen to a site. It simply means that the oxygen which does arrive is consumed as fast as it arrives and does not therefore accumulate. So what are the processes by which oxygen is consumed? Well in the presence of organic residues, decomposition by micro-organisms is a candidate. Third, if anoxia is to be relied upon for 'preservation', then some assessment of the probability of the environmental remaining anoxic in the future is essential. This is obviously much more than a biological or biochemical issue, being determined by a range of factors including hydrology, climate, land use and development. There is here, however, a link to the second factor as oxidizable carbon to act as an oxygen sink is unlikely to be inexhaustible. Therefore, for these reasons we are back to considering the question of whether the decay of particular components of the archaeological record is sufficiently slow to lead to no appreciable loss of archaeological information.

The Wareham experimental earthwork

The Wareham experimental earthwork was established in 1963 at the Morden Bog National Nature Reserve near Wareham in Dorset (southern England) in an area of heathland with acidic, sandy, well-drained, podzolic soils with the broad objective of investigating the events that occur early in the creation of the archaeological record. The Wareham earthwork is one of two earthworks each comprising bank and ditch constructions built to precise specifications at contrasting sites. The other earthwork is at Overton Down, Wiltshire (also in southern England). Full descriptions of the construction, monitoring and excavations of these earthworks can be found in Jewell (1963), Bell et al (1996) and Evans and Limbrey (1974). The experiment was set up so that sections could be excavated at intervals originally proposed as 1, 2, 4, 8, 16, 32, 64 and 128 years. One of the specific investigations was to recover at intervals and monitor over time the changes to materials of archaeological interest. The buried materials included woollen cloths, sponges soaked in blood, linen cloth, leather, goatskin, hemp rope, flax rope, charred and uncharred pieces of oak and hazel wood, human bone, cremated animal bone, glass, metals and pieces of fired clay. These were buried either in contact with turves laid on the old land surface at the base of the earthwork (the turf environment) or in the upper part of the earthwork (the sand environment). However, when the experimental earthworks project was conceived, detailed biological investigations of the buried materials were not planned. The information presented here relates to materials and investigations conducted following excavation of the Wareham experimental earthwork in 1996, 33 years (not 32 years as originally planned) and is restricted to goatskin, hemp rope, flax rope and linen cloth.

During earlier excavations at Wareham, information had been collected about the preservation of the buried materials. This showed that most of the organic materials had been subject to substantial degradation by 1980 and in some cases the only material detected was that preserved under the biocidal influence of copper from copper alloy coins with which they were originally buried. In the absence of residues of organic material recovered in 1996, biological analyses of the soil from the precise location in the earthwork where different materials were buried was undertaken and the events that occurred to selected buried materials early in the history of the earthwork were simulated experimentally. A full description of this work is provided by Lawson et al (2000) and only a small selection of the data is summarised here.

The size of the microbial community (ATP content), the basal rate of respiration and the rate of respiration of the micro-organisms when supplied with a readily metabolisable substrate (D-glucose) of the soils were measured and compared with values for a control soil. The control soil was collected from a corresponding position in the earthwork where no materials had been buried in 1963. These three parameters provided estimates of the size of the resident decomposer biomass, the rate of degradation of any remnants of the buried materials that were not visible and the latent potential for activity of the decomposer organisms that may be related to the materials originally buried. The values for all of them are shown in Fig 10.2 and all were very small compared with the normal ranges for soils. These data suggest that there was a legacy from the buried materials in the soil microbial community; particularly the potential and basal rates of activity are for the buried materials in the turf environment are compared with the controls. However, the small values limit

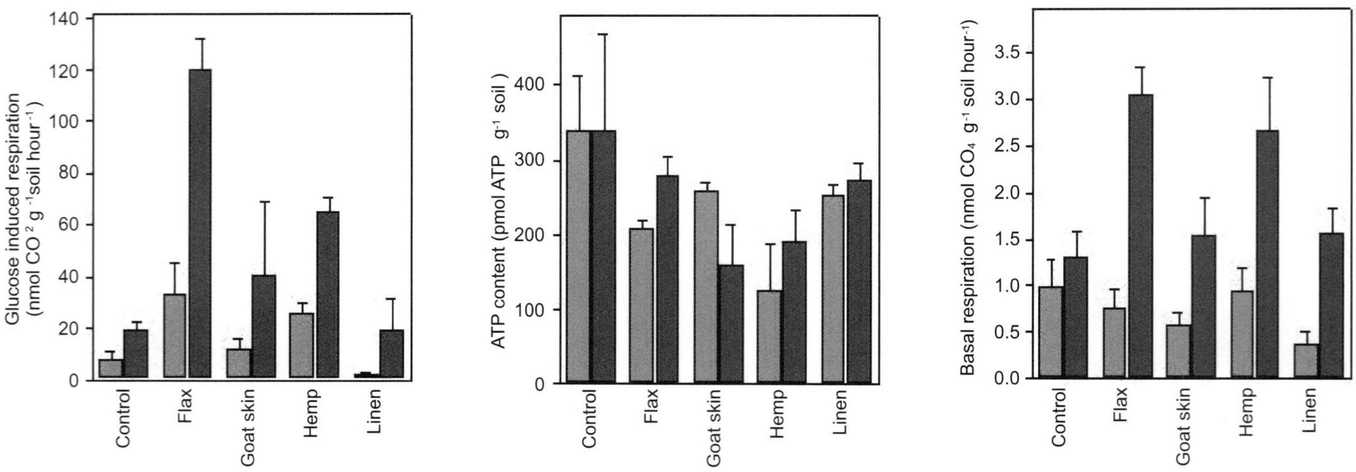

Fig 10.2 Substrate induced respiration (potential activity), ATP content (microbial biomass) and basal respiration (activity) of microbial communities in soil from the Wareham experimental earthwork associated with different buried materials (each bar is the mean of three replicates and the lines represent the standard deviations)

both the precision and reliability of interpretations that can be placed on the differences between the treatments. A more convincing interpretation is that decomposition of the organic materials had declined to a barely detectable rate of the 33-year period since burial, reflecting the fact that over this time period much of the added material had decayed. In the absence of convincing evidence that enough of the buried materials remained to be detectable, there was little point in the context of the Experimental Earthworks project in attempting to find the culprits of the decomposition. The comment of Levy (1966) made in the context of decomposed archaeological wood is apt, 'any attempt at isolating the micro-organisms present in such materials is likely to be centuries too late to find the causal organisms still active'.

The interpretation that buried materials decayed fairly rapidly in the early life to the Wareham experimental earthwork is supported by other observations reported by Lawson et al (2000) and by consideration of some of the likely environmental conditions in the earthwork. It is not possible to determine the exact conditions in the earthwork, but based on work by Harris and Birch (1990) (Fig 10.3), it is reasonable to believe that most of the soil pore volume had been oxic since its construction in 1963. Harris and Birch (1990) showed that in soil heaps constructed from sandy-textured soil for soil storage during open cast coal operations, only a small fraction of the pore volume was anaerobic for a heap 2.0m in height (not a large fraction as reported erroneously by Lawson et al 2000). The Wareham experimental earthwork is made from sandy soils, was 1.5m high when constructed and by 1996 had compacted and been eroded to a height of about 1.0m. The probability of the earthwork being predominantly oxic since 1963 is consistent with the relatively rapid disappearance of the buried organic materials due to biodegradation and the poor preservation of these materials between 1963 and 1980 reported by Bell et al (1996). Had the earthwork been made from clay-textured soil, then the prediction from Harris and Birch (1990) is that

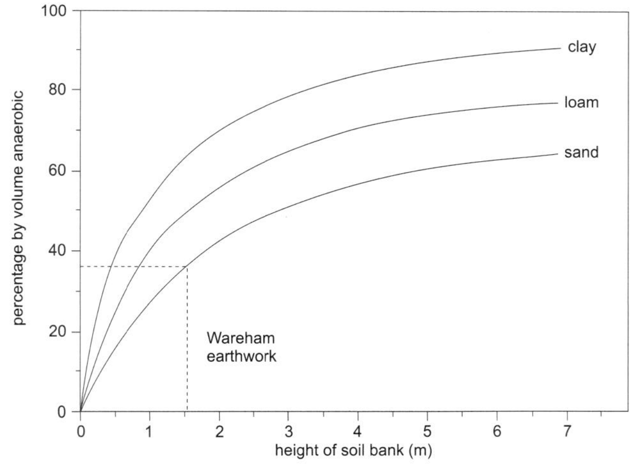

Fig 10.3 Predicted relationship between percentage anaerobic pore volume and height for soil heaps constructed of soils with different textures (adapted from Harris and Birch 1990)

about 80% of the poor volume would have been anaerobic. Perhaps conducting such an experiment should be considered.

Relevance to *in situ* preservation

Clearly the preceding discussion is really only relevant to the preservation of organic materials. The most important point to make is that attempts at preservation of organic materials need to be based on a sound understanding of the decomposition process. Nixon (nd) provides an example where alteration of environmental conditions by the introduction of oxygen and the omnipresence of decomposer organisms accelerated decomposition, whilst *in situ* preservation by reburial was being attempted. Whether this was an isolated event and whether it had serious archaeological consequences is not known. It does, however, illustrate the importance of maintaining environmental conditions in the burial or reburial environment and of the difficulty of doing this when it is perhaps only one factor was limiting decomposition and all the other factors are favourable. The work on biodegradation of buried organic materials in the Experimental Earthworks project is unfortunately very limited; in part this is because such analyses were not planned at the outset. This comment is not intended as a criticism because soil biological research and methods were still relatively primitive 30 to 40 years ago and would not have been able to provide much useful information for archaeological scientists. What it serves to do, however, is to emphasise the point that ancient materials only become ancient if they survive being young and for organic materials, whether they survive being young will be determined largely by the factors that govern decomposition.

Acknowledgements

The contributions of my collaborators Tracy Lawson and Rob Janaway are acknowledged, as are the founders of the experimental earthworks for their foresight and planning and all those involved at the outset and who have subsequently helped sustain the earthworks. The UK Natural Environment Research Council funded some of the work reported here (GR9/02751).

Bibliography

Bell, M G, Fowler, P J, and Hillson, S W (eds), 1996 *The Experimental Earthworks project 1960–1992*, CBA Res Rep 100, York

Corfield, M, Hinton, P, Nixon, T, and Pollard, M (eds), nd [1998] *Preserving archaeological remains in* situ*: proceedings of the conference of 1st–3rd April 1996*, London

Evans, J G, and Limbrey, S, 1974 The experimental earthwork at

Morden Bog, Wareham, Dorset, England: 1963–1972, *Proc Prehist Soc* 40, 619–26

Harris, J A, and Birch, P, 1990 The effects of heavy engineering and stockpiling on the soil microbial community, in *Microbiology in civil engineering* (ed P Howsam), London

Hopkins, D W, nd [1998] The biology of the burial environment, in *Preserving archaeological remains in situ: proceedings of the conference of 1st–3rd April 1996* (eds M Corfield, P Hinton, T Nixon and M Pollard), 73–85, London

Jenkinson, D S, and Rayner, J H, 1977 The turnover of soil organic matter in some of the Rothamsted classical experiments, *Soil Sci* 123, 298–305

Jewell, P A (ed), 1963 *The experimental earthwork at Overton Down, Wiltshire, 1960*, London

Lawson, T, Hopkins, D W, Chudek, J A, Janaway, RC, and Bell, M G, 2000 The experimental earthwork at Wareham, Dorset after 33 years: 3, Interaction of soil organisms with buried materials, *J Archaeol Sci* 27, 273–85

Levy, J F, 1966 The condition of archaeological wood, in *Science in archaeology: a survey of progress and research* (eds D Brothwell and E Higgs), 188–190, London

Nixon, T, nd [1998] Practically preserved: observations on the impact of construction on urban archaeological deposits, in *Preserving archaeological remains in situ: proceedings of the conference of 1st–3rd April 1996* (eds M Corfield, P Hinton, T Nixon and M Pollard), 39–46, London

11

Building for the new millennium: monitoring for the first ten years?

ELIZABETH PEACOCK and GORDON TURNER-WALKER

Elizabeth Peacock
Senior Conservation Scientist
Institute of Archaeology and Cultural History Vitenskapsmuseum
Norwegian University of Science and Technology
NO-7491 Trondheim, Norway

Gordon Turner-Walker
Guest Member of Staff with the Ancient Biomolecules Group
at the University of Newcastle
Newcastle Research Group
Fossil Fuels and Environmental Geochemistry
Drummond Building, University of Newcastle
Newcastle upon Tyne NE1 7RU, UK

Abstract

The proposed redevelopment for continued domestic use of a small residential site in the protected historic city centre of Trondheim led to the first instance of a mitigation strategy based upon long-term monitoring of the buried archaeological remains to be implemented in Norway. Monitoring of the soil moisture content and water chemistry were facilitated by the pre-construction installation of permanent access tubes and water suction samplers. Based upon the parameters being monitored, preliminary assessments can now be made five years into the project. The selected monitoring technologies have proved successfully adaptable to northern climatic conditions. The neutron probe system, although now somewhat dated, is robust and provides reliable, reproducible data. However, blind installation of access tubes for soil water samplers without prior trenching or trial pits reduces the likelihood of targeting the more degradation-sensitive pockets of the project site.

Introduction

For ethico-political and financial reasons, it has become increasingly necessary to preserve the archaeological heritage for future generations through *in situ* preservation rather than full-scale rescue excavation and reburial, or by capping of exposed archaeological sites. Throughout Europe strategies for the preservation of this heritage have moved toward adopting policies that promote minimal intervention.

In preparation for the 1000th anniversary of the founding of the mid Norwegian City of Trondheim, the city embarked upon a comprehensive renovation programme leading up to the 1997 celebrations. In particular, the city was anxious to fill numerous gaps in street frontages in the historic city centre. Of particular concern was an under-developed, vacant lot located on the street *Schultzgate* and situated directly opposite The Church of Our Lady, which dates from the 13th century.

Plans to develop this site by building modest, two-storey wooden flats for the elderly meant that a full-scale excavation of the archaeology prior to construction was financially prohibitive. A long-term programme – initially ten years – of environmental monitoring was agreed as a viable alternative, even though the knowledge and experience necessary to implement such a scheme was not well established, especially in consideration of the subarctic conditions of Central Norway.

Background

The city of Trondheim was founded in AD 997 and became an important political and religious centre during the early Middle Ages. The area around the project site is of medieval origin, and from at least this period both building plots were occupied by predominantly wood-built structures. The plan of *Schultzgate* was originally laid out following the large city fire of 1846 that destroyed much of the area. As a consequence of the fire, the city council passed a resolution requiring new buildings to be constructed in brick. Small two-storey buildings with full and half cellars were constructed, fronting onto the street. Several of these were torn down in 1972 with the demolition rubble being levelled out to backfill the cellars. The site was thereafter levelled off with gravel and used as a car park.

Schultzgate lies within the automatically protected precinct of medieval Trondheim. By law, any redevelopment within this zone must be preceded by a complete investigation of the archaeology in the affected area, and the costs involved in such an investigation borne by the developer. Between 1977 and 1991 several watching briefs and smaller archaeological excavations were conducted in the neighbouring plots. These revealed up to 3.8m of wet, cultural strata in the neighbourhood and many well preserved organic finds were recovered. In 1988 a new local authority development plan for Trondheim City Centre was adopted for this section of the city, in which the Norwegian Directorate for Cultural Heritage opened the door for redevelopment without full-scale archaeological excavation. This plan is based upon minimising ground disturbance, and construction without basements or any form of structural foundations that penetrated into the underlying archaeology.

Erecting a building on top of a protected site, and without sinking structural support into the cultural strata, was unproven in Norway as a non-destructive alternative to full-scale investigation. Planning permission was however granted for the redevelopment of *Schultzgate* and hinged upon the developer financing a long-term programme of monitoring of the status of the cultural strata beneath the new residential buildings.

Desk-based assessment

A desk-based assessment of historical source material and previous field investigations was carried out by the Norwegian Institute for Cultural Heritage Research (NIKU) in order to establish as full a picture as possible of the conditions at the site and how they related to the proposed development (Reed 1995). This information was consulted when designing the monitoring programme.

Site survey

Prior to construction, NIKU carried out an archaeological and geotechnical field evaluation of the project site (Reed 1995). The purpose of this was twofold: first, to map the geometry and near-surface condition of the foundations of the demolished buildings; second, to identify at what sub-surface depths the archaeological strata begins. Under archaeological supervision, the modern overburden of gravel sealing the demolition rubble was removed to expose the underlying made ground. The mapping proved difficult because much of the surface was covered with rubble. Moreover the cellars had been backfilled with debris of similar construction material and for the most part it was not possible to differentiate between surviving foundation walls and backfill. Only with trial pits and trenches would the foundations be precisely located and their condition evaluated. But such intervention would disrupt precisely the archaeology that was designated to be preserved *in situ*. Consequentially, this exploratory work could not be carried out. This 'Catch-22' situation highlighted the practical difficulties of incorporating existing foundations into redevelopment schemes.

Engineering a building solution

Because the project site consisted of two building plots that straddle an alley, *Munkhaugveita*, two L-shaped buildings were proposed. These were to be built at the street and alley frontages, each with interior courtyards and a back row of connecting outhouses. The outhouses were to be allocated for general storage, refuse and to facilitate the monitoring programme.

To accommodate preservation requirements a lightweight, timber construction was designed consisting of two storeys plus loft, set on a ground wall with crawl space. Beams, flooring, outer walls and clapboard were all specified in wood. In order to make use of the *in situ* remains of earlier foundation walls, the width of the new buildings was kept to that of the demolished masonry buildings. The new structures were to be set on a new, cast concrete footing that would support the outer frame of the building. For additional support another footing would be cast inside this outer footing framework, which would run parallel to the longer walls to connect the two narrower walls. After construction the footings allowed for a ventilated crawl space under the ground floor that additionally accommodated services such as mains water and wastewater. Furthermore, the crawl space provided access to the buried cultural levels should this prove necessary at some later date.

The footing would extend into the made ground 0.5–1.0m, and the made ground would be removed from the crawl spaces. Although the underlying levels had probably experienced some unloading since the 19th-century masonry buildings were demolished in 1972, they were still supporting the load from the remaining foundations and demolition rubble. The calculated additional loadings from the projected structures were however small in comparison. These would be spread between the three footing walls. Taking into account the reduced load resulting from removal of the crawl space debris it was calculated that the new buildings would not exceed the pre-construction loading.

Designing a monitoring programme

A number of technologies are available that can be installed at a saturated or waterlogged urban site of archaeological interest to monitor the sub-surface conditions and provide data on the stability of the site (Davis nd). However, the site at *Schultzgate* posed several limiting factors on the technologies that could be considered for installation.

The water table at *Schultzgate* is located deep within the natural deposits that lie beneath the site. The buried archaeology lies several metres above the water table with the consequence that there is no direct access to groundwater. The use of open-ended standpipes or dipwells to both monitor changes in the water table and to provide access points for the retrieval of water samples was therefore not an option.

The cultural levels were not exposed, and would not be exposed during construction, either by foundation trenches or by rescue excavation. The implications from this were that at no time would there be direct access to soil profiles where probe instrumentation used in conjunction with dataloggers could be installed directly. Nor would it be possible to adequately determine the optimum depths or locations for installation of monitoring equipment. Furthermore, the lack of any survey of the stratigraphic make-up of the cultural levels substantially increased the probability that some instrumentation would be installed in deposits that were not of the type one was interested in monitoring (ie water samplers may be installed in backfill rather than archaeological layers). This likelihood was further increased by the complexity and heterogeneity typically found in urban archaeological deposits.

The site's location in Trondheim's subarctic climatic zone meant that the below-surface levels would be subjected to prolonged freezing, perhaps for long periods during the winter. Consequently, access points out-of-doors would not be easily accessible and would present extreme working conditions for the researchers.

With the assistance of the environmental specialists Hunting Land and Environment two monitoring solutions were selected: measuring moisture content employing the neutron probe system; and dipwells fitted with soil water suction samplers.

Methodology

The neutron probe is designed as a field instrument for measuring *in situ* the volumetric water content of the soil and of the underlying unsaturated zone (Institute of Hydrology 1981). The measurement is made by means of a probe that is lowered into an access tube previously installed vertically in the soil profile. The permanent aluminium alloy access tubes are sealed at the lower end with a nose cone and require great care in installation to ensure intimate contact between the tube and surrounding soil. The probe contains a sealed Americium-Beryllium radioactive source from which fast neutrons are emitted into the soil. On collision with hydrogen atoms these neutrons scatter and lose speed and energy. The Boron trifluoride detector, also contained in the probe, registers the density of the resulting 'cloud' of slowed neutrons. The count rate for each depth is converted into volumetric soil moisture content by means of an appropriate calibration curve.

Each suction sampler consists of a PVC tube with a porous ceramic cup bonded to one (the lower) end (Soilmoisture Equipment Corp 1997). The other, serviceable, end of the sampler is completely sealed and two tube connectors protrude from the top. Two polyethylene access tubes are connected to these connectors and are used for pressurising and recovering samples. A vacuum is established with a simple hand pump to pull water from the surrounding soil through the porous cup into the sampler tube. The collected water is then pumped to the surface and into a container.

Installing the monitoring system

As previously described the design for both plots consisted of L-shaped structures fronting on both the street and alley with interior courtyards and a back row of outhouses. The narrow window-of-opportunity between the granting of planning permission and the onset of winter ruled out installing the monitoring system prior to development of the site. Construction needed to be able to reach a certain phase before the onset of winter; otherwise, it could not be started until the following spring. This precluded the opportunity to gather vital predevelopment, baseline soil moisture content and chemistry data for later comparison. The occasion of installation did however, provide the opportunity to visually describe and sample the underlying deposits as boreholes were sunk to accommodate the neutron probe tubes and dipwells.

Installation was carried out in November 1996 after construction work had started but before the new buildings began to take shape. For both comfort and security concerns as many access points as possible were located inside protected enclosures, ie the outhouses. Six neutron probe access tubes were installed: two at the smaller plot and four at the larger plot. One was located in each of the courtyards and the others were located in outhouses. Six pairs of suction samplers were installed at two different depths: three at each plot with some pairs located under the courtyard and others located under outhouses.

No access points were installed in areas that would be under the finished main buildings (ie crawl spaces) even though there is buried archaeology beneath the underlying older cellars. Installation would have required boring through these cellar floors. For the neutron probe system, which requires setting the probe directly on the access tube, taking readings inside private residences would be impractical. Although the water samplers can be activated some distance from their installation points, should they malfunction location beneath a building would also be impractical.

Further into the construction process, protective concrete access chambers with secure lids were built around the various access tubes. In addition, near-surface enclosed plastic channelling was installed across the courtyards to provide for running the suction sampler tubing to outhouses pre-designated for use by the monitoring project. This was to permit the collection of water samples during the winter months.

Immediately following installation, the first neutron probe readings were successfully taken. No water samples were recovered however, although the presence of air in the system indicated that the samplers had been installed successfully and were functioning as designed.

Monitoring practicalities

The commencement of regular monitoring upon completion of construction became delayed due to several factors. Production of new neutron probes ceased just after installation, and it took considerable time to acquire a certified used model. Analysis of water samples by a local laboratory proved to have a three-month turn around which, considering the necessity of 'immediate' measurement for some parameters, was not acceptable. Water samples were recovered in June 1997, and regular monitoring on a quarterly basis began in 1998, two years after installation. Teething problems, mostly climate based, have included: the freezing in place of neutron probe access covers in the winter; flooding of access pipes, and freezing of water samples in the underground discharge tubes while pumping samples across the courtyards.

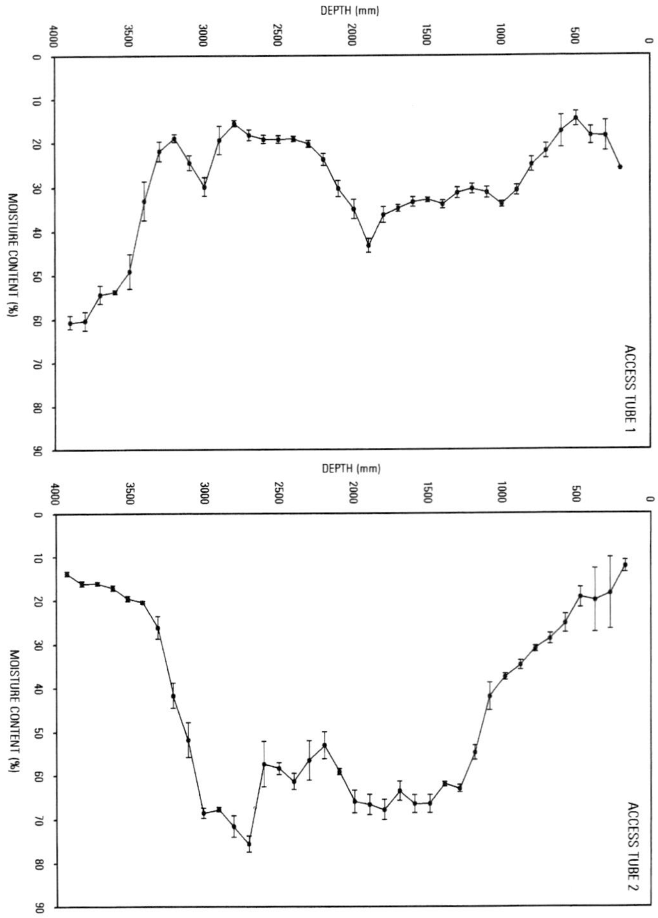

Fig 11.1 Moisture profiles for neutron probe access tubes 1 and 2, error bars indicate the variation in quarterly measurements taken over a three-year period; the distribution of wet and dry zones down the profiles is radically different between the two tube locations which lie approximately 4m apart

Neutron probe

Obtaining field data with the neutron probe is straightforward, but laborious. Readings are taken every 10cm down each access tube. The first reading is taken at the lowest point 3.8m or 3.9m below ground level and the last at approximately 0.20m. A calibration curve based upon the textural character of the deposits is used to convert the field readings into estimated volumetric moisture content. The soil texture character of the deposit profile for each access tube was determined from the contents of the borehole, but a site-specific calibration curve was not created. Instead the standard calibration curves for the three principal soil types were evaluated, and the standard 'loam' calibration curve was selected for use with the *Schultzgate* field data (Davies 1997).

The converted readings for a profile can be plotted to form a moisture profile (eg Fig 11.1). Moisture profiles for *Schultzgate* differ across each plot and between the two plots. Comparison with the soil textural characterisation for each profile, however, shows close agreement. Moreover, quarterly moisture data for each access location is consistently similar (Fig 11.1). To investigate the reproducibility of the probe, four consecutive sets of readings were taken at one of the outside access locations. The results indicate the method is highly reproducible, and that the close fit of the various moisture profiles is real (Fig 11.2). Furthermore, the sensitivity of the method was highlighted when readings, taken after a post-installation shrinkage gap in the soil around one of the tubes was filled, registered the change in soil moisture surrounding the tube. With the exception of two 0.20m pockets at one location, moisture profiles of the cumulative data show little deviation over five years of monitoring.

Soil water samplers

Unlike the straightforward readings obtained with the neutron probe, the soil water chemistry for *Schultzgate* is complicated and highly variable. It is variable in the amount of sample recovered (if any) and in the nature of the measured parameters. Of the twelve installed samplers water has been collected from nine (Fig 11.3). Two samplers do not function, another has always been empty, and three others have low volumes of collection. Of the three samplers that most frequently contain water, only one (Dipwell 1A) is consistent in delivering a full sample.

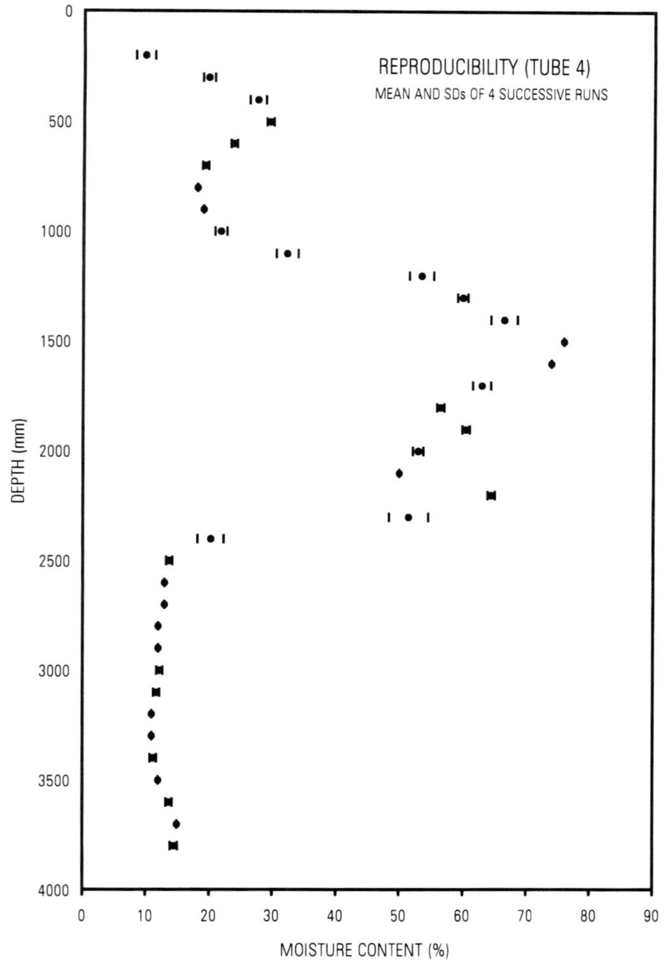

Fig 11.2 *Reproducibility of neutron probe measurement; four successive measurements of the moisture content down the profile of neutron probe access tube 4*

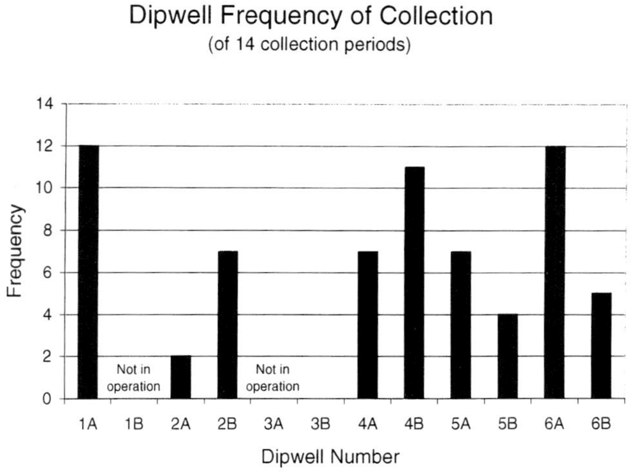

Fig 11.3 *The variation in frequency of collection of water samples from the 12 soil water suction samplers at* Schultzgate

Empty samplers arise from a low soil moisture content combined with a coarse, sandy soil, a frozen soil, or the lack of full intimate soil contact with the porous ceramic cup. Soil texture characteristics of the deposit profile for each water sampler access borehole were logged during boring to aid in targeting the samplers into wet, organic-rich, pockets. Although, the auger allowed for the recovery of core samples, these were sleeved and not readily viewable for detailed logging on site. With hindsight, the desk-based assessment should have placed more emphasis on the limited stratigraphic and geotechnical details of earlier investigations rather than the archaeological potential of the area, particularly because this information was not forthcoming in the pre-construction site survey. Closer analysis of the details of the desk-based assessment would have warned of the degree of complexity of the sub-surface stratigraphy.

Later comparison with the detailed soil texture log along with laboratory analysis on deposit samples provides limited insight into the volume, frequency of collection, and nature of the water samples. Even though many of the deposits into which the samplers have been installed have been characterised as organic cultural layers, a high frequency of included natural sands, gravel and brick debris is reported. The sampler that is operational but is consistently empty most likely is not in intimate contact with the surrounding soil. The borehole log that describes the deposit surrounding this sampler as 'made ground' supports this. The three samplers that have a very low frequency of collection are located in coarse/gravely organic sandy loam pointing to low water retention. The three samplers that have a medium frequency of collection are situated in deposits characterised as organic loam with inclusions of sand. Finally, two of the three high frequency samplers are situated in black loamy peat, while the third is in an organic sandy loam with brick debris.

Recovered soil water can be analysed for a wide variety of parameters and using a range of methods. Research, primarily in the UK, has indicated that factors such as pH, redox potential (Eh), conductivity, dissolved oxygen, temperature and selected ion concentrations are contributory to the preservation/degradation of organic archaeological materials. Conductivity, pH, redox potential, dissolved oxygen and nitrite, nitrate, sulphite and sulphate concentrations were selected for measurement of the *Schultzgate* samples. Field measurement of the soil water chemistry was ruled out based upon climate considerations. Measurement in the field is at the best of times difficult and would not be practical during the long winter months. A measurement regime consisting of in-season field measurement and out-of-season laboratory measurement would introduce an undesired discrepancy in the data. Consequently, all analysis is carried out in the conservation laboratory at Vitenskapsmuseum, a five-minute walk from the project site.

Another consideration was the degree of measurement precision. High quality, expensive, water analysis with a processing turn-around of three months was available at a local research institute. The *Schultzgate* monitoring programme seeks to map trends in the soil water chemistry, and for this it was concluded that high quality analysis was not needed. Moreover, the importance of immediacy in carrying out measurement once samples were collected was prioritised. Consequently, the analytical regime is a combination of standard benchtop meters with selective electrodes and a reflectrometric test strip system.

Laboratory measurements on the soil water samples have been difficult to interpret, even following three years of quarterly monitoring. The lack of a pre-construction and construction baseline has added to the difficulty. Plotting the Eh against the pH does provide some insight into characterising the microenvironments of the pockets from which the water samples are being extracted (Fig 11.4). The majority of measurements cluster together in the mildly oxidised 'WET' zone, and are slightly more basic than rainwater. The exception is one group, all of which are from Dipwell 1A, which lies in the more reducing 'WATERLOGGED' zone. What this suggests is that much of the soil water being recovered and analysed is recent rainwater that has percolated down through the soil profile and that Dipwell 1A is situated in a wet, preserving pocket. This correlates well with the nature of the site as one in which the buried archaeology lies well above the water table and the chemical interaction with the burial environment being through the soil water which is subject to climatic conditions.

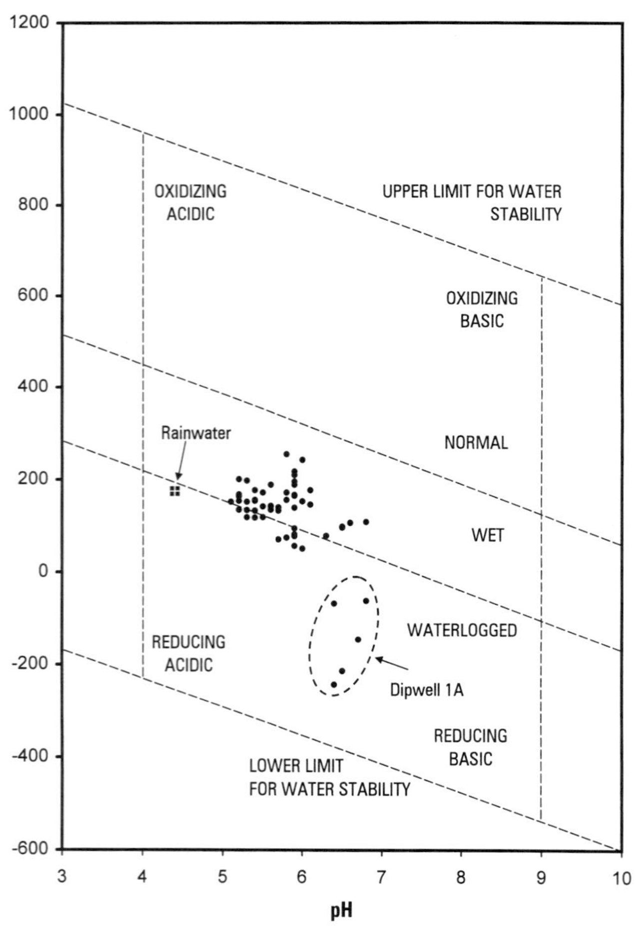

Fig 11.4 Eh-pH diagram for the Schultzgate dipwells

Appropriateness of selected monitoring systems

The soil moisture profiles at *Schultzgate* are clearly not changing, although there has been some slight reduction in the uppermost 0.10–0.20m in some tubes since the building phase was completed in the winter of 1996–7. The variability of both the recovery of water samples, and their chemistry, highlights the necessity of frequent regular monitoring over several years before it is possible to develop a baseline of knowledge about the site. This is in contrast to other reported studies that have operated on a much shorter, eg one-year, time frame. However, much of this other work has focused on non-urban, rural sites where the sub-surface stratigraphy is much less complex than that found in an urban context, and where the cultural layers lie close to or within the water table.

The neutron probe system provides soil moisture data down a profile and is well-suited to monitoring an urban site such as *Schultzgate* where there has not been pre-construction access to determine the optimum depths or locations for installation of monitoring equipment. However, such a starting point dramatically reduces the potential success rate of the water suction sampler system. This system does not provide for a series of soil water samples down an access tube, but rather one sample at a pre-selected, fixed depth. If no water is recovered at that depth for whatever reason, there is no soil water data from the entire access tube. The total amount of data feeding into the monitoring database is potentially less than projected, compromising both the comprehensiveness of the monitoring programme and the accuracy of site stability assessments made based upon the soil water chemistry data.

Soil water samplers are less suitable for a monitoring programme such as that at *Schultzgate*. *Schultzgate* is a site that firstly, has only pockets, not large areas of anoxic wet burial environments; secondly, there is a known but unmapped complex made ground/cultural layer sub-surface stratigraphy; and thirdly where there is no access to detailed soil profiles to ensure appropriate placement of suction samplers in deposits that have archaeological organic materials preserved in them. The project also highlights that for sites where there is no pre-construction access to the buried archaeology, it is even more imperative that monitoring equipment be installed and activated prior to the start of construction so that the monitoring record has a pre-construction baseline.

Conclusion

Though ultimately any damage to this site cannot be assessed until the new buildings come down, the archaeological monitoring permits confidence that, in this case, the mitigation strategy is successful and the site remains well preserved. Furthermore, if this site is excavated in the future, the present artefact archive representing 30 years of urban excavations in the city will provide a material preservation baseline with which artefacts recovered from these deposits can be compared.

A small residential development in a redevelopment-protected historic city centre introduces mitigation concerns of a different nature than those encountered with large commercial property. Commercially it is arguable as to whether even a monitoring solution is financially viable for a small domestic development. The cost of the ten-year monitoring programme at *Schultzgate* is approximately 10% the estimated price of a full-scale archaeological investigation. Still, had it not been for the fact that the developer in this case is one of the largest construction companies in the country, with in-house resources to develop alternative building strategies, the site may well not have been developed.

Bibliography

Davis, M, 1997 Monitoring of archaeological deposits at Schultzgate 3–7, Trondheim, Norway, unpub Hunting Technical Services rep R1169 305395:NOR-02

Davis, M, nd [1998] *In situ* monitoring of wet archaeological environments: a review of available monitoring technologies, in *Preserving archaeological remains* in situ: *proceedings of the conference of 1st–3rd April 1996* (eds M Corfield, P Hinton, T Nixon and M Pollard), 21–5, London

Institute of Hydrology, 1981 *User's handbook for the Institute of Hydrology neutron probe system*, rep 79, Wallingford

Reed, I, 1995 Prosjektbeskrivelse. Schultzgate 3, 5 og 7, Trondheim: overvåking av middelalderske kulturlag, unpub NIKU rep

Soilmoisture Equipment Corp, 1997 1920F1/1920F1K1 pressure-vacuum soil water sampler operating instructions, http://www.soilmoisture.com/opinst.htm

12

Advances in research into the *in situ* management of historic shipwreck sites

IAN OXLEY

Ian Oxley
Assistant Inspector of Ancient Monuments with Historic Scotland
Centre for Environmental Resource Management
Department of Civil and Offshore Engineering
Heriot-Watt University, Edinburgh
Scotland EH14 4AS, UK

Abstract

This paper considers recent advances in the quantification of processes acting on historic shipwreck sites and the ways the data can be incorporated into management schemes (with their assessment, monitoring and intervention procedures), data management and manipulation in geographical information systems (GIS), deterioration matrices, and site formation modelling which might be predictive. Such developments will serve to underpin the success of strategies of *in situ* management that seek to maximise the value of shipwreck heritage resources such as impact mitigation, raising of awareness, promotion of wider access, and the fostering of research initiatives.

Introduction

Historic shipwreck management is a complex subject that is developing globally, driven by the increasingly obvious fact that many important ships are submerged worldwide and that management *in situ* may be the only viable strategy to retard their eventual disintegration. It is also clear that we are experiencing a rapid rise in high profile initiatives (in whatever environment they lie: local, national and international; or in extreme depths and far from land). These projects often specifically target individual sites with strong historic associations such as the *Marie Celeste*, HMS *Hood* and the SS *Carpathia*, (the ship that rescued survivors from the SS *Titanic*). These developments have been accelerated by improvements in search, survey and intervention technology particularly deep water positioning, and by the growth of recreational diving (with emphasis on ever deeper experiences requiring mixed gas breathing systems and 'technical' training). Most of these advances are out of the normal reach of archaeological organisations.

However, some initiatives foresee a time when we might have a consistent and integrated approach to historic shipwrecks across environmental, administrative and jurisdictional boundaries. Discussions are underway regarding the protection of shipwreck resources in international waters, currently being addressed by the *Draft UNESCO convention on the protection of the underwater cultural heritage* (Dromgoole and Gaskell 1999), and shipwreck recording initiatives are becoming integrated into maritime heritage resource management developments on a national scale (eg for Ireland see Breen and Forsythe 2001).

Management options vary from complete recovery after comprehensive on-site assessment (ie the CSS *H L Hunley*, Murphy 1998), partial recovery and stabilisation (on the *Monitor*, *Monitor* National Marine Sanctuary 1997), and individual site stabilisation (Pournou et al 1999). Nevertheless the variety of factors involved in deterioration processes, as well as the complexity of their interrelationships, are not widely understood or accepted (Gregory 1999), and it is highly likely that management strategies that do not fully take into account the actual factors governing the causes of change to a site will inevitably fail.

Furthermore, the recommended conditions for protecting one type of archaeological material or context will not necessarily be conducive to preserving another and the real effects of even low technology stabilisation strategies (such as sand-bagging) are not yet fully understood. Such mitigation strategies to deal with observed impacts have often been implemented as short-term, stopgap measures as reported in earlier reviews of historic shipwreck management *in situ* (Oxley nd).

Management is not synonymous with preservation and it cannot eliminate change completely but only promote procedures that might reduce the detrimental effects resulting from recognised impacts. Management of historic shipwreck resources can be seen as an ongoing, proactive necessity backed up with quantified environmental information, ie data collected to a standard and using techniques acceptable to the wider disciplines of marine science. Management approaches based on a high level of marine environmental knowledge are preferred in many countries, and formal policy and proposal documents (often called 'Management' or 'Conservation' Plans) based on comprehensive desk-based and field assessments are becoming common (*Monitor* National Marine Sanctuary 1997; Smith 1998; Breen 2001).

As shipwrecks need their own set of theoretical and methodological approaches (Gibbins 1990), so should their management be approached specifically as they have been deposited into the archaeological record by special processes. Furthermore, the medium they are submerged in is far more dynamic than air or soil from the point of view of post-depositional processes.

Finally, in the same way as our natural marine heritage, shipwrecks are subject to growing threats of destruction given that global marine and coastal development is increasing rapidly. This promotes a parallel rise in the popularity of marine recreational pursuits such as underwater tourism, but it also increases the rate of discovery of previously unknown wreck sites whilst exposing the heritage resource to heightened risk of damage or loss. In addition, there are more opportunities for the implementation of appropriate management structures. Nevertheless the rate of discovery of new sites will inevitably slow in the future, most of the readily accessible sites will be discovered, but will the sites be in any condition to serve as valuable heritage resources? This paper illustrates recent research work on historic shipwreck management in Scotland on a national and regional scale (including the remains of the German High Seas Fleet in Scapa Flow), shipwreck assessment and monitoring on the USS *Arizona* in Pearl Harbor, field experiments in Belfast Lough to improve the quality control of geophysical survey, and advances in individual site management research on the *La Surveillante* in Bantry Bay, Ireland.

Historic shipwreck management research

There are many interrelated factors involved in the *in situ* management of historic shipwrecks and a research programme has been devised to overcome some of the deficiencies in the management of Scotland's shipwreck heritage caused by this complexity (Oxley 2001a). The research is approached at three levels: national, regional and local, and it is deliberately linked to a highly developed knowledge of the marine environment in and around shipwreck sites. Within the context of a current understanding of the range, quantity and condition of the shipwreck archaeological resource in Scotland advances in the quantification of processes acting on such sites, in the ways the data can be incorporated into management schemes (with their assessment, monitoring and intervention procedures), and the ways that such data can be managed and manipulated in GIS, deterioration matrices, and formation and predictive modelling, are being addressed.

The Scapa Flow Maritime Archeology Project (ScapaMAP) initiative

The principal regional case study involves Scapa Flow Maritime Archeology Project (ScapaMAP) a multi-disciplinary, multi-institution, international project involving government agencies, industry and the academic community concentrating on Scapa Flow in Orkney (Oxley 2001a). The study area, a large semi-enclosed sea area, contains the remains of the scuttled German High Seas Fleet (1919) which form a unique underwater record of one of the great periods of British and German maritime history. Seven wrecks of major warships remain, up to 25,000 tons and over 200m in length, and between them lie concentrations of other wreckage associated with salvage activities on the existing wrecks and the vessels that were raised and scrapped.

The vessels and wrecks have seen radical changes in their perceived value over a short period of time, from weapons of mass destruction (1918–21), salvage resource (1923), diving amenity (1960s onwards), to present-day national historic and archaeological heritage (1990s). These wrecks are of particular significance and in May 2001 Historic Scotland scheduled them under the Ancient Monuments and Archaeological Areas Act, 1979. This is the first time in the UK that this legislation has been used to protect submerged archaeological sites and is an example of a government agency developing and following through policies aimed at an integrated approach to heritage management in all environments (Historic Scotland 1999).

In early 2001, the ScapaMAP project employed a dynamically focused, very high-resolution multibeam echosounder to generate the most accurate data of the wrecks and surrounding area to date. The survey was aimed at providing baseline bathymetric data for presentation in a variety of applications including 3D visualisation software and GIS. One example is a plan view of the battleship SMS *Markgraf* (Fig 12.1). The extensive, post-scuttling salvage damage is clearly visible in both images and work is underway to test the use of such data-sets for establishing monitoring programmes. It is hoped that periodic, high precision bathymetric surveys together with targeted video mosaicing will provide immediate feedback on deterioration processes such as the collapse of ship structure. In this way condition data can be used to generate baselines by which change can be monitored and measured, thus providing critical parameters for tracking the survival of archaeological resources, such as the Features of Interest proposed on the *La Surveillante* wreck site below.

Management framework development: *La Surveillante*

It is clear that complex frameworks contribute to maintaining and/or restoring conditions on shipwreck sites which are conducive to the long-term aims of shipwreck management *in situ*. Guided by the overriding aim of non-net loss of any part of the archaeological resource, effective management should be responsive and adaptive, working with local interests in a way that builds support for conservation objectives (Kelleher 1999). To achieve these aims managers should adopt a systems approach, use inter-disciplinary teams, and follow a clear sequence of decision-making. Best practice in the management of marine protected areas (see SNH et al 1997) provide a useful framework for the development of an

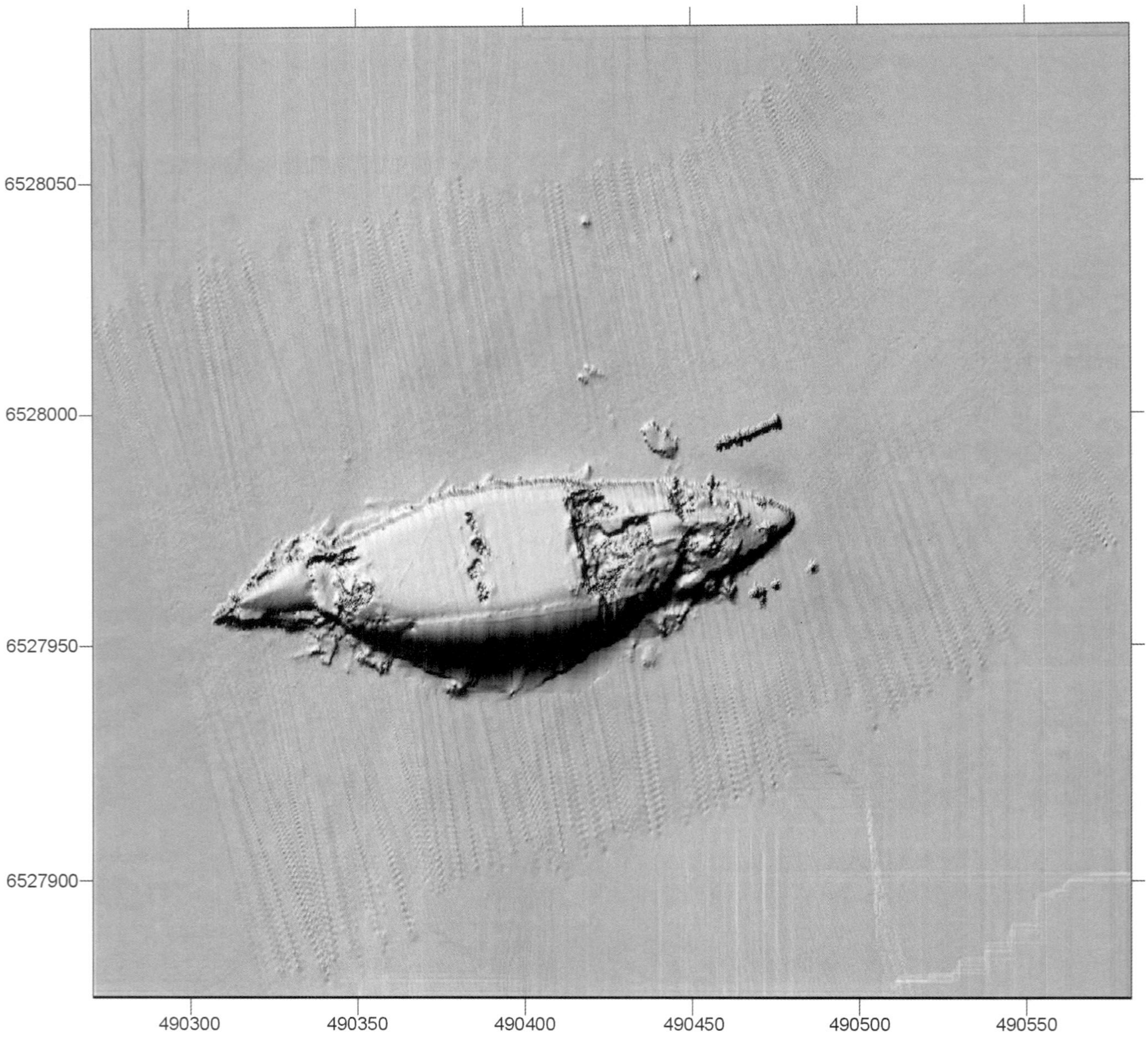

Fig 12.1 Plot, using Surfer, showing post-scuttling salvage damage of the bathymetry over the German High Seas Fleet battleship SMS Markgraf (Konig Class, over 25,000 tons, around 175m in length) in Scapa Flow; the survey system consisted of a Reson 8125, TSS POS/MV 320 attitude and navigation system, and a Racal Landstar DGPS (courtesy of the Archaeological Diving Unit, University of St Andrews)

historic shipwreck Management Scheme for *La Surveillante* (Breen 2001).

La Surveillante is an integral part of the local marine environment and must clearly be managed as such. Similarly, because of the highly fluid and connected nature of the sea, which efficiently transmits substances and processes for long distances, a protected area will rarely succeed unless its management is embedded in an integrated ecosystem management regime.

Management research on *La Surveillante in situ* describes a recommended approach and by outlining the next steps in designing, implementing and maintaining a formal Management Scheme (Oxley 2001b). The work carried out to date on the wreck site goes a very long way to fulfilling the first part of the Management Scheme process – that of assessing and understanding the current situation regarding the site. From this the condition of the site can be summarised, and the nature, extent and rate of negative processes estimated.

Conservation Objectives help identify management needs, confirm existing measures or identify the need to introduce new measures to maintain or restore the features of the site. Such objectives should be specific, attainable, measurable and regularly reviewed because they should provide the standard against which the success or failure of site management could be judged. The primary Conservation Objective for the *La Surveillante* might be the preservation of the current physical archaeological evidence on the site.

The next stage in the Management Scheme development process would be the synthesis of the assessment data to analyse the nature and effects of current use and activities to reach common understanding of those processes that may have a beneficial, neutral or harmful impact. Features of interest present on the site could then be identified together with their attributes that would be monitored over the medium to long term using accepted and testable methodologies. Such monitoring would track trends in factors such as the stability of the visible wreck structure, the integrity of inboard deposit, and the condition of the surface concretion on the anchor and guns, the condition of the debris field immediately associated with the wreck.

Assessment and monitoring: Pearl Harbor

In the United States the National Park Service is carrying out research into the assessment and monitoring of the battleships USS *Arizona* and USS *Utah* in Pearl Harbor, Hawaii (Lenihan 1989). This work, led by the NPS Submerged Resources Center (SRC), uses multiple lines of inquiry to determine exactly what forces are acting on these historic ships and what can be done to ensure their long-term preservation (L Murphy and M Russell, pers comm).

Efforts include structural monitoring using advanced Global Positioning System (GPS) technology coupled with traditional historic building monitoring devices; electrochemical and microbiological characterisation of the steel hull's corrosion; analysis of oil leaking from the hull and trapped in interior spaces to determine internal leak-points as indicators of interior structural collapse; and interior exploration of *Arizona* using a small Remotely Operated Vehicle (ROV) capable of taking scientific measurements and collecting samples for analysis. SRC is gathering existing data, site plans, and engineering drawings into a GIS format that will be the basis for future computer modelling, including a Finite Element Model, of *Arizona*'s hull.

In situ corrosion measurements have been taken systematically along *Arizona*'s hull, including pH and corrosion potential (Ecorr). The instrument used has underwater probes deployed by diving archaeologists and metallurgists connected to a surface readout by a 100-foot cable. Underwater communications gear provided a link between the divers and the topside recording station. Other parameters such as dissolved oxygen were also recorded and samples collected for analysis by a microbiologist to determine the role of microbes in the overall corrosion process.

Quality control in geophysics: the Belfast Lough acoustic test facility

Research into the quality control of prospecting and survey techniques used in the marine geophysical seabed mapping programme carried out by Dr Rory Quinn based within the Coastal Studies Research Group in the School of Environmental Studies at the University of Ulster, Coleraine. The work is focused in four main areas: the application of marine geophysics to wreck-site investigations and site-formation processes; the role of marine geophysics in multi-disciplined investigations of maritime cultural landscapes; classifying marine geophysical signatures of anthropogenic materials ('cultural anomalies'); application of marine geophysics to mapping offshore natural resources and understanding offshore processes.

Initial results of the marine geophysical investigation, involving a multi-disciplinary team of geophysicists, archaeologists and geomorphologists from the University and Environment and Heritage Service (DoE NI) indicate that the approach adopted is successful in providing information on the archaeological resource and geomorphology of the inshore coastal waters of Northern Ireland. For example, a side-scan sonar and Chirp sub-bottom profiler of Belfast Lough highlighted the maritime cultural resource of the Lough including the imaging of seventeen wreck sites (Quinn et al 1999).

In the summer of 2001 a joint research project with the Archaeological Diving Unit (University of St Andrews) involved the construction of an acoustic Test Facility in the Lough. The experimental site is laid out along two baselines: a 70m long northern baseline and a 50m long southern baseline. The material laid down includes: ceramics; green wood; waterlogged wood; leather; gravel; aggregate; glass; skeletal remains; fish basket; amphora; range of aluminium tubes and iron rods (200–60mm diameter, horizontal, vertical and at various inclinations); car tyres (horizontal and vertical). Analysis of the sonar signatures of the various materials using different survey configurations and instruments will aid our understanding of the capabilities of geophysical survey of marine archaeological sites.

Future directions

Despite these developments it is clear that the fact that the submerged archaeological resource cannot be easily accessed by researchers, archaeological practitioners and decision makers without specialist skills, techniques and equipment. Such requirements can be readily acquired by recreational divers, but they are not generally available on the required scale to government agencies apart from specific cases such as the Ministry of Defence in relation to HMS *Royal Oak*, or the organisations which call on the technical support of the Archaeological Diving Unit. Paradoxically, Maritime and Coastguard Agency monitoring of the SS *Richard Montgomery* (Maritime and Coastguard Agency 2000), Department of the Environment, Transport and the Regions investigations into the loss of the MV *Derbyshire* (DETR 2000), forensic examination of the SS *Titanic* (Garztke et al 2000) and recent enquiries into unexplained fishing boat losses in recent decades indicate that stronger links with other organisations investigating shipwreck sites would be beneficial in developing assessment and monitoring methodologies.

Furthermore, significant improvements in other types of access exist which can be managed in an appropriate way. Examples include: 'telepresence' using developments in information technology; adventurous tourism (eg recreational visits to the SS *Titanic* site are now common); personal submersibles are on the market; glass-bottomed or glass-panelled vessels are common world-wide (eg Kyle of Lochalsh, Scotland); a tourist Remotely Operated Vehicle in Scapa Flow focusing on the German High Seas Fleet wrecks is very popular; and there is a rise in media interest in historic shipwrecks (eg Channel 4's recent sponsoring of the investigating of the *Bismarck*/HMS *Hood* sites). Ironically, some multi-national shipwreck or 'historic' salvage organisations have espoused interests in investigating historic shipwrecks for their media value, seeking returns from marketing the search rather than selling off objects in an often unpredictable market place.

The requirement for an integrated approach to the UK's submerged heritage is apparent and all marine archaeological sites should enjoy parity of esteem and treatment with terrestrial archaeological remains as referred to by a preliminary policy statement produced by English Heritage (2000). The same document makes further points about the UK situation: the density of shipwreck remains in UK territorial waters is likely to be amongst the highest in the world with the greatest number in English waters; the overall resource is poorly surveyed, not well understood and, as a result, has poorly developed research frameworks; very few of the sites have been ground-truthed or assessed in terms of their preservation and susceptibility to impacts, by an archaeologically-competent authority; in marine archaeology generally in the UK the professional framework is poorly developed and supported, and amateur archaeologists have a more central role than they do on land; the number of recreational divers per length of coastline and proximity to the coast, are possibly the highest in the world, the number of sport dives is similarly significant in terms of the world proportion of dives. Many of these dives are on wreck sites.

A major problem is government will and motivation which is further weakened by legislative anomalies. The Protection of Wrecks Act 1973 confers no responsibility on any organisation for funding the management of sites of 'national' importance (including survey, excavation, conservation and publication) (Fenwick and Gale 1998) and in many cases throughout the UK, Designated Historic Wreck Sites are deteriorating through natural and anthropogenic processes (Oxley nd).

The necessary linkages between marine sciences, heritage bodies and their technical/conservation experience (whose experience is largely terrestrial) are seriously underdeveloped. This situation is as bad in the UK as anywhere else. If a 10–20,000 ton structure dating to World War I or World War II, complete with the interred remains of hundreds of people killed in military action, unchanged apart from natural degradation, existed on land there would be considerable support for appropriate protection and management. Similarly, if a large, wooden structure from 1703, housing a full complement and all equipment and stores was found, largely intact; we would be doing something to protect it. There is such a case in English waters where the legally protected Designated Historic Wreck Site of the warship the *Stirling Castle* on the Goodwin Sands is being actively destroyed, yet its survey, monitoring and rescue is being carried out principally by local, voluntary, self-funded efforts (Peacock 2000). Data about the processes destroying this nationally significant heritage resource is sparse; central government resources cannot be used for the support of the rescue efforts; we have little knowledge about specific stabilisation possibilities, costs and chances of success.

As more and more areas of the seabed become accessible the challenge of balancing the promotion of appropriate access with ensuring protection will become more acute. Opportunities for cooperation and integration with other marine sciences and organisations should be pursued, together with the raising of public awareness and the development of schemes promoting 'holistic' wreck diving for recreational divers, and for involving amateur archaeologists in monitoring and reporting initiatives. Finally significant gains will be made from extending data management initiatives (such as maritime archaeological GIS) from site and area mapping to the intra-site analysis of spatial data relating to condition, stability and exposure to impacts (Mather and Watts 1998).

Conclusions

Shipwrecks are monuments of significant archaeological and historical importance with the potential for providing widespread benefits through education, research and amenity to present and future generations. Assuming that the preferred option is to manage the site *in situ*, such benefits will only be accrued through the development of effective management methodologies.

We need to move on from a traditional response characterised by individual shipwreck investigations, carried out in a context of increasing diving access which often involves trophy-hunting, usually in the unforgiving glare of the public eye. Innovative approaches could also include local stewardship or management partnerships, promoting cooperative projects based on strong inter-disciplinary links. New finds should be approached using a background of steadily increasing knowledge of the whole shipwreck archaeological resource and incorporating a carefully considered implementation of a range of legislative and administrative tools.

Some advances in *in situ* management research, taking advantage of developments in marine environmental resource management, have been described here together with investigations of the monitoring and quantifying of processes that affect the condition of sites. Such initiatives will provide much needed confidence in understanding impacts and agents of change on a site, and offer the chance of neutralising or reducing negative trends.

Despite these few advances we do not understand the processes occurring on historic wreck sites with any degree of confidence and, as yet, heritage regulatory organisations do not have the resources to acquire or maintain effective access to most of the submerged heritage on a national, local or global scale, or to carry out effective mitigation of impacts when they get there.

Acknowledgements

The assistance of the following persons and organisations is gratefully acknowledged: Dr Hamish Mair and Bobby Forbes (Department of Civil and Offshore Engineering, Heriot-Watt University); Dr Brian Calder (Center for Coastal and Ocean Mapping and Joint Hydrography Center, University of New Hampshire); Rich Lear (Reson Offshore Ltd, Aberdeen); Martin Dean and Mark Lawrence (Archaeological Diving Unit, University of St Andrews); Dr Rory Quinn and Colin Breen (Centre for Maritime Archaeology, University of Ulster); Larry Murphy and Matt Russell (Submerged Cultural Resources Unit, US National Parks Service).

Bibliography

Breen, C (ed), 2001 *Integrated maritime investigations on the historic shipwreck* La Surveillante, Coleraine

Breen, C, and Forsythe, W 2001 Management and protection of the maritime cultural resource in Ireland, *Coastal Management* 29(1), 41–51

DETR, 2000 Department of the Environment, Transport and the Regions, *MV* Derbyshire *re-opened formal investigation*, London

Dromgoole, S, and Gaskell, N, 1999 Draft UNESCO Convention on the protection of the underwater cultural heritage 1998, *Int J Marine Coastal Law* 14(2), 171–206

English Heritage, 2000 *Taking to the water: a preliminary policy statement on the management of marine archaeology in England*, London

Fenwick, V, and Gale, A, 1998 *Historic shipwrecks: discovered, protected and investigated*, Stroud

Garzke, W H, Foecke, T, Matthias, P, and Wood, D, 2000 A marine forensic analysis of the RMS *Titanic*, in *OCEANS 2000 MTS/IEEE conference: where marine science and technology meet* (ed MTS/IEEE), 673–90, Piscataway

Gibbins, D, 1990 Analytical approaches in maritime archaeology: a Mediterranean perspective, *Antiquity* 64, 376–89

Gregory, D, 1999 Monitoring the effect of sacrificial anodes on the large iron artefacts on the Duart Point wreck, 1997, *Int J Nautical Archaeol* 28(2), 164–73

Historic Scotland, 1999 *Conserving the underwater heritage*, Historic Scotland Operational Policy Pap HP 6, Edinburgh

Kelleher, G, 1999 *Guidelines for marine protected areas: world commission on protected areas, best practice protected area guidelines*, ser 3, Cambridge

Lenihan, D J (ed), 1989 *USS Arizona Memorial and Pearl Harbor National Historic Landmark: submerged cultural resources study*, Southwest Cultural Resources Center Professional Pap 23, Santa Fe, New Mexico

Maritime and Coastguard Agency, 2000 *Report on the wreck of the SS* Richard Montgomery, Southampton

Mather, I R, and Watts, G P Jr, 1998 Geographic information systems for submerged cultural resource management and site specific investigations, in *Underwater archaeology* (eds L E Babits, C Fach, and R H Harris), 3–13, Tucson, Arizona

Monitor National Marine Sanctuary, 1997 *Charting a new course for the* Monitor: *comprehensive, long range preservation plan with options for management, stabilization, preservation, recovery, conservation and exhibition of materials and artefacts from the Monitor National Marine Sanctuary*, Newport News, Virginia

Murphy, L E (ed), 1998 H L Hunley *site assessment*, Cultural Resources Management Professional Pap 62, Santa Fe, New Mexico

Oxley, I, nd [1998] The *in situ* preservation of underwater sites, in *Preserving archaeological remains in situ: proceedings of the conference of 1st–3rd*

April 1996 (eds M Corfield, P Hinton, T Nixon and M Pollard), 159–73, London

Oxley, I, 2001a Towards the integrated management of Scotland's cultural heritage, examining historic shipwrecks as marine environmental resources, *World Archaeol* 32(3), 413–26

Oxley, I, 2001b Towards an effective management scheme for the *La Surveillante* historic wreck site and its management *in situ*, in Breen 2001, 103–17

Peacock, R, 2000 Seadive, *Nautical Archaeol* 2, 12

Pournou, A, Jones, A M, and Moss, S T, 1999 Monitoring the environment of the Zakynthos wreck site, in *Art 99: 6th international conference on non-destructive testing and microanalysis for the diagnostics and conservation of the cultural and environmental heritage, May* 17–20, 1999 (ed Associazone Italiana Prove Non-Distruttive), 2001–18, Rome

Quinn, R, Cooper, A, and Williams, B, 1999 Marine geophysical investigation of Belfast Lough, *Archaeol Ireland* 13(1)

Smith, T, 1998 Project AE2: investigation of the HMA AE2 submarine wreck site, unpub New South Wales Heritage Office rep

SNH, EN, DOE(NI), CCW, JNCC, 1997 *Natura 2000: European marine sites: an introduction to management*, Scottish Natural Heritage, English Nature, Environment and Heritage Service, Countryside Council for Wales and Joint Nature Conservation Committee, Perth

13

Sutton Common (South Yorkshire): a monument at risk in England's wetlands

ROBERT VAN DE NOORT

Abstract

The work on the Iron Age site at Sutton Common, South Yorkshire, UK, has provided both inspiration and a testing ground for the development of English Heritage's strategy for wetlands. This paper concentrates on the non-technical aspects of the developing conservation management of the site, which includes *in situ* preservation of selected waterlogged remains, and summarises the main results of the Monuments at Risk in England's Wetlands project, the new strategy for which it formed the basis.

Introduction

This paper concerns the wider, non-technical, issues surrounding the *in situ* preservation of archaeological sites, alongside the technical aspects of preserving waterlogged remains. The principle site-specific issues of *in situ* preservation of archaeological sites containing waterlogged remains have been widely discussed elsewhere (eg Coles 1995; Kenward and Hall 2000; Coles and Olivier 2001). The need to keep the group water table high, in order to maintain an anoxic environment that inhibits microbial activity, requires that such an archaeological site cannot be preserved *in situ* without a certain degree of control of the hydrology of the site. In practical terms this means that wetland sites cannot be preserved in isolation as 'monument islands', but that a successful conservation management plan must involve consideration of the hydrological catchment of the wider region. As the legal framework does not provide mechanisms to enforce control of the hydrology, the agency wishing to achieve a successful conservation management of monuments in wetlands must therefore consider alternative methods.

English Heritage, the principal agency responsible for the management of archaeological sites in England, has in the last three decades commissioned a number of wetland projects, for example at Flag Fen and Seahenge, as well as survey programmes in the Somerset Levels, the Fens of East Anglia, the mires of the north-west of England and in the area around the Humber (recently summarised in *Current Archaeology* 172). On completion of the final survey, that of the Humber wetlands, English Heritage could look back at a programme of academic research that in terms of scale has few, if any, parallels in Europe. Nevertheless, the need for further work on the future management of these wetlands was required with particular reference to the conservation management of wetlands and wetland sites. In 2000, English Heritage commissioned the Monuments at Risk in England's Wetlands (MAREW) project from the University of Exeter. This programme was to provide a general picture of the condition of England's wetland archaeological resource and the risks it faces, and to create a benchmark against which future changes in England's wetlands could be monitored. It was to pay

Robert Van de Noort
Senior Lecturer in Archaeology
Department of Archaeology
University of Exeter
Exeter EX4 4QE, UK

particular attention to the effect of hydrological changes on the waterlogged organic archaeological and palaeoenvironmental remains in wetlands. It was also to form the basis for a new English Heritage strategy for wetlands.

The active conservation management strategy of the Iron Age site of Sutton Common, in South Yorkshire, started in 1997 and this enigmatic site with extensive waterlogged remains has acted both as an inspiration to many aspects of the new English Heritage strategy for wetlands, and as a testing ground for a range of core aspects of the conservation management approach advocated in the strategy. It is therefore valid, in this paper concerning the wider issues of *in situ* preservation of wetlands sites, to discuss the main aspects of the work at Sutton Common, before discussing the main results form the MAREW project and the key features of the English Heritage strategy for wetlands.

Sutton Common

The name 'Sutton Common' is used by the Ordnance Survey for three fields south of the small town of Askern (centred around National Grid Reference SE 563122) in the parish of Sutton, South Yorkshire. Sutton Common is dissected by the palaeochannel of the Hampole Beck, which is now completely drained. Until its enclosure in *c* 1850, the area was wet and peat formed the soil of the Common. Two enclosures of prehistoric date (South Yorkshire scheduled monument 291) form a pair of features on opposite sides of the palaeochannel. Both enclosures are situated on 'islands' of sands and clay of the '25-foot drift'/Lake Humber clays. The larger of the two enclosures is situated on the eastern side of the former Hampole Beck, the smaller of the two enclosures on the west side (Fig 13.1). Various suggestions have been made as to what the site's function was, ranging from that of a Roman camp or camps (eg Whiting 1936), to a 'lowland hillfort', refuge camp, a site with extensive grain storage facilities, an elite settlement, a site with ideological connotations signified by its locations between various wetland landscapes or possibly a combination of these (eg Parker Pearson and Sydes 1997 and elsewhere), but none of these suggestion can be proven to be more valid than others.

The prehistoric enclosures at Sutton Common have received considerable attention by archaeologists. In the 1860s, the site was mapped by Scott Surtees, and later, in the 1930s, Charles Whiting undertook the excavation of a number

Fig 13.1 Aerial photograph of Sutton Common, showing the remains of the two enclosures straddling the palaeochannel of the Hampole Beck (photo Neil Mitchell; copyright CCT/AP UK)

of trenches across both enclosures (1936). He described the site as a Roman camp (or two camps), which showed well-preserved archaeological remains, including parts of a ladder, and structures, including upstanding wooden remains. No further archaeological fieldwork was undertaken here until after the bulldozing of the larger enclosure in 1980. In 1987, the South Yorkshire Archaeology Unit, with support from English Heritage, undertook the assessment of the waterlogged deposits of the smaller of the two enclosures. This was the first of a number of assessments aimed at determining the state of preservation and the rate of desiccation of the organic remains. Work with similar objectives followed in 1988, 1992 and 1993, the latter by the University of Sheffield. The information on these assessments were collated and discussed in Parker Pearson and Sydes (1997).

In 1997, following the transfer of land to the Carstairs Countryside Trust (CCT) with financial support from English Heritage and the Heritage Lottery Fund, a conservation management strategy could be developed and implemented with the full support of the new landowner. A high-resolution digital survey of the area was undertaken, which showed that the elements of prehistoric landscape were caused by the variable desiccation of organic and inorganic sediments at Sutton Common, indicating that wet-preserved deposits still existed across the site (recently summarised in Chapman and Van de Noort 2001). Further work in 1998 reassessed the state of preservation and rate of desiccation of the site. It concluded that the survival of features, some organic, inside the larger enclosure beneath the ploughsoil was good, although the 'occupation layer' had not survived the recent agricultural activities. Opportunities for long-term *in situ* preservation of archaeological features of the area exist, and have since formed the basis for the work undertaken by CCT, English Heritage and the other agencies involved in the work at Sutton Common (see the Acknowledgements for a fuller description of the organisations involved).

On the basis of the work undertaken in 1998, it was recommended that a programme aimed at (re-)creating a wetter environment would not only place the monument in an environmental context which is more akin to its contemporaneous surrounds, where its functions and meanings can be much better understood and explained, but additionally, enhancement of *in situ* preservation of important elements of the waterlogged archaeological resource may be achieved. Moreover where desiccation has advanced considerably, preservation by record rather than preservation *in situ* should be considered and any programme aimed at creating a wetter environment should be accompanied by a programme of monitoring of the groundwater hydrology and its effect on the archaeological deposits. Hydrological data suggest that permanent waterlogging of the primary deposits within the ditches of the smaller enclosure and of parts of the palaeochannel of the Hampole Beck can be achieved; but the possibilities for *in situ* preservation of organic archaeological and palaeoenvironmental material in the larger enclosure following the creation of a wetland are more problematic.

Since 1998, additional work at Sutton Common has been based on these recommendations. A programme of monitoring the water table and its impact on the burial environment has been instigated through the CCT sponsorship of a research studentship (currently being undertaken by James Cheetham at the University of Hull; eg Van de Noort et al 2001a and 2001b). In 1999, additional work was commissioned by English Heritage to assess the extent and state of preservation of archaeological remains in the interior of the larger enclosure, and to excavate the entranceway to the larger enclosure and a small part of the causeway linking both enclosures across the Hampole Beck. This resulted in a much improved understanding of the site and, in particular, the manner in which the two enclosures were linked in the Iron Age. However, the research also confirmed earlier concerns that the long-term *in situ* preservation of waterlogged remains within the larger enclosure is not feasible. In order to address this issue English Heritage has commissioned the University of Exeter to develop an excavation and research project.

A series of 'town meetings' in the Askern Miners Welfare engaged local people with the project, and have provided the springboard for discussion about how the conservation management strategy could be incorporated in economic development of this former coal-mining community. The need to engage the young people with the work at Sutton Common formed the basis for the project 'Rubbish and Archaeology' undertaken in a partnership between CCT, English Heritage and the University of Exeter (Van de Noort and Panter 2000). This project involves experimental archaeology and scientific research into the deterioration of organic and inorganic remains through an experimental burial project with archaeological researchers and local schoolchildren participating side by side. In September 2000, 152 school children buried objects after analysis. These were to be re-excavated during the subsequent Nation Science Week whereupon the objects would be analysed again (unfortunately this had to be postponed following the Foot and Mouth disease outbreak). The project is intended to provide the children with some insight into the creation of the archaeological record, with hands on experience of some simple and some more advanced scientific methods, and to involve them closely with the archaeological work undertaken in the past and in the future at Sutton Common.

In terms of policy, the work at Sutton Common confirmed considerable shortfalls in the legal framework when it comes to the protection of wetland sites. This includes the limited definition of what constitutes an 'archaeological site' and the exclusion of the palaeoenvironmental archive from protection through scheduling. It also includes the wide issue of responsibility for management of the drainage system, and the integration and delegation of responsibilities among a range of governmental and non-governmental organisations involved in the environment.

Monuments at risk in England's wetlands

The reasons for the Monuments at Risk in England's Wetlands (MAREW) project included the growing realisation within the archaeological community that despite the enormous progress made in archaeological wetland research, relative little progress has been made on archaeological wetland management. Notwithstanding the publication of *Wetland management; a survey for English Heritage* (Coles 1995), few examples of the successful *in situ* management of wetland archaeological sites exist in this country. Furthermore, the efficacy of the management and mitigation of the main threats to the wetlands in England, including drainage and peat extraction, through either the scheduling process or planning guidance (eg PPG16) is not clearly understood and, finally, research into the effectiveness of *in situ* preservation of wetland sites is poorly developed.

The aim of MAREW was the collation and assessment of information that was required to underpin the future strategy for the archaeological management of wetlands. It aims to provide quantitative data on the extent of the wetland heritage and the rate of its destruction and deterioration. The majority of data were obtained through analysis of existing literature and data, principal among these the 101 Sites and Monuments Record (SMR) offices in England and the monographs of the English Heritage funded wetland surveys, and through liaison with a large number of governmental and non-governmental agencies which hold data and information on wetlands – their destruction and conservation. One questionnaire, sent to all local authority archaeological officers, tested the awareness of wetland-specific issues, in particular the impact of changes of the hydrology on waterlogged archaeological and palaeo-environmental remains outside areas of development. A second questionnaire tested the awareness of the cultural heritage in wetlands amongst nature conservation and wildlife organisations.

MAREW concentrated on three categories of wetlands: upland peatlands; lowland peatlands and alluviated lands below 10m OD, totalling *c* 1.1 million ha of land (Fig 13.2). Several categories of wetlands, as defined by the Ramsar Convention on Wetlands, had to be omitted from the study, that is the rivers and their floodplains above 10m OD and the coastal wetlands. These areas were considered insufficiently researched in the field to allow for confident quantification of their archaeological potential. For the same reason the waterfronts of towns such as London and York and the 'urban wetlands' were omitted from the study. MAREW recommended that these areas should be the subject of separate studies.

The key results of MAREW can be summaries as follows (Van de Noort et al 2002). The identifiable archaeological resource of England's wetlands comprises 13,400 monuments, including 1800 monuments in upland peatlands, 4200 in lowland peatlands, and 7400 in alluviated lowlands. In addition, a significant number of unidentifiable monuments remain deeply buried, mainly in the alluviated lowlands and beneath upland peat. We must also recognise the importance of the extensive body of palaeoenvironmental source material, forming either matrices of archaeological sites, or separate 'archives' that hold information on landscape and climate change (eg Burton and Hodgon 1987).

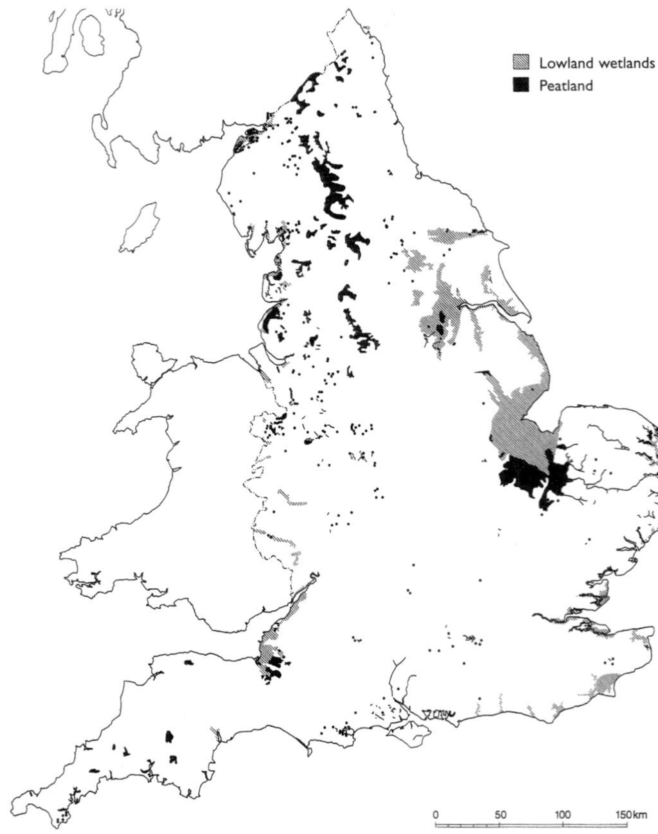

Fig 13.2 Map of the main wetlands in England, as defined for the purpose of the Monuments at Risk in England's Wetlands project (Mike Rouillard)

Seven key-causes of monument destruction in England's wetlands have been recognised. These are: drainage; water abstraction; conversion of pasture into arable land; peat wastage; peat erosion; peat extraction and urban and industrial expansion onto wetlands (Fig 13.3). Of these, the effect of drainage and the associated change of land use has the greatest impact either directly, through the lowering of the water table, or indirectly, for example through 'peat wastage'. Lowland peatlands have suffered more than other wetland areas, and MAREW estimates that at least 50% of the original extent of lowland peatland has been lost in the last 50 years alone. In the immediate future, the pressure of these threats on our wetlands is likely to be aggravated by the impact of climate change, especially sea level change (Bergkamp and Orlando 1999).

Over the last 50 years, the number of wetland monuments that have suffered from wholesale destruction is estimated at 2950. This includes mainly sites in lowland peatlands that have been destroyed through peat wastage (2020 monuments), peat extraction (230 monuments) and urban and industrial expansion onto wetlands (estimated at 700 monuments). Over the same period, the number of wetland

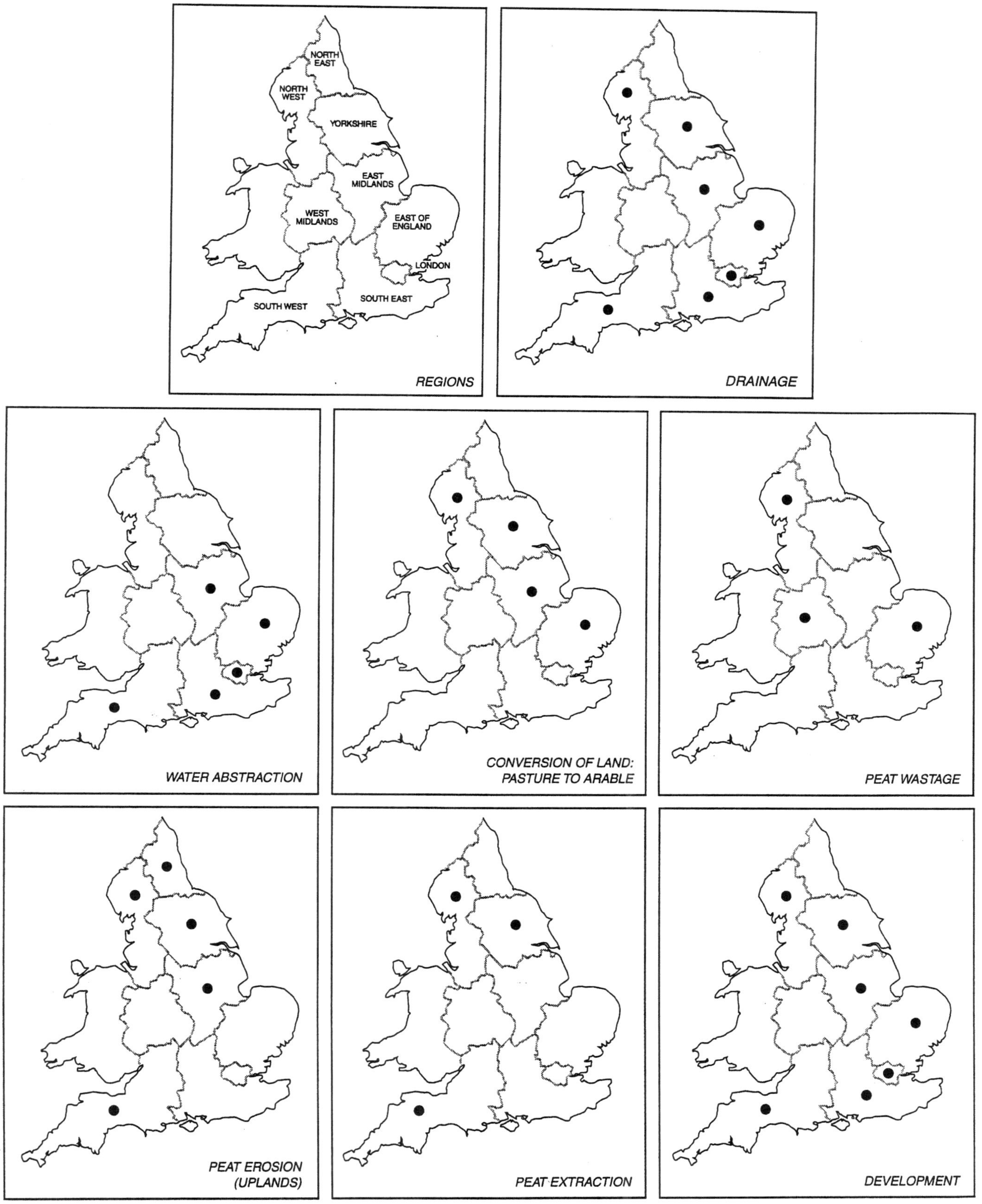

Fig 13.3 Characterisation of the threats to England's wetland regions as devised by English Heritage (Mike Rouillard)

monuments that have suffered from damage, desiccation and partial destruction is estimated at 10,450 (with some suffering more than one type of damage). This includes mainly sites in alluviated lowlands and lowland peatlands that have been damaged through drainage (9020 monuments), and ploughing (1650 monuments suffering from drainage and desiccation are now also under arable instead of pasture land use), while 360 sites are no longer protected by upland peat. An additional number of wetland sites survive more or less intact, buried beneath upland peat (estimated at 1440 monuments), lowland peat (estimated at 2940 monuments) and alluvial sediments in the lowlands of England (no estimate attempted). These sites are likely to be well preserved, although a number of key causes of change in particular water abstraction, drainage and urban and industrial expansion onto wetlands threaten this resource.

Appreciation of the wetland resource in the archaeological community was assessed by a short questionnaire sent out to all local government-based curatorial archaeologists. On the basis of the responses (64 out of 87 questionnaires were returned, 74%), 43 archaeologists (72% of respondents) stated that wetlands existed within their local government area, and 40 (67%) recognised that these wetlands included an archaeological resource or potential. Only six of the 43 local authorities with wetlands (14%) had policy procedures in place to deal with wetland-specific matters, although the type of procedure ranged from detailed statements that were taken into account in the planning process including the issues the impact of hydrological changes beyond the limit of the development (eg Avon, Cambridgeshire and Somerset) to additional but limited requirements in briefs for archaeological work.

In its recommendations to English Heritage, MAREW drew on a range of 'best practice' examples from across the country and beyond (eg Coles 1995), and the experience at Sutton Common provided considerable inspiration. The management strategy at Sutton Common involved a range of practical mechanisms to protect the waterlogged remains, including the use of dams and changing the efficacy of the existing drainage system, in close cooperation with the local Internal Drainage Board. Alongside the practical work, the monitoring programme is ongoing. The need to consider the wider hydrological context of the Sutton Common site and the need to protect palaeoenvironmental archives had resulted in 'interagency' cooperation on the ground, with English Heritage, English Nature and Countryside Agency and others. The work at Sutton Common also highlighted a number of deficiencies in the existing policies and legislation. The outreach to the local community and education projects undertaken with local schools has contributed considerably to the appreciation and awareness of the archaeology of Sutton Common of the public. Finally, and most importantly, at Sutton Common we recognised that not all waterlogged remains could be preserved *in situ*. These remains will therefore be excavated as part of an academic research project aimed at unravelling the Sutton Common enigma (Van de Noort and Chapman 2001).

English Heritage strategy for wetlands

Considering the extent of the cultural heritage resource of our wetlands, and the present and future threats it faces, the future conservation and management of the monuments in England's wetlands cannot be undertaken by the archaeological community alone. However, the cultural heritage value of wetlands sits alongside nature conservation and recreational values. MAREW recognised that the future of monuments in England's wetlands can only be addressed through cooperation with other organisations, and that this must form the cornerstone for any strategic approach.

English Heritage's strategy for wetlands is not only based on the management of individual sites within their wider landscape setting, but also the wider issues of outreach and education, procedures and policy and research.

Management strategies

To promote practical mechanisms to conserve and protect the cultural heritage by developing guidance and best practice for the integration of cultural heritage and nature conservation in wetland management.

Outreach and education

To promote and disseminate understanding of the cultural heritage of wetlands as an essential precondition for the development of successful management policies.

Procedures and policy

To promote the cultural heritage interests of wetlands in the work of local authorities, national, international and intergovernmental agencies.

Research

To promote applied research to underpin and inform the management of wetlands.

An implementation plan has been developed, which will address many of the issues raised by MAREW. In terms of management strategies, this will involve the production of an inventory of the most important wetland monuments in England, and the design and implementation of site-specific conservation management strategies. The development of mechanisms for a selection of these sites to function as 'beacon sites', ie platforms for interagency cooperation forms a key component of disseminating 'best practices'. The development of partnership with other agencies' conservation management strategies for wetland landscapes forms the basis for the successful preservation of wetlandscapes.

As for outreach and education, this will involve the furthering of cooperation and effective liaison between English Heritage and statutory organisations relevant to wetland management. It will also include the dissemination of information through the popular media and the provision of access to wetland archaeological research during excavations through 'open days', publicity and participation.

Concerning procedure and policy, English Heritage will develop, in consultation with Association of Local Government Archaeological Officers (ALGAO), a geographical information systems (GIS) resource for curatorial authorities to be used in the planning process. This GIS resource will incorporate map data on the extent of the various types of wetland landscapes linked to an assessment of archaeological potential and contextual mitigation policies. It will also involve development and training of local authority archaeologists, managers and consultants on the hydrological context of archaeological sites in the planning process for consideration within the existing PPG16 framework. In consultation with ALGAO English Heritage will enter negotiations with the Peat Producers Association (PPA) on the adoption of a uniform archaeological policy in areas under commercial peat extraction and 'best practice' guidance on the cultural heritage for use by the peat industry, conservation bodies, landowners and farming and wildlife groups will be produced. Finally, English Heritage will continue to consider changes in legislation (eg the Culture and Recreation Bill), policy and planning regulations at all levels, where possible with its partners, including the Europae Archaeologiae Consilium (EAC) (Coles and Olivier 2001).

In terms of research, English Heritage will develop pilot surveys of the upland peatlands in England, linked to an assessment of the erosion of upland peat and develop assessments of the rivers and their floodplains above 10m OD, the intertidal wetlands and urban waterlogged resource. It will develop research programmes that provide qualitative assessments of wetland monuments that remain deeply buried by peat and alluvial deposits, develop research on the causes of peat wastage and develop research into the impact of statutory designation on archaeological monuments. The need for the development of research into wetland-specific prospection techniques is recognised. English Heritage will also develop research into the preservation of archaeological remains *in situ*, including the effect of the chemical oxidants on wetlands under pasture, the effect of bioturbation of different wetland tree and plant species and the dynamics of hydrology in various wetland landscapes and its impact on *in situ* preservation.

Conclusion

The future of wetlands in England and beyond benefits from its recent reappraisal and revaluation, mainly on the basis of the threatened flora and fauna, through the work of the Ramsar Convention on Wetlands and other organisations. The recent recognition by Ramsar that the cultural heritage of wetlands forms a vital part of the long-term sustainability of wetlands, has lead to this being the theme for World Wetlands Day 2002 (Davidson in Coles and Olivier 2001).

The future of wetland archaeology rests essentially on the successful cooperation of organisations and individuals who have an interest in wetlands. Only through interagency cooperation can a substantial part of what remains of our wetlands be protected. Only through 'whole wetland' management, for example as promoted by the National Trust at Wicken Fen in Cambridgeshire, can the sustainable *in situ* preservation of archaeological wetland monuments be delivered. Further research on a range of issues is required to play a constructive part in this process, and to ensure that the archaeological remains receive the optimal benefit from wetland conservation projects.

The future of Sutton Common will be assessed and the waterlogged archaeological remains here will be preserved *in situ* if we can restore the landscape to what it was before 1850. Whether we will succeed depends on a range of factors, not all under the control of those involved, varying from rainfall to the long-term impact of the agricultural fertilisers and nutrients added to the soil before 1997. Nevertheless, with the forthcoming excavations of the larger enclosure, the completion of the 'Rubbish and Archaeology' project and the study of the impact of the practical mechanisms to restore the Common, we expect the Sutton Common project to deliver a range of results that will benefit wetland archaeology for some time to come.

Acknowledgements

The 'Sutton Common project' is spearheaded by the Trustees of the Carstairs Countryside Trust (CCT), who own the land, in partnership with English Heritage, English Nature, Countryside Agency, the Universities of Exeter and Hull and Grantham Brundell and Farran. The project, which includes land acquisitions, wildlife and landscape enhancement, archaeological and palaeoenvironmental evaluations, research and conservation, and engineering works to raise groundwater levels has been made possible through the cooperation of the Sheard Family Trust and financial support from: English Heritage, Heritage Lottery Fund, English Nature, Countryside Agency, Darrington Quarries (Landfill Tax Credits) through Waste Recycling Environmental (WREN), James Goodhart, The Pilgrim Trust, the University of Hull and DEFRA's Countryside Stewardship Scheme. Help has also been given by Doncaster Naturalists Society and Crombie, Wilkinson Solicitors and Carter Jonas, Property Consultants, both of York. This innovative project forms one of the trial schemes in the Humberhead Levels 'Value in Wetness' Land Management Initiative. Participation with the Askern Community Partnership over public access and enjoyment or the site seeks to contribute to the environmental and economic regeneration of this Coalfields area in South Yorkshire.

The MAREW report was commissioned by English Heritage and co-authored with William Fletcher, Gavin Thomas and Ian Carstairs. The English Heritage Strategy for Wetlands was co-authored with Adrian Olivier. I am grateful to all for their cooperation in the past years.

Bibliography

Berkamp, G, and Orlando, B, 1999 Wetlands and climate change; exploring collaboration between the convention on Wetlands (Ramsar, Iran 1971) and the United Nations Framework Convention on Climate Change, background paper IUCN, The World Conservation Union

Burton, R G O, and Hodgon, J M, 1987 *Lowland peat in England and Wales*, Soil Survey in England and Wales, Harpenden

Chapman, H P, and Van de Noort, R, 2001 High-resolution wetland prespecting, using GPS and GIS: landscape studies at Sutton Common (South Yorkshire), and Meare Village East (Somerset), *J Archaeol Sci* 28, 365–75

Coles, B, 1995 *Wetland management: a survey for English Heritage*, Exeter

Coles, B, and Oliver, A, 2001 *The heritage management of wetlands in Europe*, Europae Archaeologiae Consilium Occas Pap 1, Wetland Archaeol Res Project 16, Brussels and Exeter

Kenward, H, and Hall, A, 2001 Decay of delicate organic remains in shallow urban deposits: are we at a watershed? *Antiquity* 74, 519–25

Parker Pearson, M, and Sydes, R, 1997 The Iron Age enclosures and prehistoric landscape at Sutton Common, South Yorkshire, *Proc Prehist Soc* 63, 221–59

Van de Noort, R, and Chapman, H P, 2001 Sutton Common, Sutton, South Yorkshire, Project Design SCOM001–03 (v 2), unpub Univ Exeter rep

Van de Noort, R, and Panter, I, 2000 Rubbish and archaeology; an educational project *Past, the newsletter of the Prehist Soc* 36, 3

Van de Noort, R, and Chapman, H P, and Cheetham, J L, 2001a In situ preservation as a dynamic process: the example of Sutton Common, UK, *Antiquity* 75, 94–100

Van de Noort, R, Chapman, H, Cheetham, J, 2001b Science-base conservation and management in wetland archaeology; the example of Sutton Common UK, in *Enduring records: the environmental and cultural heritage of wetlands* (ed B A Purdy), 277–86, Oxford

Van de Noort, R, Fletcher, W, Thomas G, Cartsiars, I, and Patrick, D, 2002 *Monuments at risk in England's wetlands: final report*, Exeter (www.ex.ac.uk.marew)

Whiting, C E, 1936 Excavations on Sutton Common, 1933, 1934 and 1935, *Yorkshire Archaeol J* 33, 57–80

14

Deposit monitoring in Tønsberg, Norway

IAN REED

Abstract

The archaeological deposits in Norway's medieval towns have been heavily depleted over time, with as much as 50% now destroyed. During recent years the cost of archaeological excavations has risen dramatically. This has given rise to a desire for alternative building methods that will preserve the archaeological deposits *in situ* and at the same time allow for the sustainable development of the town centres.

Following a limited depth archaeological excavation of a *c* 500m² site on the waterfront in Nedre Langgate, Tønsberg, a hotel was built over the remaining archaeological deposits. Prior to the construction of the hotel, monitoring instruments were installed on the north side of the site to record any possible changes in the water content and temperature in the remaining highly organic deposits. In addition, a series of dipwells was installed along the southern side of the site, with an additional one to the north-east of it. These monitor the changes in the levels of the groundwater and the groundwater chemistry. The monitoring programme started in November 1999.

Introduction

There are eight medieval towns in Norway. In addition there were a number of other smaller population nuclei and market centres which had an urban character, although these lost their importance by the end of the Middle Ages. Of the eight towns only four have substantial areas of intact waterlogged deposits. In each case, these are mainly, though not exclusively, located in the immediate vicinity of the urban waterfront. The archaeological deposits in the medieval towns are non-renewable resources which have been greatly depleted during the last 30 years due to the fact that the towns are living environments. During this period the aim of the Norwegian Cultural Heritage Act has been to preserve these deposits as ancient monuments. Permission to remove protected deposits can be regarded as the exception rather than the rule, and is only granted when special grounds exist. For the medieval towns, however, applications to penetrate the archaeological deposits have been denied in only exceptional cases. As a result of this, a conservative estimate indicates that on average at least 26% of the area occupied by archaeological deposits has been removed, while in the case of some individual towns it could be as much as 50%.

Furthermore, during the last ten years the costs of archaeological excavations have become relatively high. It is not unusual for excavations to cost between £375,000 and £750,000. The most expensive excavation to date cost in excess of £2,250,000. At the same time, the heritage management authorities have adopted a more restrictive attitude with regard to granting permission to excavate these protected archaeological deposits. As a result of this there are today a number of undeveloped plots within the medieval

Ian Reed
Researcher/archaeologist
Norwegian Institute for Cultural Heritage Research
Kjøpmannsgate 25
7013 Trondheim, Norway

towns. In order that these plots do not remain undeveloped forever, and in order to enable a sustainable development of the town centres, there is a need to develop building methods that meet the developers' wishes while at the same time maintaining heritage management's interests. In addition to this, the services (water and sewage pipes) in most towns are reaching considerable age and are ready for modernisation. There is a similar need for developing new methods which will allow this to be undertaken while at the same time limiting the scale of intrusion.

The site

The site in Nedre Langgate 40 lies near the present-day waterfront at a height of between 2.0 and 4.0m above sea level (asl), in an area known to encompass the high- and late-medieval waterfront development. In this area the shoreline displacement has been about 100m over the last 1000 years. Core sampling carried out during the assessment of the project and prior to the building work showed that the depth of the deposits here varied between *c* 4.0m and 8.5m. Some of the late medieval deposits were removed by 19th- and early 20th-century development. Further deposits were removed in connection with the laying of new sewers in 1976 when a *c* 1.20m wide trench was excavated archaeologically across the north-eastern end of the site.

The site developer wished to keep the costs of archaeological excavation to a minimum, so the building was planned without a basement and without the use of piled foundations. The Norwegian Directorate for Cultural Heritage (Riksantikvaren) gave permission for the plate foundations to be laid on top of the archaeological deposits, but insisted on the archaeological excavation of the deposits down to the required level. A limited depth excavation of an area of *c* 500m^2 took place in 1999. Following the excavation, permission was given to insert eight piles between 15m and 25m from the medieval waterfront. These were inserted using the odex drilling method; core samples being extracted from the deposits during the work.

The earliest structures uncovered during the excavation were large earth-filled timber foundation structures dated to the 13th century. These box-like foundation structures were constructed on the foreshore and formed the foundations for warehouses and quays. The rows of buildings constructed on these had their gable ends orientated towards the sea and faced onto a *c* 3.5m wide wooden paved passage which ran between them.

As one of the conditions for granting permission to develop the site, Riksantikvaren demanded that the site should be monitored and that this should be paid for by the developers. A monitoring project was designed by the Norwegian Institute for Cultural Heritage Research (NIKU) with the aim of looking at changes in both the groundwater level and groundwater chemistry, as well as changes in the moisture content of the archaeological deposits. In addition, the project is monitoring the possible effects of the new building and excavation on a listed 18th-century customs house immediately adjacent to the southern side of the site. The building is being monitored for possible structural damage that may be caused by subsidence linked to possible changes in the groundwater table or to possible changes in the archaeological deposits. The project is designed to run for at least five years, which corresponds to the guaranteed life expectancy of the soil moisture cells. The total cost for the five-year programme is approximately £100,000. Riksantikvaren is, however, willing to insist on the developer financing the monitoring for a period of up to 15 years should this be deemed necessary.

The monitoring programme

It was originally planned to install six dipwells to give a cross section starting *c* 65m to the north-east of the site and running down to its south-west corner. Unfortunately, due to the depth down to the water table, it was not possible to install the furthest point, and the first point in the cross section was therefore placed *c* 22m to the north-east of the site. Three dipwells were installed along the southern side of the site, and a fifth well was installed on the northern side of the site. The height of the groundwater was measured at weekly intervals from September to December 1999, and at monthly intervals subsequent to this. At the same time, water samples are drawn off with the aid of a vacuum pump, and the pH value, redox potential (mV) and temperature are measured using a Hanna Portable pH Meter type HI 8424.

In addition to the dipwells, a series of 6 Delta-TML2 soil moisture sensors and 6 Campbell Scientific Ltd – type 107 temperature gauges were installed in a vertical section on the northern side of the site in November 1999. These are connected to a CR 10X datalogger from Campbell Scientific Ltd. The datalogger collects information on the temperature and moisture content once a day.

During the installation of the probes, samples of the deposits were taken for laboratory analysis in order to characterise the material. Generally, the deposits had a high organic content, mainly wood chips, while those around point 5 (where a fire deposit was located) contained quantities of charcoal. The samples were analysed by the Norwegian Institute for Soil Research (Jordforsk) for water content, organic content, pH and conductivity (Table 14.1).

Water content

The probes where installed in a vertical section, as mentioned previously. Five of these were placed at intervals between 0.00m above sea level to 1.82m above sea level, while the sixth was placed below wooden paving in the section to the north-east at approximately the same level as point 4. The water content at the lowest point, point 1, shows more irregular fluctuations than the others, something that is

Table 14.1 Results of the analysis of the soil samples taken during the installation of the monitoring equipment

Sample	Height (m asl)	Water content %	Organic content %	pH	Conductivity
1	0.00	61.6	29.6	6.7	1.63
2	0.59	65.5	55.5	6.6	0.84
3	0.92	75.5	60.5	6.0	1.35
4	1.30	70.7	52.5	5.8	2.45
5	1.82	47.8	17.8	2.8	9.34
6	1.29	65.3	32.5	6.0	1.36

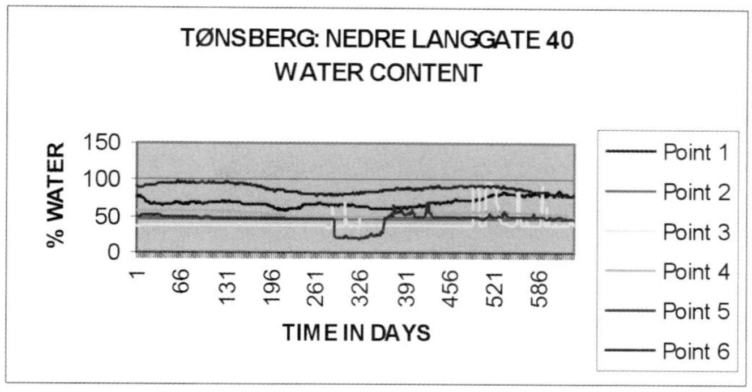

Fig 14.1 Daily measurement: water content

probably the result of fluctuations in the water table, a phenomenon possibly linked to tidal variation (Fig 14.1). The others, with the exception of point 5, show fluctuations that can probably be related to climatic conditions, amongst other things. The marked peaks in the measurements at point 3 are the result of short-term saturation of the deposits during a period of exceptionally high rainfall in August and September 2000. The exceptionally high peaks at the same point in March and April 2001 are not readily explainable, as there was very little rainfall at this time.

Point 5 shows a marked and rapid drop in the water content between the end of August and the beginning of November 2000. This was due to the probe being uncovered during some unscheduled digging by the developers who lowered the ground surface by about 0.7m. The probe was later reinstalled and is now registering just below the surface.

It is interesting to note the differences in the water content at points 4 and 6. These points are at approximately the same level and are located about 1.0m from one another, but display considerable differences in water content. At point 4 the water content varies between c 50% and c 60%, while at point 6 it varies between c 80% and c 98%.

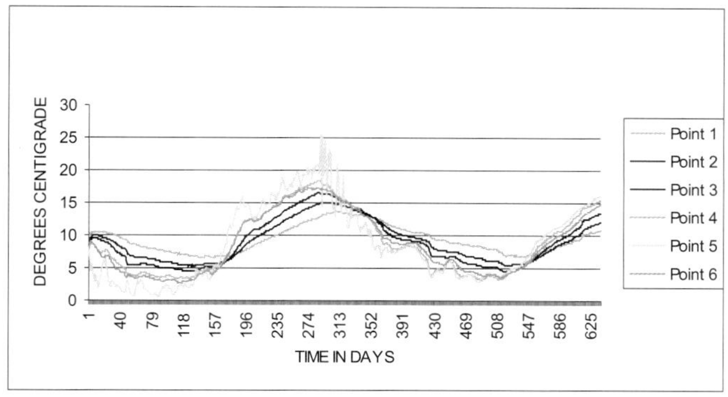

Fig 14.2 Daily measurements: soil temperature

Temperature

The temperature is being monitored (Fig 14.2) in order to document whether covering the deposits causes an increase in temperature as has been suggested is the case in Sweden (Larsson 1995, 15). It is probable that this increase in temperature could lead to increased microbial activity (Borg 1993, 49; Hopkins nd, 80). It is clear that the observed fluctuations in temperature, irrespective of depth, mirror the seasonal changes in temperature. It is obvious from the figure that the closer to the surface the measurements are taken, the greater are the fluctuations. It is clear, however, that there has not been a rise in temperature since the monitoring area was covered in 2000. Indeed, the temperature of the deposits shows a drop of between 1°C and 3°C, dependent on the depth below the surface, when one compares the same point in time in both 2000 and 2001. It is also interesting to note that while points 4 and 6 show marked differences in their water content, the temperature shows only minor differences.

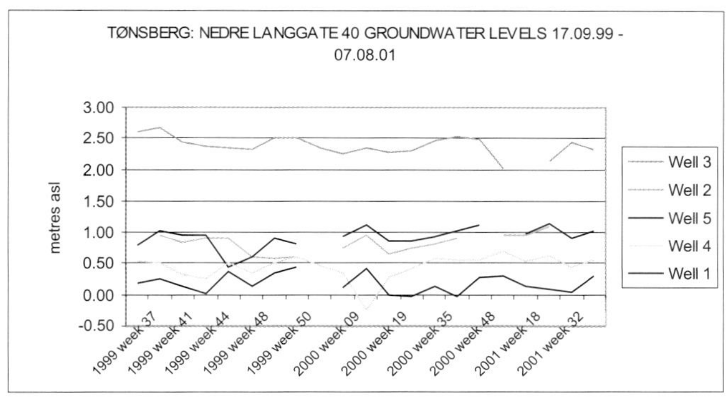

Fig 14.3 Variations in recorded groundwater levels

Groundwater levels

There have been some suggestions that deep excavations and the removal of archaeological deposits can cause the lowering of the water table around new buildings (Larsson 1995, 14), and that this can cause dewatering of the surrounding areas. The dipwells have been inserted in order to test this and to see what effects, if any, this development has on the groundwater level in the area adjacent to the site (Fig 14.3). The levels in these will undoubtedly be affected by barometric pressure and possibly also by the amount of precipitation, although it is likely that much of this will run off into drains and will not affect the water table noticeably.

During the winter months it has been difficult to take readings from the dipwells due to the caps covering the wells being frozen solid, or due to large amounts of snow or ice covering them. As a result of this the measurements are not continuous. Generally the wells display local variations, with wells 2 and 5 being the most similar, which is perhaps not surprising since they are more or less equidistant from the present-day waterfront. Well 3, which is furthest from the site, shows some variation in the water level, with a marked drop registered at the end of 2000. Unfortunately we are missing data for a five month period during the winter 2000/2001 and therefore cannot say anything about what happened during this period of time. There is a slight decline in the water level in well 3, which could be indicative of the occurrence of dewatering, but the variations may just as possibly reflect the natural variations here.

Redox potential and pH values

The redox potential and pH values are both taken using the same instrument. This has not been without its faults, and has been sent in for control and re-calibration on a number of occasions. Consequently it has not been possible to adhere to the original time schedule for monitoring. At the same time, while the monitoring of the pH values has been relatively straightforward, this has not been the case with the redox potential data. This data has shown considerable variation within each well, and indeed it has been difficult to get an exact reading in the field. The redox value in one case started at -273,6mV, followed by a steady climb to -172mV without stabilising. On average, the redox potential in the different dipwells varies between 160mV and 200mV. Caple and Dungworth (1998) have pointed out the problems regarding the relationship between the quality of dipwell water and the surrounding waterlogged burial environment, with the dipwell water yielding higher redox values than the *in situ* measurements. If this is the case then it is probable that *in situ* measurements here would give lower values.

With regard to the pH values, these have generally been in the range of 6.4 to 6.91 but with values on two occasions of 7.07 and 7.52, and in one case as high as 10.3. With the exception of the few high values, it is probable that these measurements do give a correct picture of the level of acidity of the groundwater.

Turning to the correlation between the redox potential and the pH values, given that there are problems with the validity of these measurements, they do, however, appear to indicate that the conditions are conducive to the preservation of organic material. The Eh-pH diagram (Fig 14.4) is based on the average of the last 12 measurements for each well.

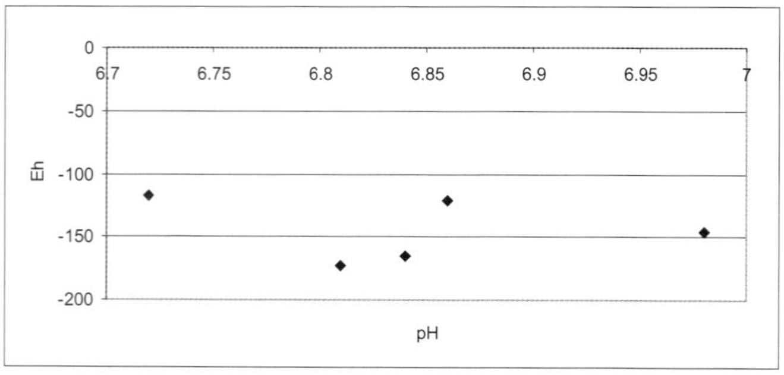

Fig 14.4 Eh-pH diagram based on the average of the last 12 measurements for each well

Conclusions

The monitoring project has been running for nearly two years, and both the reliability of the probes and their ability to monitor sudden changes have been demonstrated on several occasions. The data recorded so far by the probes would seem to indicate that conditions in the buried environment below the new hotel are at present stable, and that this building does not at present seem to have had any adverse effect on the deposits preserved below it. In fact, with regards to the ground temperature, it has had a positive effect insomuch as it now acts as a form of insulation. As a result of this, the soil temperature has been lowered and is more stable.

It is interesting to note here that the sewer trench originally excavated in 1976 was reopened during the course of the archaeological excavation, and the sections documented. There was little or no evidence of degradation of the organic deposits, and features recorded on the original section drawings were readily identifiable. The trench had been backfilled with the excavated soil, and only at the very top was gravel laid as a bedding for asphalt.

The monitoring project has also been a fruitful learning process, particularly with regard to the practicalities of monitoring on a construction site. The project was designed not to come into conflict with any part of the new building, but as the construction process progressed, plans were changed and we were not always informed of the changes. One result of this was the uncovering of one of the monitoring probes. Similarly, the dipwells along the south side of the site were originally placed well outside the limits of the original planned construction, but as time progressed two of these ended up within the construction area. As a consequence of these experiences, Riksantikvaren will in future insist on a detailed construction plan showing all areas needed to be excavated for the building and necessary services before any new monitoring project is planned.

Bibliography

Borg, G C, 1993 Hotet mot medeltida kulturlager ur naturvetenskaplig synspunkt. *Medeltidsarkeologiska föreningen* 93(2), 44–57

Caple, C, and Dungworth, D, 1998 Waterlogged anoxic archaeological burial environments, unpub Ancient Monuments Laboratory Rep 22/98

Hopkins, D W, nd [1998] The biology of the burial environment, in *Preserving archaeological remains* in situ: *proceedings of the conference of 1st–3rd April 1996* (eds M Corfield, P Hinton, T Nixon and M Pollard), 73–85, London

Larsson, S 1995 *Nedbrytningen av urbana kulturlager*, Arkeologiska rapporter från Lund 10, Lund

15

Monitoring methods in mires and meadows: five years of studies at Nydam Mose, Denmark

HENNING MATTHIESEN, DAVID GREGORY, BIRGIT SØRENSEN, TANJA ALSTRØM and POUL JENSEN

Henning Matthiesen
Scientist

Birgit Sørensen
Head of Section

Tanja Alstrøm
PhD student

Poul Jensen
Senior Scientist

Department of Conservation
National Museum of Denmark
PO Box 260
DK-2800 Lyngby, Denmark

David Gregory
Senior Scientist
National Museum of Denmark
Centre for Maritime Archaeology
Havnevej 7
DK-4000 Roskilde, Denmark

Abstract

Present-day Nydam Mose is an 11ha waterlogged meadow. During the Iron Age the site was a small lake into which weapons and ship parts were deposited on several occasions between AD 200 and 500. Archaeological excavations in Nydam were begun in the 1850s and later resumed in the 1990s by the National Museum of Denmark. Due to the volume of finds and resources required for conservation excavations were stopped in 1997 to investigate the feasibility of preserving the site *in situ*. Over the last five years a programme of investigation to examine the deterioration of archaeological material has been ongoing. Three approaches have been taken: examination of the environment in Nydam using chemical, electrochemical, geological and hydrological techniques; study of corrosion rates on modern model materials; and examination of artefacts by X-ray and microscopic methods. Selected results of these investigations are presented in this paper paying special attention to metallic artefacts. The effects of a fluctuating water table will be discussed as well as the future *in situ* preservation of Nydam Mose.

Introduction

This paper deals with an ongoing feasibility study into preserving the site of Nydam Mose *in situ*. At the beginning of the study in 1997 the aim was to determine the physical, chemical and biological nature of the environment and what effects these parameters would have on the metal and wooden artefacts in the meadow (Sørensen and Gregory 1998). The methodologies developed and employed will be described, along with the results obtained. To complement the monitoring, studies of modern test materials as well as conserved and unconserved archaeological artefacts are ongoing. The first results from studies of wooden artefacts have already been published (Gregory et al 2001). In this paper results from the studies of metallic materials are presented.

Background to Nydam

Nydam Mose is an 11ha water meadow located in a small valley in Southern Jutland, Denmark (Fig 15.1). It is surrounded by moraine hills with intensive farming right to the bank of the meadow.

During the Iron Age Nydam Mose was a shallow freshwater lake into which sacrificial offerings were deposited on at least five or six different occasions during the period from approximately AD 200 to 500. The artefacts that were deposited appear to have been military and nautical paraphernalia. This includes two large rowing boats and another boat, which had been broken into pieces prior to its deposition. All kinds of weapons such as swords, arrows,

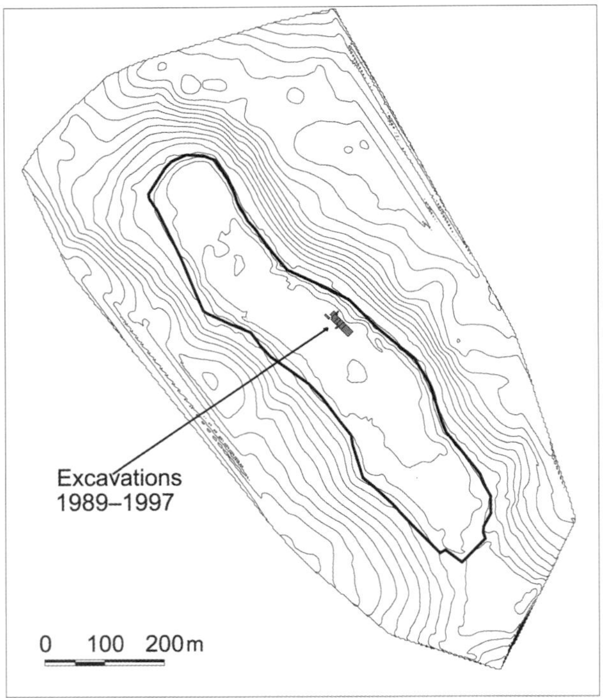

Fig 15.1 Topographic map of Nydam Mose showing the area excavated during the 1990s, the outline of the meadow is indicated by a thick line (contour interval 1.0m)

spears, richly decorated shields, scabbards with precious metal fittings and personal equipment such as knives and tweezers were thrown into the lake and, in some cases, swords and lances were stabbed into the lake bed.

Around AD 500 the lake became overgrown and evolved into mire. In the 16th century a small ditch was dammed to create a millpond. In the 20th century the water supply to the ditch was disturbed and drainage was established, turning the mire into a meadow. In 1993 legislation was passed so that the archaeological remains in the meadow would be protected from agricultural soil treatment and driving with heavy machinery.

The changing conditions in Nydam Mose have resulted in a layered structure of peat and gyttja, described by Christensen and Kolstrup (1998). The stratigraphy of the excavated area (Fig 15.1) is 0.5–1.0m of Iron Age gyttja, overlain by 1.0–1.2m of peat. In some places the peat has been mixed up by peat digging over the centuries. The finds in Nydam are deposited in two discrete layers approximately 0.80m and 1.25m below the ground surface, respectively.

Archaeological finds at Nydam Mose were discovered in the middle of the 19th century, where Conrad Engelhardt carried out excavations and found the Nydam Boat. Since then several excavations have been carried out, the latest by the Institute of Maritime Archaeology of the Danish National Museum, who re-excavated Engelhardt's old pits and their surroundings during 1989–1997 (Rieck et al 1999). The re-excavation covered an area of 600m^2 and yielded approximately 15,000 wooden artefacts and 1500 metal artefacts (Fig 15.2).

Due to the volume of finds and resources required for

Fig 15.2 A weapon heap discovered in 1990

conservation, excavations were stopped in 1997 to investigate the feasibility of preserving the site *in situ*. As only a minor part of the area has been excavated (Rieck et al 1999), it is predicted that many artefacts of wood, iron, bronze, precious metals, bone, leather, amber, and glass still remain on site.

In situ preservation

The 1997 decision to stop the excavations was – and still is – a matter of debate, as it is thought by many that the artefacts are at risk of accelerated deterioration. It has been stated by archaeologists that those artefacts excavated in Nydam during the 1990s are in a worse state of preservation than those excavated in the 19th century. In other countries, similar statements have been supported by statistical evidence during recent years (eg for bronzes in Sweden, (Fjaestad et al 1998) and metals in Germany, (Scharff et al 2000)). However, these studies were made in environments quite different to Nydam Mose.

A research programme was started in 1997 in order to gain a better understanding of the environment in Nydam – its variation in time and space – and of possible deterioration mechanisms for the artefacts. Amongst the questions raised were those concerning whether the find layers dry out during dry summers and what are the possible effects on the artefacts; whether archaeological excavations threaten the surrounding finds and whether anything can be done to improve the environment at Nydam. The central question was whether the 1997 decision of preserving the site *in situ* was a sound decision. Despite not all the questions being answered, some of the methods and selected results are presented here.

Monitoring methods

Water level

From 1997 to 1999 a system of 20 perforated plastic tubes (dipwells) with a length of 1.25m and a diameter of 30mm

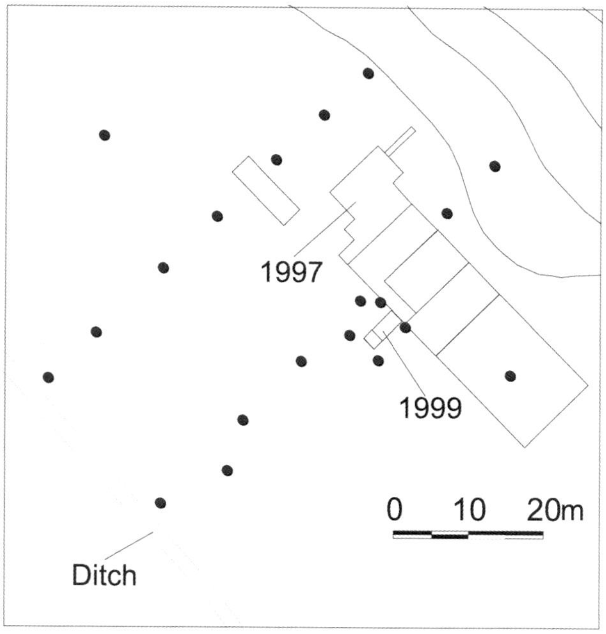

Fig 15.3 Positions of dipwells in Nydam Mose in relation to the excavated areas (1989–1997 and 1999)

were inserted on the site (Fig 15.3). The tubes still remain on the site.

The positions were selected in order to gain information on the general variation in the water table on the site and also to monitor the effects of pumping during the 1997 and 1999 excavations. The positions of the tops of the pipes were measured accurately. Subsequently the water levels in the pipes were measured at intervals during the period 1997 to 2001.

Pore water composition

The chemical parameters monitored in Nydam include pH, oxygen, sulphate, sulphide, ammonium, nitrate, nitrite, phosphate, potassium, chloride, iron, and manganese. The parameters were chosen because either they affect the deterioration of the archaeological material directly (pH, oxygen, chloride, sulphate), serve as nutrients for micro-organisms (ammonium, nitrate, phosphate), or show the redox conditions in the area (sulphate, sulphide, nitrate, iron, manganese). Besides these, gold and platinum electrodes have been employed in the area to test their reliability in measuring redox potential, but these data have not been evaluated yet.

From 1997 to 2000 most measurements were made on water samples from the inserted pipes. The procedure included withdrawal of the water with a syringe, centrifugation (if necessary), filtration, and measurement, using standard procedures.

However, using water from the pipes has its limitations, as it will not reveal possible vertical differences in the water chemistry. For example considerable differences are possible between the peat and the gyttja layers in the meadow. Further, the water in the pipes might be influenced by atmospheric contact, especially by oxygen in the pipe.

A new measuring technique with microelectrodes mounted in a hollow steel spike penetrating the different soil layers gives insight into the vertical variation and avoids atmospheric contact (Alstrøm 2001). To date, electrodes for oxygen and pH measurement have been used and this technique will be used in Nydam in 2001.

Corrosion measurements on finds

Twenty-eight lances from the 1993 and 1994 excavations and 75 lances from the 1991 excavation have been X-rayed in order to measure the corrosion depth. These lances have all been conserved and exterior corrosion products have been removed. The X-ray shows the original surface and the contours of the preserved iron core. The corrosion depth is measured in 10–15 positions at 5mm intervals on each side of the picture using a digital sliding gauge. If the whole socket or part of it is fully converted to corrosion products a default value (1.75mm) for 'totally corroded' was used.

In 1999 three 'fresh' lances were excavated and kept in ethanol without conservation. The purpose of the small 1999 excavation was specifically to obtain unconserved artefacts to analyse the state of preservation. The analysis of these lances is still pending.

Corrosion measurements on test plates

Twenty-eight pre-weighed iron test plates were placed at 0.80m and 1.45m below ground surface in October 1997. The test plates consisted of a 10 x 35 x 0.5mm iron plate fixed to the end of a plastic tube. Inside the tube a cable was fixed to the test plate to allow potential measurements. These test plates were retrieved in May 1999 and reweighed after removal of the corrosion products. The corrosion rate (in $\mu m \cdot year^{-1}$) could then be calculated using the density and surface area of the test plates as well as the exposure time.

In November 1999 and August 2000 new pre-weighed iron, copper, and bronze test plates were placed on the site 1.00m below ground surface, this time only connected to the surface by a thin thread. The first of the iron test plates was excavated in June 2001 and reweighed.

Results and discussion

Water level

The water level is without doubt the most important single parameter when it comes to preservation of waterlogged archaeological material *in situ*. The water level in Nydam Mose compared to the position of the archaeological remains has been intensively monitored during the last few years and modelling the results is ongoing. Fig 15.4a shows results from April 1997, where the water level was approximately 0.20m below the meadow surface, with a slight decline in the water table towards the ditch in the central part of the meadow (lower left side of picture).

During May and June 1997 an archaeological excavation took place in the area shown on Fig 15.3. The water level in the excavation site was kept below 2.80m through pumping. The water level in the surrounding area was measured in the pipes in the weeks up to 27 June, when the pumping stopped and the excavation site was backfilled (Fig 15.4b). The figure shows decreased water levels as far as 35m away from the small excavation site, so that at least one-third of a hectare of the meadow was influenced. It was observed, that the area affected by pumping gradually increased during the excavation period; the effect was not immediate.

After the excavation the water level remained low in the dipwells (1 July: 3.09m–3.61m; 19 July: 3.17m–3.65m) with a clear decrease towards the excavation site. A dry summer gave extremely low values at the start of September (1 September: 2.92m–3.34m), with the same decreasing trend. This was still above (but fairly close to) the known find layers (2.40m–2.80m). In the next season in October 1997, heavy rainfall had returned the area to a condition similar to Fig 15.4a (13 October: 3.50m–3.67m).

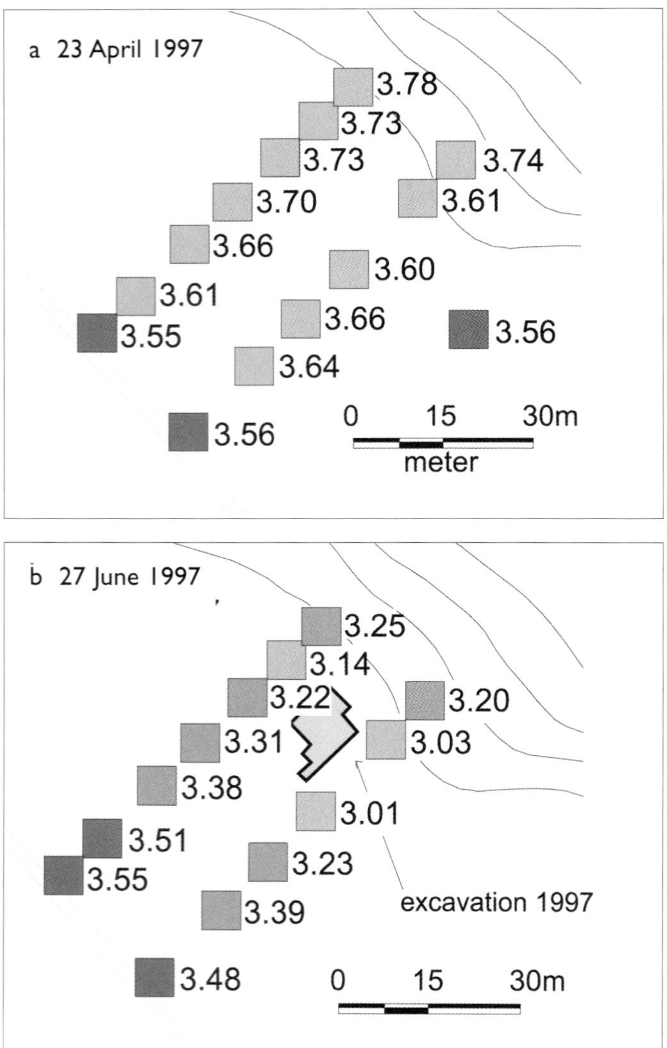

Fig 15.4 *a – water levels (m above sea level) measured 23 April 1997 (before start of excavation); b – 27 June 1997(at the end of excavation); for comparison, the surface of the meadow is 3.70–4.00m above sea level, and the archaeological remains are found 2.40–2.80m above sea level*

Monitoring water levels in this way is rather time consuming, and definitely not possible in all archaeological excavations. To this end Fig 15.5 is a predictive model of the water level in the meadow as influenced by excavations. It shows how the water table decreases as a function of time and distance to the excavation site. The water table measured on 23rd of April (17 days before the excavation started) has been used as 'background' and subtracted from the water heights measured during the excavation. The results show a water table development very similar to what is expected from a standard hydraulic model for pumping (Freeze and Cherry 1979). The water table decreases in the immediate vicinity of the excavation site (giving a log shaped curve), and later the effect spreads out (giving a more linear curve). In practical terms this means, that a short excavation in Nydam (eg for one week) will only have very local effects, whereas longer excavations will influence a much bigger area.

A preliminary fitting of the data in Fig 15.5 to a pumping model gave a hydraulic conductivity within the normal range for undecomposed peat (0.001–0.1cm/s; Clymo 1983). Our next step is to try to measure the hydraulic conductivity directly at Nydam and at other sites. With a good model it should then be possible to estimate the effects of future excavations and give advice about pumping (depth and intervals), the use of sheet pilings, and possibly re-watering of a site after excavation and backfilling.

Water chemistry

Up to 12 different readings have been measured in 15–20 dipwells during ten fieldwork seasons 1997–2001, so it is

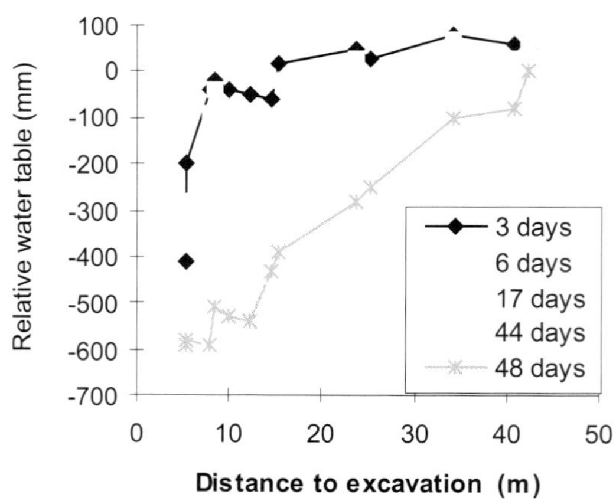

Fig 15.5 *Decreasing water table as measured during an excavation from 10 May to 27 June 1997; the different curves show results from 13 May (after 3 days), 16 May, 27 May, 23 June and 27 June (after 48 days) and the water table is given relative to values measured on the 23 April, which is the last measurement before the excavation*

only possible to present a very small selection of the available data. Averaging the results is not always possible (or wise), as some of the measured readings show a high degree of spatial and temporal variation through the meadow. Keeping this in mind, some of the general concentration levels measured in the dipwells during undisturbed conditions are shown in Table 15.1 to give an impression of the environment.

The neutral to slightly acidic conditions and rather high concentrations of nutrients and dissolved salts are in accordance with a classification of Nydam Mose as a rich fen (Heathwaite and Göttlich 1993). Highly variable results have been obtained for ammonium, nitrate, phosphate, sulphate, and dissolved iron without any unambiguous correlation to, for example, time of year or water level. This variation indicates a highly dynamic environment, where not only oxygen, but also sulphate, nitrate and iron (III) are quantitatively important as oxidants. It is still too early to conclude, what the 'typical conditions' in Table 15.1 mean to the deterioration rate of archaeological artefacts of different materials.

Table 15.1 Typical concentration levels measured in dipwells at Nydam between 1997 and 2001; excluded are the results measured during 'extreme' conditions (eg during the 1997 and 1999 excavations). Preliminary results from the steel spike and microelectrodes used in June 2001 indicate, that the oxygen levels measured in dipwells might be higher than in the surrounding soil pore water

Readings	Concentration level
PH	6.4–7.0
Sulphate	5–100mg SO_4/l
Chloride	20–60mg Cl/l
Sulphide	<0.03–1mg S/l
Dissolved iron	1–50mg Fe/l
Dissolved manganese	< 0.5–4mg Mn/l
Potassium	<10 mg K/L
Nitrate	<2–20mg NO_3/l
Nitrite	0.1–0.5mg NO_2/l
Ammonium	1–15mg NH_4/l
Phosphate	0.2–10mg PO_4/l
Oxygen	<0.1–2mg O_2/l
Conductivity	400–1000µS/cm

Effect of drainage

During the excavations in 1997 and 1999 the water level was lowered in the surrounding dipwells. These events are well suited to model the chemical effects of drainage, where the water level is lowered permanently, or more natural fluctuations, where the water level is lowered, such as during extreme dry summers.

Fig 15.6 shows the results from measurements of sulphate on 29 June 1997, just at the end of the 1997 excavation, and on 14 October 1997, three and a half months later. The results show increased sulphate levels near the excavation site. Other effects of the excavation are indicated in Table 15.2.

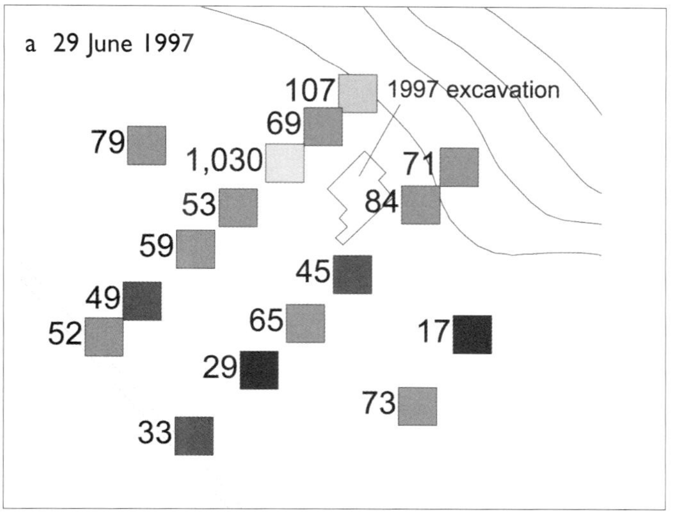

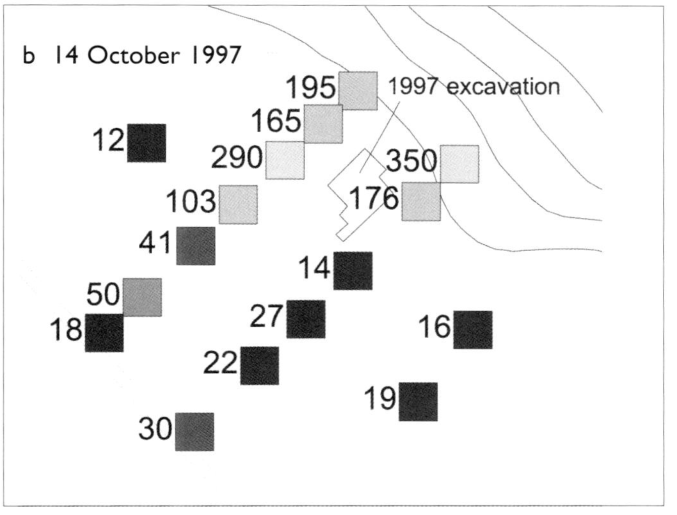

Fig 15.6 Concentrations of sulphate (mg/l) measured in the dipwells: a – 29 June 1997 (at the end of the excavation); b – 14 October 1997 (after the water table had reached its normal level)

Table 15.2 Range of concentrations measured in dipwells during and after drainage in 1997 (results from 27 May and 14 October 1997); only readings influenced by the excavation are shown; for comparison the normal range (from Table 15.1) is shown and some extreme ranges measured on 9 September 1998. All units are mg/l, except for pH (unitless)

Readings	Normal range	During and after drainage	'Natural' extreme (09/09/1998)
pH	6.4–7.0	5.7–6.4	6.0–6.8
Sulphate	5–100	20–1000	5–160
Potassium	<10	4–240	not measured
Nitrate	<2–15	3–90	2–40
Nitrite	0.1–0.5	0.1–14	0.1–0.3
Ammonium	1–15	2–130	1–300

The sulphate most probably stems from an oxidation of sulphur and sulphide compounds in the peat caused by the lowering of the water table during the excavation (Fig 15.4). Oxidation of reduced sulphur compounds gives rise to a production of H^+ and thus a falling pH, a phenomenon well known from for example drainage of mires (Heathwaite and Göttlich 1993) and also observed in this study (Table 15.2).

Most of the produced H^+ reacts with the peat in an ion exchange reaction, where H^+ is bound by the peat and base ions such as K^+, Ca^{2+}, Mg^{2+} and Na^+ are released (Clymo 1983). Of these, only potassium was measured at Nydam, and indeed a high release was found during the excavation (Table 15.2). By binding H^+, the peat serves as an effective pH buffer. It can be calculated, that without this buffering effect a sulphate production of say 100mg/l would cause pH to drop down below 3, which could be fatal to the archaeological artefacts. The high buffering capacity of peat (100–150μmol·g^{-1}, Heathwaite and Göttlich 1993) is therefore an extremely important feature.

The increased levels of nitrate, nitrite and ammonium shown in Table 15.2 are probably due to a strongly increased decomposition of peat during aeration.

On 5 May 1998, ten months after the excavation had finished, increased sulphate levels of 12–90mg/l were found in dipwells near the excavation site compared to a background level of 5mg/l further away. This shows, that even if the find layers were not at any time drained themselves, the archaeological remains may still be influenced by the excavation and a subsequent dry summer, through changed water composition in the area. It also shows, that such effects are observed in the area a long time after the excavation has finished.

The environment indicated in Table 15.2 'during and after drainage' is more hostile to iron artefacts than the 'normal' environment. For example, the high sulphate concentrations can enhance corrosion of iron by destabilising protective corrosion layers (Mattson 1992) or by enhancing microbial corrosion (Hilbert 2000). Also, the higher concentrations of dissolved ions can have an enhancing effect (Mattson 1992).

It is important to note that the dry summer following the excavation in 1997 has probably increased the effects on the environment significantly, and there are likely to be other 'natural variations' of the environment which cannot yet be explained (Table 15.2, extreme ranges from 9 September 1998).

Corrosion measurements on test plates

The effects of drainage on the water chemistry were distinct, as shown above. But can these changes influence the artefacts in the meadow? The corrosion rate measured on test plates can give an indication. In Fig 15.7a the calculated corrosion rates are shown for iron test plates buried at 0.80m (2.90m–3.20m above sea level) in October 1997. At that time, the water table had reached its normal level following the 1997 excavation. The test plates were retrieved in May 1999, before the small 1999 excavation. Six test plates buried at a depth of 1.45m (2.25m–2.55m above sea level) generally showed lower corrosion rates (3–29um/y), with a single exception (data not shown). For comparison the first results from iron test plates buried in November 1999 are also shown (Fig 15.7b), five months after the excavation, at a depth of 1.00m (2.79m–3.00m above sea level).

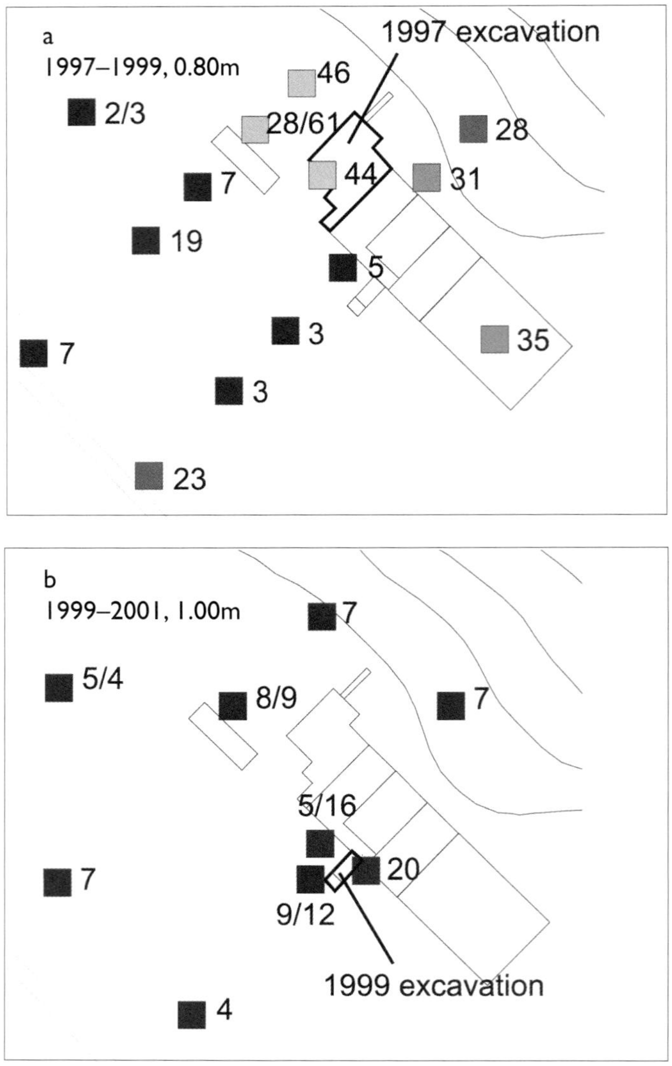

Fig 15.7 Corrosion rates on test plates of iron; a – one set of test plates were buried in October 1997 at 0.80m and retrieved in May 1999; b – the other set was buried in November 1999 at 1.00m and retrieved in June 2001 (all numbers are in μm·y^{-1})

Corrosion rates from 1997 to 1999 are characterised by large variations. Still, the results indicate that there is an effect from the changed water chemistry at the site, as the highest corrosion rates are found in and around the excavation site. This cannot be an effect of direct aeration of the test plates as they were first buried after the water table in the area had reached its normal level of 3.50m–3.70m above sea level.

The corrosion rates for the period 1999–2001 are more homogenous, and considerably lower than the rates from 1997–1999. But also there seems to be an effect from the small 1999 excavation, even though the test plates were first buried five months after the excavation.

From November 1999 to June 2001 no major changes in water table have been registered during the four fieldwork seasons (lowest level measured was 3.48m). If we assume that the corrosion rates measured away from the excavation (5–10μm·y^{-1} for a fresh metallic surface) are representative of the conditions in Nydam, the archaeological iron artefacts are not acutely endangered under normal conditions with a high water level. The artefacts are protected by a thick layer of corrosion product, so their corrosion rate is probably considerably lower than for test plates.

Corrosion measurements on finds

The first results from corrosion measurements on the finds (Sørensen and Gregory 1998) showed an average corrosion depth of 0.58mm on lances excavated in 1991, and an average depth of 1.47mm on lances excavated in 1993–1994. As the two populations have been excavated at different sites and different depths, they cannot be directly compared. It is unlikely that the difference between the populations is exclusively due to the longer exposure time of the 1993–1994 finds – in that case the average corrosion rate in the three year period should be in the order of $300 \mu m \cdot y^{-1}$ which is way above the highest corrosion rates measured on modern unprotected iron test plates.

In order to compare the two lance populations properly, we are in process of registering all the lances on a geographical information system (GIS). In such a system all data are identified by an exact geographical location, which is important in areas with high horizontal and vertical variance. The GIS allows easy comparison between different parameters and has already shown itself useful for evaluating the environmental data, all maps shown in this article are produced by this system.

We are also in process of investigating the corrosion layers from 'fresh' archaeological lances, excavated in 1999. We hope that these studies will give us further insight to the deterioration and protective mechanisms for iron artefacts, so that we can end up with a more realistic estimate of the current corrosion rate.

Preliminary conclusions and future for Nydam

Interpretation of the results from the environmental monitoring is still ongoing, as is the development of the methodology. We cannot yet conclude from the measured parameters that the archaeological artefacts in Nydam are at great risk. That is, as long as the site is left undisturbed with only moderate fluctuations in water level.

A low water level, whether caused by an excavation or a dry summer, has a quite drastic effect on the environment. The concentration of several ions increases, and the pH decreases, even after the water level returns to normal. These environmental changes increase the corrosion rate on modern test plates placed at the same depth as the archaeological find layers. The increased corrosion rate is not an effect of direct atmospheric exposure of the test plates, as they were first introduced in the meadow after the water level had returned to normal.

It is still open to question whether the archaeological artefacts remaining in Nydam have been affected by the excavations – we do not claim that this is the case. Rather, we see the results as a warning signal of what effects a fluctuating water table might have. This means that it is not wise to ignore the water level in the meadow just because the archaeological find layers are never dried out – changed environmental conditions might reach deeper than the water level. The preservation conditions can probably be improved by stabilising the water table at a higher level to avoid strong fluctuations, as for example during dry summers. Also, pumping during excavations should be planned to minimise its effects on the surroundings.

Depending on negotiations and funding, the water level in Nydam will be increased and stabilised from 2004. First, however, a new onslaught is to take place at Nydam, this time against the small plant 'Horsetail' which is deteriorating the wooden artefacts with its long roots (Gregory et al 2001).

Bibliography

Alstrøm, T, in prep The use of micro sensors for measurements of environmental parameters in wetland areas, *Proceedings of the 8th WOAM conference, June 2001*, Stockholm

Christensen, C, and Kolstrup, E, 1998 Nydam Mose - en jernaldersø med krigsbytteofre, *Geologisk Nyt* 6, 6–9

Clymo, R S, 1983 Peat, in *Mires: swamp, bog, fen and moor* (ed A J P Gore), 159–224, Amsterdam

Fjaestad, M, Ullén, A G, Tronner, K, Borg, G C, and Sandberg, M, 1998 Are recently excavated bronze artefacts more deteriorated than earlier finds? in *Proceedings of the international conference of metals conservation* (eds W Mourey and L Robbiola), 71–9, Draguignan-Figanières

Freeze, R A, and Cherry, J A, 1979 *Groundwater*, New Jersey

Gregory, D, Matthiesen, H, and Björdal, C, in prep *In situ* preservation of artefacts in Nydam Mose: studies into environmental monitoring and the deterioration of wooden artefacts, *Proceedings of the 8th WOAM conference, June 2001*, Stockholm

Heathwaite, A L, and Göttlich, K, 1993 *Mires: process, exploitation and conservation*, Chichester

Hilbert, L R, 2000 Monitoring microbially influenced corrosion, unpub PhD thesis, Technical Univ Denmark

Mattson, E, 1992 *Elektrokemi och Korrosionslära*, Korrosionsinstitutet, Stockholm

Rieck, F, Jørgensen, E, Petersen, P V, and Christensen, C, 1999 ...som samlede Ofre fra en talrig Krigerflok. Status over Nationalmuseets Nydamprojekt 1989–1997, *Nationalmuseets Arbejdsmark*, 11–34

Scharff, W, Arnold, C, Gerwin, W, Huesmann, I, Menzel, K, Pötzsch, A, Tolksdorf-Lienemann, E, and Tröller-Reimer, A, 2000 *Schutz archäologischer Funde aus Metall vor immissionsbedingter Schädigung*, Konrad Theiss Verlag, Stuttgart

Sørensen, B and Gregory, D, 1988 *In situ* preservation of artefacts in Nydam Mose, in *Proceedings of the international conference of metals conservation* (eds W Mourey and L Robbiola), 94–9, Draguignan-Figanières

16

Design and decision on five development sites in London: Governor's House, Millennium Footbridge, Alder Castle, Plantation Place and Park Lane, Croydon

16a

The reuse of pile locations at Governor's House development site, City of London

RICHARD HUGHES, BRIAN COLES and RICHARD HENLEY

Richard Hughes
Consultant Archaeologist

Brian Coles
Geotechnical Engineer

Richard Henley
Structural Engineer

Ove Arup International
13 Fitzroy Street
London W1T 6BQ, UK

Abstract

This paper demonstrates how addressing archaeology early in the design process and consultation between different interested parties including the developers, architects, engineers, English Heritage, Corporation of London Planning Department and the archaeologists can lead to the implementation of a policy resulting in extensive preservation *in situ*. The strategy in this instance protected three Scheduled Ancient Monuments and the vast majority of the archaeological resource, but at the same time allowed limited evaluation and excavation of parts of the site and led to significant archaeological results. It is a good case history demonstrating how new piling is set in predetermined locations where archaeological contexts had already been totally truncated.

Introduction

The Governor's House redevelopment scheme was realised through the unique contribution to integrated design by the consultant archaeologist, structural and geotechnical engineers, and building services engineers of Ove Arup and Partners. The archaeological and engineering timetable for the groundworks is shown in Table 16.1.

Up to some 10m below the site lie the remains of what is believed to be the Roman Provincial Governor's Palace. The international archaeological importance of this site has meant that the teams were heavily constrained in what they could do. The design team had to find a way to construct the new building without causing harm to this Scheduled Ancient Monument. Their priority was to develop a design philosophy robust enough to evolve intact through the design stages and on to practical and safe construction. The principal objective was to minimise the loss of archaeology while maximising the client's commercial objectives. The scheme required there to be a well-determined set of archaeological research objectives where impacts into the fill and alluvial soils could not be avoided.

Through a range of engineering solutions most of the archaeological resources were fully preserved *in situ*. This case study, now several years on from the completion of the development, stands out as a 'first' and still one of only a few examples of reusing pile positions in an urban setting and where the geotechnical profession is now concerned with past piling effects on ground conditions and the limitations of locating new piles. There is now a growing literature on reusing piles and pile locations and this will have a growing potential to aid with mitigating development effects on important archaeological sites.

Table 16.1 The archaeological and engineering timetable for the groundworks

Task	Date
Archaeological desk study	May–August 1994
SI and archaeological evaluation	September–December 1994
Scheduled Monument Consent, based on up-front and well-honed engineering and archaeological mitigation designs	January–May 1995
Archaeological site works	May–September 1995
Piling contract	June– August 1996

The site

The Argent Development Consortium Limited development site lies in a central City of London location just west of London Bridge and north of Upper Thames Street, in an area with occupation dating back to the early Roman period. Most of the site (the western three quarters) lies within a Scheduled Ancient Monument, the Roman Governor's Palace. The scheduling legally protects all the archaeological remains on the site and in the immediate setting, of whatever age and depth. No development disturbances were allowed to the ground, from ground level downwards, without Scheduled Monument Consent and to gain this required very early complete engineering and archaeological designs, comprehensive construction method statements and integrated programmes. The Corporation of London also protects the whole site from uncontrolled disturbance of the archaeological resource by their statutory obligations referenced in the City's Unitary Development Plan of 1994.

Site history

The Governor's House development site is located centrally in the Roman City of Londinium, with first phase development dating to the mid 1st century AD (Fig 16.1). In the Roman period the site was located to the south-west of the Forum and Basilica complex and east of the River Walbrook. The former was the main administrative and cultural centre of the city, and perhaps of England, and the latter was a main semi-navigable waterway traversing north up through the city. The site was dramatically located on the descending terrace slopes overlooking the River Thames foreshore, located along what is now Upper Thames Street.

From the mid 19th century, the general area was known to have contained substantial Roman remains. In a jigsaw fashion, the plan of a magnificent building complex has been pieced together, and this has been interpreted as the Roman Governor's Palace. Substantial Late Saxon, medieval and post-medieval remains were also known to be on site, but partially disturbed by the basements and foundations of the previous building.

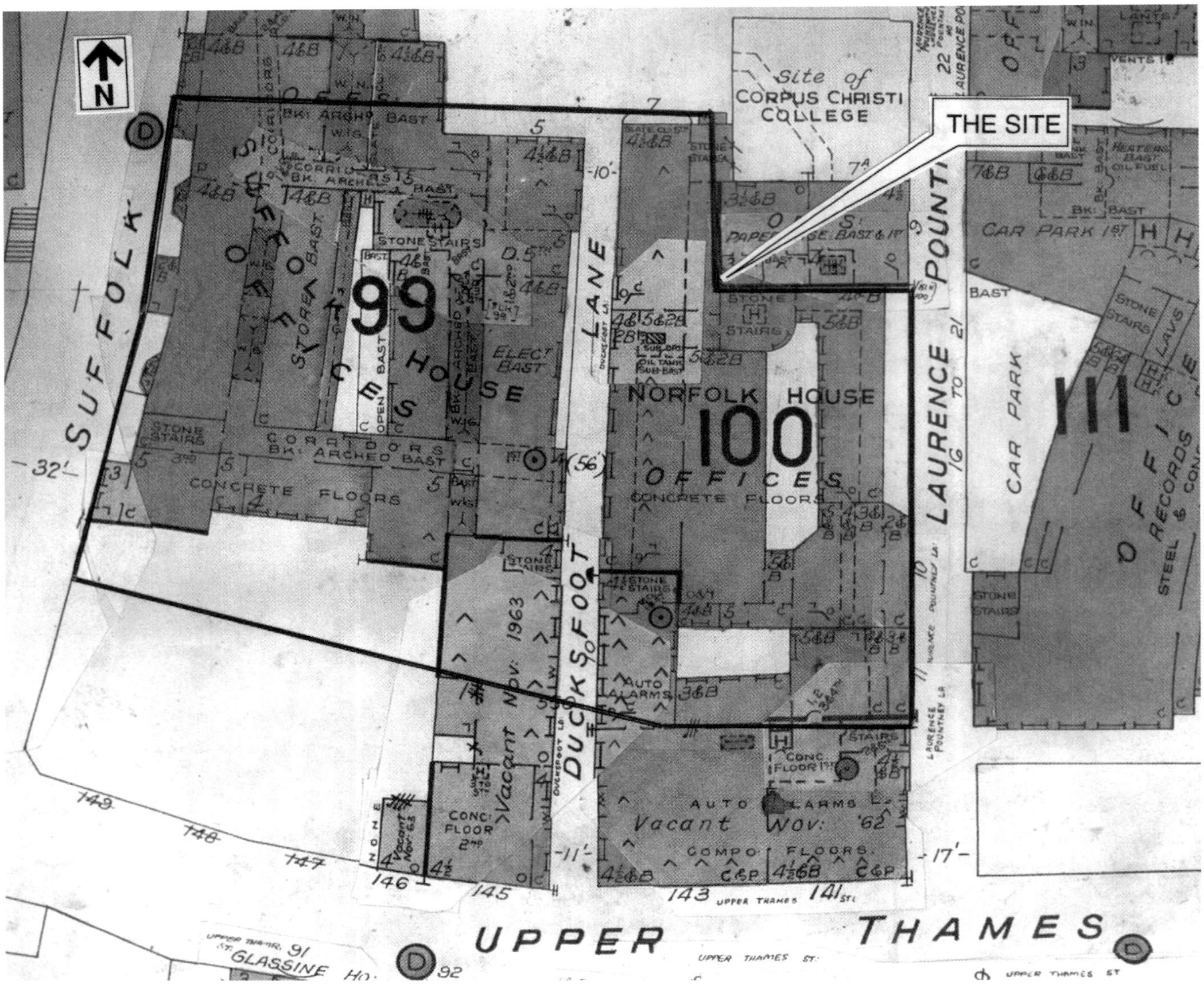

Fig 16.1 Site location shown on a Goad Fire Insurance Plan of 1959, and which was used with other data for predicting the locations where old foundations had penetrated through archaeological resources and therefore providing good potential locations for new piles

The archaeological works necessitated by the development confirmed the extent and quality of the archaeological resources. The early Roman riverfront revetment and reclamation dumps were located and analysed and these were found to be overlying marshy soil formation of far greater antiquity. Many features were identified of the building complex on the east side of the Roman Governor's Palace.

Archaeological mitigation strategy

The basis of the Governor's House mitigation strategy was to develop a groundworks scheme that had minimal impact on the buried archaeological resource (providing maximum *in situ* preservation and protection). This was achieved by fully integrating a consultant archaeologist into the design team, evolving a foundation and substructure design that tried to totally avoid the archaeological resource, and investigating the location and quality of the resource in a non-damaging/full documentation process (and just as importantly, investigating the ground where research indicated the resource should be absent). There was a three phase Scheduled Monument Consent and planning approval process which was obtained as a result of regular design meetings and discussions with English Heritage and the Corporation of London.

Foundation solutions

Ground investigation

Between October 1994 and January 1995 ground investigation was carried out, consisting of 5 boreholes (maximum depth 70m) and 12 trial pits. This investigation allowed the stratigraphy of the ground to be defined and also provided data on the engineering properties of the soil through which the piles were to be built. The trial pits were archaeologically excavated. Some were located where major Victorian load-bearing walls had been and where their foundations were still expected to be and some where new piles were to be installed. The aim was to find locations of deep modern disturbances and therefore places of no archaeological resource (Fig 16.1). This would then aid obtaining Scheduled Monument Consent for the final foundation design. The stratigraphy was as shown in Table 16.2.

Table 16.2 Details of stratigraphy as recorded in the boreholes and trial pits

Strata	Thickness (m)
Fill and alluvium	5–10
Terrace gravel	1–5
London Clay	36–43
Woolwich and Reading	18
Thanet Sands	1+

Pile design

To maximise the load carrying capacity of the piles to be built and therefore minimise their dimensions, additional data on the strength of the London Clay was obtained from three load tests on piles built on the site in 1968. This test data plus in situ strength testing and laboratory testing of samples recovered from the 1994–1995 ground investigation resulted in a pile design line for undrained shear strength in the London Clay as shown in Table 16.3. Between each level the undrained shear strength is considered to increase linearly with depth.

Table 16.3 Undrained shear strength of samples recovered from the 1994–1995 ground investigation

Level (m OD)	Undrained shear strength (kN/m≈)
2.5	86
-1.0	94
-12.5	200
-21.5	200
-38.0	310

The piles were designed with the other following design parameters:

Straight shafted piles
Overall factor of safety = 2.5
Shaft adhesive factor = 0.6
Bearing capacity factor, Nc = 9

Under-reamed piles
Factor of safety
Shaft = 1.5, Base = 3.0
or
Overall factor of safety = 2.5
(whichever gave the minimum capacity)
Shaft adhesive factor = 0.5
Bearing capacity factor, Nc = 9

The final pile design was carried out by Bachy who then built the piles.

Piling

The new building is founded on piles of 1050mm, 1135mm and 1800mm diameter, but slightly larger taking account of the casing sizes. Pile loads are typically between 5000kN and 19500kN. The piles are founded in the London Clay and are a mix of straight shafted and under-reamed piles, all founded above *c* -30m OD. The pile layout maximises symmetry around the perimeters that gives the most opportunities for future reuse of the pile. To aid with this possibility the piles designs and full construction records have been lodged with the archaeological archive.

Because of the design mitigation philosophy to minimise the loss of archaeological resource, two-thirds of the new piles are located over existing piles. This was an innovative solution developed between Arup and Bachy and the process has now been applied on several other Arup projects. It was therefore necessary to bore out many of the existing piles prior to forming the new ones. The existing piles which were to be removed were 600 and 750mm diameter, *c* 20–25m long and lightly reinforced.

The new piles required for the development were 1050, 1350 and 1800mm diameter, up to 40m long and substantially reinforced, both straight shafted and under-reamed. The total number of new piles was 25 of which 12 are straight shafted and 13 under-reamed to increase the bearing capacity at the base. Sixteen old piles were removed and replaced. The piling contract period was nine weeks, this taking into account the precise location of piles in relationship to the known archaeological resources and the extra time needed for drilling out the old piles. Six of the larger diameter piles were pre-excavated for a 9.0m depth through the fill and alluvium. In these few locations no previous foundations were located and therefore the Scheduled Monument Consent allowed for a programme of archaeological excavation by the Museum of London Archaeology Service (MoLAS) set against predetermined research objectives.

The piling processes were monitored full time by a resident engineer and, as the pile construction went through the fill and alluvium formations, by the archaeological consultant as well (Ove Arup). During the piling operations there were weekly meetings with English Heritage and the Corporation of London.

The pile replacement method comprised: locating the old

pile according to past engineering drawings; locating the pile cap and removing it to confirm the location of the top of the pile; oscillate down the casing (Fig 16.2) with a cutting edge immediately around the existing pile to its construction depth, with the cutting edge removing the odd bulges in old pile sides and allowing for the old pile being off vertical (typically 750mm diameter). The old pile and the reinforcement was then drilled out as a single operation and with the old pile material coming up looking like a very finely chopped up spaghetti bolognaise (Figs 16.3–16.4); the hole was filled in with weak concrete and bentonite mix, the casing extracted and the mix allowed set. Then the new casing was drilled down to the required sealing off level in London Clay (typically 1200–1500mm diameter) and the new pile augered out, including the concrete-bentonite mix designed to have a strength slightly in excess of the London Clay but still 'drillable', and an annulus of fill and subsoils to the base level of the original pile and continue to the new pile depth. The new pile was then cast and the casing progressively extracted; the pile top trimmed to form the connection to the pile cap/ground beam.

On site the working method related very much to the quality sensitivity of the piling rig operator (Bachy) and support crew and these were fully briefed about archaeological concerns. The archaeological monitoring took into account and helped to establish a consistent method: tracking of a drilling rig on a piling mat designed with knowledge of the underlying fill soils and shallow archaeological resources; precise location of the pile casing to ± 50mm maximum; ensuring the pile casing did not twist off location as it slowly rotated in and sealed off in London Clay; ensuring pile verticality to better than 1 in 200. The aim was to limit vibrations of the casing insertion oscillations and pile boring, especially when working through concrete and dense gravel soils. Accelerometer instrumentation was not used but the aim was not to feel vibrations through site boots at 2.0m from the pile location. This represents a ground acceleration of less than 1mm per second and site experience confirms that this is small and non-damaging. The method also enabled limiting the insertion of pile casing with 'muddying-in' water and repetitive raising and dropping of the casing; providing a

Fig 16.3 Details of the auger head being fitted with special cutting teeth for drilling out old piles

Fig 16.2 The Bachy piling rig oscillating down casing around an old pile

Fig 16.4 Augering out an old pile showing details of the extracted concrete and reinforcement

spillage sand 'sponge' around the rig and right up to the edge of the casing – additional to the piling mat; no drilling-in of spare auger heads, into the surround fill near the rig; measuring the pile depth and verticality and the cleanliness of the pile base; sampling and testing of the concrete; limiting the height of the concrete 'head' when extracting the casing through the fill. The casing had to be pulled out slowly and precisely through the fill so there was no side collapse and concrete penetration into the fill; a reaction plate was used for the casing extraction callipers, so the surrounding ground was not compressed and the trimming-down of the pile top was confined to be within the thickness of the piling mat. The site operations went very smoothly and to schedule and the process has subsequently been used on other Arup contracts.

Ground beams

The substructure consists of deep ground beams. In summary these have spans of approximately 15m between perimeter piles. They carry an eight-storey structure and are all set above archaeological resources.

The systems of ground beams located around the external and internal perimeter of the building support the superstructure columns and transfer the load to the piles. On the west and north perimeters, because of the height of the archaeological resource, the ground beams are located between the lower ground and ground floor levels. Along the east and south perimeters the archaeological resource is low enough for the ground beams to be positioned below the lower ground floor level to allow for the windows and doors of the proposed office spaces. On the internal perimeter, around the proposed atrium, the ground beams are located below the lower ground floor slab within the construction footprint of the pre-existing basement.

The positions of the new piles were chosen to minimise the eccentricity between the superstructure loads, the ground beams and the piles. It was not possible to introduce additional cross beams which would, on the west part of the site, very likely conflict with any remaining archaeological resource or compromise the lower ground floor car park. The ground floor slabs resist eccentricity on the west and north ground beams. The south and east beams are designed for torsion. Adopting a superstructure frame without internal columns further eliminates the need for cross beams.

The ground beams and floor slabs are fully above the top of archaeological deposits as determined by the archaeological evaluation and pile excavations. This strategy also means that the post-medieval, medieval, mid to Late Saxon structures and soils are all preserved *in situ* and act as the buffer zone to the underlying resource of the Roman Governor's Palace.

For general interest some facts and figures are presented in Table 16.4 illustrating the size of the groundworks undertaking.

Table 16.4 Size of the groundworks

Groundwork description	
Pile caps	140kg/m^3
Main Ground beams	150kg/m^3
Annex ground beams	105kg/m^3
Columns	185kg/m^3
Overall average (including slabs, walls etc)	130kg/m^3
Total reinforcement	270t
Quantity of concrete	2125m^3

The superstructure

The archaeology and the resulting constraints imposed on the substructure led to a superstructure that generally spans 15m from the central atrium across to the perimeter columns that are generally at 4.5m centres. As a result of the same constraints, there was a requirement to minimise the superstructure loads and hence the pile sizes and lengths. The self-weight of the floor plates was therefore an important consideration. The parameters for the design of the floor structure were: overall structural services, ceiling and raised floor zone is typically 1250mm; the raised floor is (typically) 150mm; horizontal distribution of the services to be accommodated through the structural zone. Ceiling and lighting zone run below the beams; deflection of the structure to be within appropriate limits; floor vibration to be considered and fan coil positions to be coordinated with the structure.

A composite steel frame can be achieved in a number of ways and the most suitable option for this project was Cellform beams. The cellular beam solution, with 590 holes in the beam webs, offered good flexibility for service routing, good vibration and deflection characteristics and avoided the need for any propping. Transfer beams at ground floor and at the high level set-backs, and primary beams at the corner of the building are plate girders or plated beams as the cellular beams cannot accommodate the resulting large shear forces.

Vierendeel frames around the north provide resistance to horizontal loading to the main building, east and west sides of the atrium, and at the rear of the lift shafts. The Vierendeel frames around the atrium are formed in *in situ* concrete between ground floor level and foundations, and in steel above ground floor level up to the roof. The frame to the lift shafts is in steel from lower ground floor level up to the roof, and is supported by a concrete transfer beam over the existing basement.

The concrete floor slabs act as rigid diaphragms to transfer the horizontal loads to the vertical stability frames. Stability to the five-storey annexe is provided by the Vierendeel frames in the main block, with an additional frame in the east wall of the annexe, adjacent to Laurence Pountney Hill.

For general interest some facts and figures are presented in Table 16.5, illustrating the scale of the superstructure.

Table 16.5 *Scale of the superstructure*

Description	
Cellform beams	380t
Plate girders	320t
Rolled sections (floor beams)	380t
Rolled sections (columns)	425t
Roof steelwork	35t
Typical floor weight (3rd)	85kg/m^2
Overall floor weight	110kg/m^2

Conclusion

The Governor's House project demonstrates the first case history where old pile removal and replacement has been successful in improving on the amount of *in situ* preservation of archaeological resources that can be achieved on a major urban development site. It is now a proved technology and has been used on other Arup development sites and is available for further application and development.

It is important to note that the success of this development depended on the enthusiasm of Argent, the developer, to a 'partnering' approach and therefore the full integration of the archaeological consultant and geotechnical engineer with the architects, structural engineers and groundworks contractor. This was fundamentally important here where a complex full design was required prior to gaining Scheduled Monument Consent and planning consent. We typically find on most development sites the foundation design comes late in the process and often as a 'design and construct' package by a subcontractor. The Governor's House site illustrates the advantages of the integrated approach. As more archaeological sites are deemed worthy of *in situ* preservation we see a growing trend to innovative foundation solutions and where the archaeologist can help drive their design.

16b

In situ preservation of medieval waterfront archaeological resources at the north abutment of the London Millennium Footbridge

RICHARD HUGHES

and JOHN SEAMAN

Abstract

Significant archaeological excavations took place at both the Southwark and City bridgeheads of the new London Millennium Footbridge, as part of the early site works for the massive cable restraining foundations. The archaeological programme offered a unique opportunity to compare phased historic riverside reclamation and development on both sides of the River Thames at the same time.

The archaeological programme had to respond to considerable pressures resulting from the tight construction programme and budget constraint. In this respect, the local authority planning requirements for *in situ* preservation whenever possible was supportive of a strategy of partial excavation only and so deadlines could be met.

On the Southwark side of the Thames the very deep pile cap and large diameter piles passed through the full depth of archaeology and underlying alluvial soils. However, on the north side the engineering works were only to be half through the archaeology and therefore it was possible to design for *in situ* preservation of resources. A strategy was developed beforehand for assessing during excavation what should be retained in place and where and what would be better excavated.

This paper will discuss the elaborate and innovative team based process for preserving archaeological resources, and the successes of excavation to recover a full site 'story'. The paper will also report on the re-excavation of the famous 1970s Trig Lane excavations (and the determination of the diggers' rubbish preservation status), new findings from ground penetrating radar on a waterfront site (in respect of technique and predictive capabilities), and the capping techniques for the deposits retained in the ground.

Introduction

Background

The London Millennium Footbridge is the result of an international competition organised by the Financial Times Newspaper in 1996. Of the 226 teams who entered, five were short-listed to the second stage and the competition was won by a team from Ove Arup and Partners, architects Foster and Partners, the sculptor Sir Anthony Caro and cost consultants Davis, Langdon and Everest. Half of the funding came from the Millennium Commission; the rest was made up with contributions from the London Borough of Southwark, the Corporation of London, and various private and corporate sponsors.

The first new bridge in London for over one hundred years, the London Millennium Footbridge traverses between the London Borough of Southwark and the City of London, providing a pedestrian link between the Tate Gallery of

Richard Hughes
Consultant Archaeologist

John Seaman
Associate Director

Ove Arup International
13 Fitzroy Street
London W1T 6BQ, UK

Modern Art and St Paul's Cathedral. The bridge also forges links echoing the bimodal history of the river over some 10,000 years. Comparing the urban history of riverside development, on both banks and through the river at the same time during the construction process, became a unique archaeological objective for London. Fig 16.5 shows the general location of the London Millennium Footbridge.

Fig 16.5 General view of the London Millennium Footbridge showing its location in front of St Paul's Cathedral

The abutment structures

The abutment structures of the footbridge comprise two very large pile caps approximately 18 x 30m and some 3.0m thick and supported off groups of piles. The structures provide anchorage to the bridge cables and their very significant lateral forces. The Southwark pile cap is set below the base of the fill and the City pile cap is founded approximately half way through the fill, with the piles penetrating well down into London Clay.

The archaeological mitigation strategy

Strategies for deciding what resources should be excavated or retained, and how, were the subject of many discussions and meetings. Only the pile caps would be fully archaeologically excavated and priority was given to maintaining a clear set of archaeological objectives in developing the excavation programme (Ove Arup and Partners 1998c).

Below the pile caps preservation of archaeological deposits was the preferred option. On the Southwark side the pile cap base just coincided with the base of archaeological formations. The base of the pile cap in the City was estimated to lie about halfway through the archaeological formations. Here *in situ* preservation was a viable option if the overall archaeological excavation objectives could be satisfied. For engineering design reasons the strategy for preserving the archaeological resources on the City side was set in place before the excavations. This strategy was updated and fine-tuned on a weekly basis through the early phases of the archaeological/engineering groundworks programme.

Bridge design

The superstructure

The overall span of the bridge is 320m. Two groups of cables span from bank to bank over two piers placed 144m apart in the river. The dip of the cable over this central span is 2.3m, far shallower than a more conventional suspension bridge structure, and the cable profile necessarily results in large cable tensions (Fig 16.6).

Fig 16.6 Detail of the London Millennium Footbridge showing the general character of the structure

The bridge dead load is around 2 tonnes/m, resulting in a cable tension of 22.5MN. The live load adds an additional 8MN under extreme conditions. The design exploits this tension, which provides sufficient stiffness for the deck so that no additional structure is required, such as a conventional horizontal truss under the deck, to resist lateral wind load. A damping system has subsequently been installed on the bridge to prevent excessive movements occurring when it is crowded.

The abutment foundations

The most significant loads applied to the abutment foundations are the horizontal forces, some 30MN combined dead and live load, and the resulting large overturning moment. The foundation piles resist the horizontal loading by lateral loading of the soil. The moment is resisted partly by frame action but mainly by vertical push-pull axial loading of the piles. The design of the foundation system is principally constrained by the requirement to limit horizontal deformations at the cable anchorage point due to lateral movement and rotation of the pile cap.

An additional consideration was the need to limit the effect of ground movements on the adjacent structures. The horizontal loading causes some soil pressures to be applied to the piles supporting the neighbouring buildings and the river walls, which they must be able to sustain without distress.

The City substructure is formed from a 3.0m deep pile cap, which is integral with the two walls connecting to the cable anchorages. The foundations comprise a group of 12 concrete piles 2.1m diameter, bored and cast *in situ*, 28m long. The size of the piles and reinforcement has been selected primarily to provide the desired stiffness response of the foundation. As with the north abutment, the loading on the Southwark foundations is predominately horizontal. The foundations here comprise a group of 16 concrete piles 2.1m diameter, bored and cast *in situ*, 28m long with a 3.0m deep pile cap.

The abutment piles tend to apply a pressure to the back of the river walls as they are pulled forward by the force from the cables. This can lead to adverse effects on the walls unless specific protection measures are taken. A system was developed during discussion between Arup, the piling subcontractor Bachy-Soletanche and the main bridge contractor, Monberg Thorsen / Sir Robert McAlpine, whereby the front two rows of piles are surrounded by an annulus filled with a weak, cement-bentonite slurry. The slurry is sufficiently dense and strong to support the soil, but only transfers a limited amount of load to the ground as the piles deform. The pile caps are separated from the soil along the face by the river by Cellcore, a compressible grillage formed from expanded polystyrene, again to avoid loading the river wall. The other three sides of the pile caps are separated from the soil by 200mm thick expanded polystyrene board (Fig 16.7).

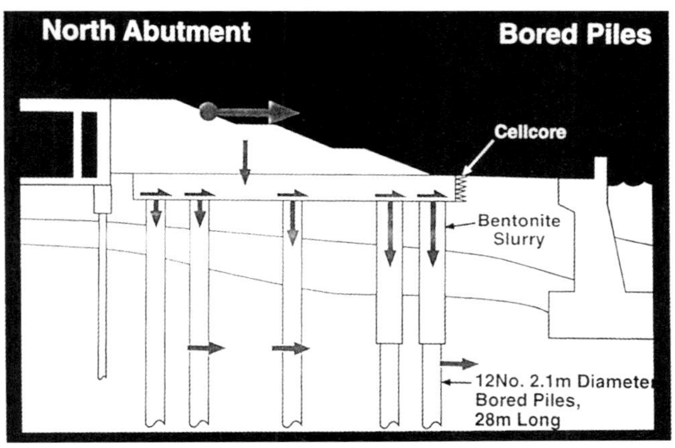

Fig 16.7 North–south cross section through the north abutment

Summary of the archaeological programme

In general, the archaeological works for the Millennium Footbridge consisted of a four-phased programme. Phase 1 comprised the archaeological excavation to the base of the pile caps in both Southwark and in the City. Here the new engineering structures totally displace the archaeological resources. Phase 1 started with a meeting of all concerned parties prior to implementation of temporary works and archaeological interventions. The archaeological excavation works were then undertaken and those related to the north abutment are outlined in this paper. Weekly site meetings occurred throughout the excavations between all concerned parties including the client, Arup, QS, Museum of London Archaeological Service (MoLAS), Balfour Beatty Construction Ltd, (BCCL, the enabling works contractor) and both local authority archaeological planning officers.

Phase 2 consisted of the field documentation of the exposed resources at the base of pile cap level in terms of their archaeological potential at greater depths, taking into account their location compared with proposed engineering groundworks. This was followed by the production of a briefing report by MoLAS which described the archaeological findings, based on the results of the evaluation. The report was then reviewed with the archaeological planning officers. From this work a strategy report was then produced by Arup for the archaeological works at deeper depths. BBCL were then able to design temporary works to allow for the implementation of the archaeological strategy. In further meetings with the archaeological planning officers the future archaeological programme was then agreed, based on the Arup and BBCL designs. At the same time MoLAS were able to redefine their archaeological research objectives.

The archaeological Phase 3 programme 'cleared the site', by archaeological excavation, at the location of the deeper engineering works and supervised resource reburial where *in situ* preservation was to occur.

The post-excavation archaeological works, Phase 4, conforming to the original contract specification and the agenda defined in MoLAS's method statement completed the programme. This is now fully achieved with publication scheduled for 2002 (MoLAS 1999; Ayre and Wroe-Brown 2002).

Details of the archaeological mitigation strategy for the two bridge abutments

Responding to the bridge design, the Arup strategy report (Ove Arup and Partners 1998e) recommended that for the north abutment the archaeological excavation would be taken to the base of the pile cap. Below this there would be a programme of excavation of some pile locations, coring of other pile locations and sampling of environmental materials. Large areas of resource would be left *in situ* below the pile cap. In Southwark the strategy proposed excavation down to the base of the fill to just below the pile cap, and excavation of some pile locations and sampling of alluvial soils. *In situ* preservation was not regarded as a useful option since little resource would be left after excavation and consequently it would be far better to recover a maximum amount of data. Arup further defined archaeological intervention activities relating to resources of various types of riverfront structures being encountered and also described engineering constraints on the City side imposed by the natural and man-made settings (Ove Arup and Partners 1998a). These included various difficulties: excavating substantial areas towards the bottom and below the base of the temporary works retaining walls; excavating below the basement level of the adjacent Swiss Bank Building; archaeological excavation (and engineering breakout of obstructions) in the pile locations next to the king post retaining wall system; the stability of very soft and very loose soils and finally archaeological and engineering excavations below water table levels.

The 'strategic' archaeological objectives aimed to minimise excavation at pile locations and around the edge of the temporary works for safety reasons. Archaeological works would concentrate in central areas where temporary works excavation stability concerns are minimised. This recognised the need to carry out sufficient work for the archaeological 'story' of the whole site to be thoroughly determined yet leave in place significant quantities of archaeological resources that would not be directly affected by engineering impacts. Some, but not all, pile locations would be archaeologically excavated, particularly those where the engineering contractor needed to break out obstructions, and those deemed to be located in key archaeological objective locations.

The evaluation of the potential resources below the base of the pile cap (+1.0m OD in the City and +1.6m OD in Southwark) was of fundamental importance for establishing the archaeological options and then implementing the best 'combination' of intervention measures below the pile caps. The evaluation included a series of undertakings which were agreed, as part of the design strategy, with the local authority archaeological planning officers. These comprised an examination of exposed partial and whole archaeological features and of soil surfaces and local removal of modern intrusions to reveal the cross section of deeper potential resources. Remote sensing was to be used (one or more techniques appropriate to the site conditions) selecting that which offered the best opportunity to identify buried features in difficult waterfront conditions. For this site GPR was the preferred method and was successful despite the salty grounds conditions. In addition it was agreed to hand auger 'core' samples – these cores to be used for identification of buried features and soils, for aiding with revising the mitigation and later used as part of the environmental research work. Local small exploration pits were excavated. Finally expert judgement systems were to be used for resource prediction modelling.

For strata below the pile caps additional guiding principles were established. The depth of excavation would be determined by the quality of soils, their stability and the overall safety of the temporary works. In the City and in Southwark there would be no excavation of a pile location if it coincided with modern features or was sampled for environmental analysis (by coring and taking monolith tins for the recovery of undisturbed samples). Small area excavations would aim to be central to the pile cap and taken down from the underside of pile cap into underlying alluvial soils; this would partially respond to 'edge' safety requirements. Partial excavation would occur at pile locations in the City from +1.0m OD to -1.0m OD, if structural features were present with elements of considerable archaeological value. In the City, there would be excavation of one or more local areas, as representative sampling of archaeological resources in adjacent 'edge' piles.

In the City and in Southwark, where there were no structural features present, it was proposed to excavate one or two sampling trenches. These trenches would be 1.0–2.0m long and 1.5m wide (with shoring as required for safety) for extraction of bulk samples during excavation and undisturbed sampling from trench sides with monolith and micro-morphological 'tins'. After excavation there would be more 'coring' and monolith tin samples taken of cohesive natural soils for laboratory analysis.

Detailed account of the City (north) abutment archaeological undertakings in phases 2 and 3

The mitigation strategy represented minimum intervention that could have been anticipated, allowing for significant *in situ* preservation of archaeological resources and at a time when the engineering design had been 'frozen'. Given the character of the deeper fill and alluvial soils and high water levels, the mitigation proposal was also deemed to be the most technical feasible solution to achieve expediently. The excavation/preservation strategy was reviewed weekly, as the archaeological excavations progressed, to ensure the right balance between the archaeological and engineering objectives.

Site preparation

The enabling works contractor was appointed to provide full support services, including removal of all modern obstructions and extraction of low-grade archaeological soil resources under archaeological supervision (eg mass dumps). Standing modern concrete structures were then removed, with no need for archaeological supervision. The modern concrete surface was then broken-out with part-time archaeological supervision. There was then a programme of 'limited' ground reduction to provide appropriate site conditions for insertion of temporary works and this was followed up with the construction of a 'king-post wall' structure. The king-post wall comprised 'H' section steel columns plunged into bore piles, drilled at intervals around the site boundary. The piles were cast to a level below that of the excavation. *In situ* concrete was cast between the 'H' sectioned columns and excavation proceeded.

The king-post structure was formed on three sides of the pile cap excavation, with the east side being the existing basement wall of the adjacent Swiss Bank building. The king posts required drilling 0.6m diameter holes into London Clay. The drilling operation was monitored with an archaeological watching brief and visual observations were made of archaeological deposits being passed through. The archaeological loss resulting from the insertion of the king posts was deemed to be acceptable because they were principally installed to allow for the archaeological excavations. This system also reduced the need for deep and expensive temporary works, required for safe archaeological access, particularly below the water table.

The observations during piling proved useful for providing information about the character and depth of the underlying archaeological resources, especially structural remains and dump reclamation soils. The walling planks were cast from time to time, as archaeological excavation progressed. This technique allowed for the trench faces to be archaeologically documented in a safe working environment. The technique is generally the best for archaeological purposes, where trench face stability is required down to a water table. The technique at lower depths does permit significant water ingress and this can adversely affect archaeological digging techniques and programmes below the water table.

The pile cap excavation programme

From +3.5m to +1.0m OD all archaeological features were fully excavated, with areas of modern spoils, obstructions, etc being removed by the contractor under archaeological supervision. Key upstanding archaeological features were left in place to be considered for retention (eg if a timber river wall ran through the area with a top at +1.5m OD the base of the pile cap could be remodelled and strengthened above).

Major archaeological structures were found in the north abutment excavations and were then the subject of trying to preserve some of them (Fig 16.8). The excavations were able to show that in the 12th century the whole area was still within the foreshore. The site was then reclaimed and the main significant findings were of six phases of 13th- to 15th-century timber riverfront revetments and with some exceptional high quality carpentry and some massive tie-backs. These structures showed a close relationship to Boss Alley, a medieval lane that was progressively extended south from Upper Thames Street and separating properties to the east and west. In the 15th century the riverscape changed again with the construction of a substantial stone wall supported by timber piles and complex jointed sleeper beams and creating part of a large dock. There was substantial evidence of post-Great Fire industrial activities. Throughout the excavations spectacular assemblages of metal and leatherwork were discovered.

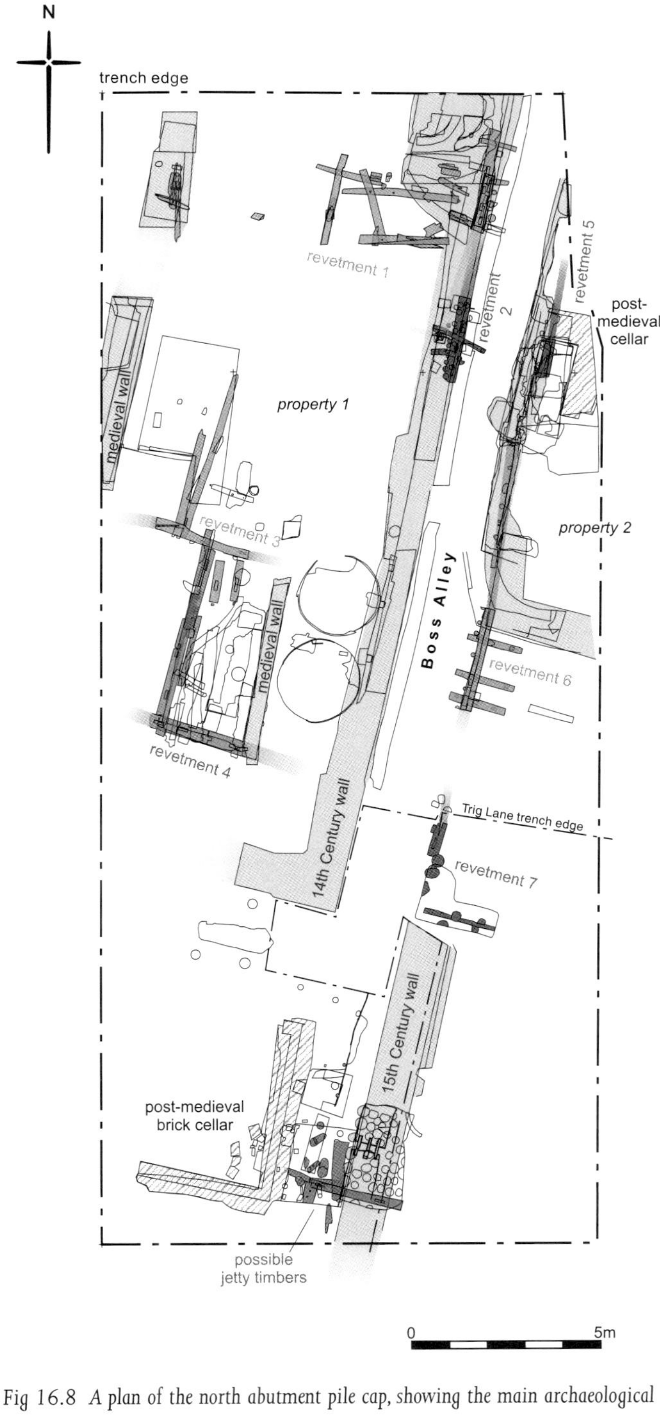

Fig 16.8 A plan of the north abutment pile cap, showing the main archaeological structures (courtesy of MoLAS)

Excavation/preservation below the City pile cap

The excavation/preservation strategy below the pile cap involved a complex decision-making process (Fig 16.9; Ove Arup and Partners 1998b). An assumption was made that about half of the pile locations would be fully excavated and this proved to be the case. Some pile locations were excavated, if the uppermost excavated surface illustrated important occupation/river front structures at deeper levels. Other pile locations were partially archaeologically excavated or cored, and accompanied by sampling of environmental soils and features. Pile locations were not excavated where archaeological excavation at higher levels showed deep modern disturbances or thick reclamation soils containing no complex archaeological features. Below 0.0m OD the assumption that there would be no 'area' archaeological excavations also proved to be the case. There was then a phase of excavating specific local features to their base, those found to lie within foreshore silty-sandy gravel deposits and down to *c* -2.0m OD maximum.

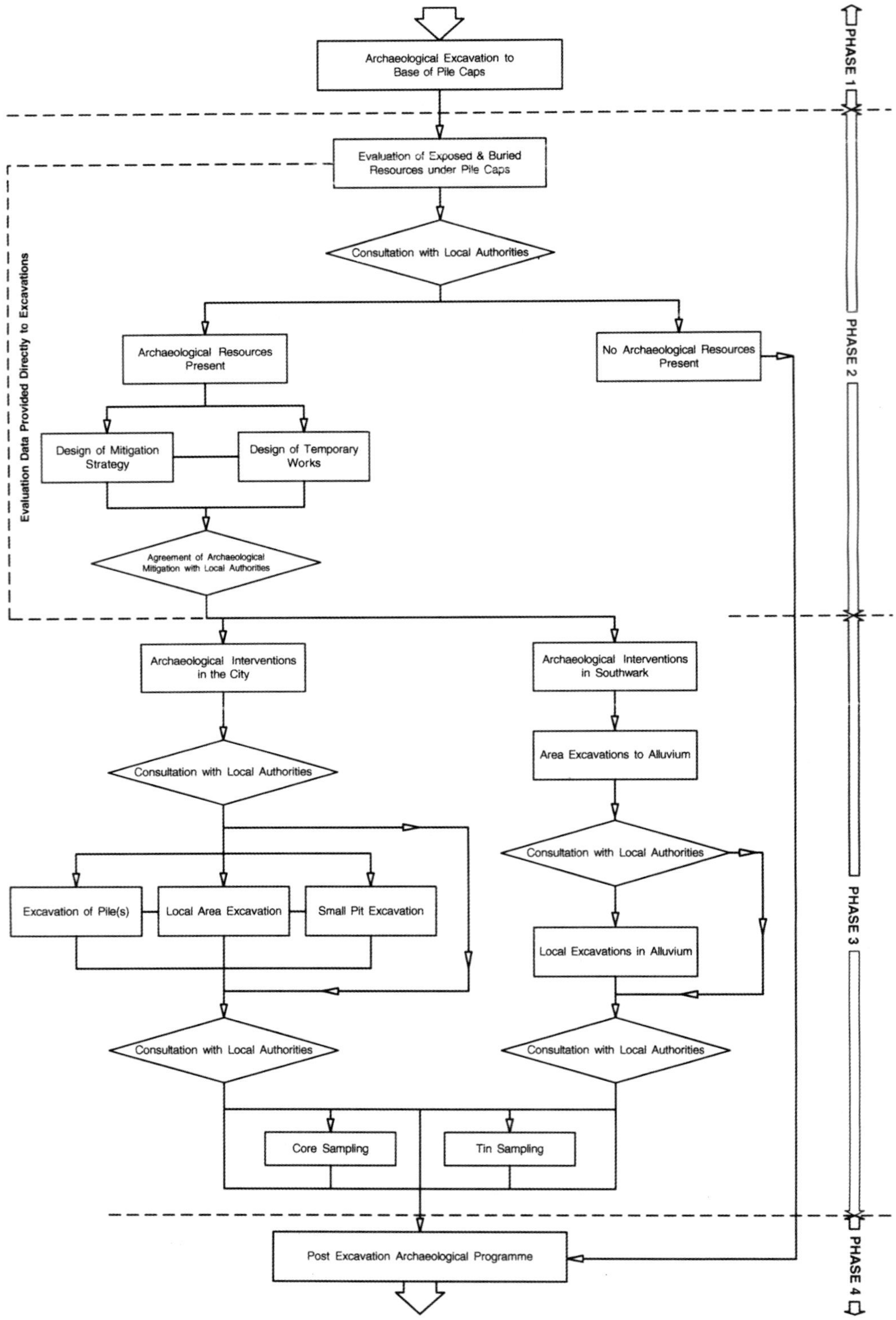

Fig 16.9 *The process of designing and enacting archaeological mitigation and allowing for in situ preservation below the north abutment pile cap*

Other objectives

During the course of the excavations other archaeological concerns were addressed. Areas not excavated or needed for engineering works, and where an archaeological resource or alluvium were encountered, were protected prior to backfilling and piling. In consultation with English Heritage and the local authority archaeological planning officers, the reburial strategy was to use compacted London Clay not the usual geotextile and sand. By this time the excavated organic soils had been removed off site. The strategy was therefore to produce a low permeability encapsulation, thereby maintaining a physical protective buffer zone and a saturated-low oxygen regime in an organic soil setting. Modern artefacts and timber structures left in the reburial environment of the 1974 Trig Lane archaeological excavations were recovered for assessment of their condition. It was not considered necessary to instrument the site, or to determine the preservation status from time to time, given the good level of preservation found in the Trig Lane archaeological excavation and the clay backfill capping

Conclusions

The archaeological excavation and *in situ* preservation programme was an extremely complex undertaking and has been rightly praised for having been well executed and precisely conforming to the overall project programme and allocated budget. The archaeological objectives of this waterfront site were fully achieved and have considerably added to the findings of the previous Trig Lane archaeological excavations. The success relied on well-planned and then controlled mitigation strategies, fully integrating archaeological and engineering site works. The dedicated 'partnering' of a large project team was instrumental in this success.

Acknowledgements

Thanks are given to the considerable support given to the developers archaeological programme throughout the project by: The Millennium Bridge Trust; The London Borough of Southwark; Museum of London Archaeological Service; Corporation of London and English Heritage.

Bibliography

Ayre, J, and Wroe-Brown, R, 2002 *The London Millennium Bridge: excavation of the medieval and later waterfronts at Peter's Hill, City of London, and Bankside, Southwark*, MoLAS Stud Ser 6, London

MoLAS, 1999 London Millennium Bridge, Bankside and Peter's Hill London SE1 and EC4: an archaeological interim report, unpub MoLAS rep

Ove Arup and Partners, 1998a, The London Millennium Footbridge: archaeological excavation strategy addendum report, October, unpub rep

Ove Arup and Partners, 1998b, The London Millennium Footbridge: archaeological mitigation below the abutment pile caps, November, unpub rep

Ove Arup and Partners, 1998c, The London Millennium Footbridge: specification for archaeological excavations, November, unpub rep

Ove Arup and Partners, 1998d The Millennium Footbridge: archaeological desk study, February, unpub rep

Ove Arup and Partners, 1998e The Millennium Footbridge: strategy report for archaeological excavations, September, unpub rep

16c

Archaeology and preservation at Alder House, Castle House and Falcon House

RICHARD HUGHES and

JONATHAN BUTLER

Abstract

Alder House/Castle House redevelopment site is located in the City of London at the external south-west corner of the Roman Cripplegate Fort, at its junction with the city wall. Resulting from engineering site investigations and an archaeological evaluation, the mitigation strategy was determined to be one of maximum *in situ* preservation – in accordance with Corporation planning conditions and a Scheduled Ancient Monument protection requirement.

This paper will demonstrate how the new structure's basement foundations and services were redesigned to successfully retain all the archaeological deposits. This was possible due to the considerable engineering and archaeological effort to locate and understand several phases of old foundations and site the new foundations through them. The engineering contractor's site skill of piling through old heavily reinforced concrete pad foundations made a significant contribution to not damaging archaeological deposits.

At the same time, a flexible strategy developed with English Heritage and the Corporation allowing for a highly successful 'bonus' archaeological programme. This was kindly supported by the developer. Here, the careful removal of old shallow brick and concrete penetrations into the ground enabled archaeological structures and sequences to be technically examined. A major improved archaeological understanding resulted, related to the insertion of the Roman fort/city wall and ditches, and then the multi-phased medieval ditches and a city wall bastion.

Introduction

During 1997–1999 Argent Group Investments plc acquired a major redevelopment site at Alder/Castle/Falcon House, 1–6 Aldersgate Street, City of London. The scheme was for commercial offices fronting both Aldersgate Street to the west and Noble Street to the east (Fig 16.10).

Previous archaeological investigations in 1922 on the site had revealed the city wall and a hollow bastion to be present along the southern boundary of the development area (RCHME 1928, 90). Further investigations in 1939 in a tunnel from the south-west corner of the site under Aldersgate Street observed the western part of Aldersgate itself (Maloney 1983, 110–11). Finally, Professor Grimes in the post-war period observed the remains of Cripplegate Fort with a corner turret and later city wall immediately to the east of the site in Noble Street (Grimes 1968, 21, 47–8). These observations led to the scheduling of the three ancient monuments identified: County Monument Number 26D, the city wall bastion; County Monument Number 26R the Roman, medieval and post-medieval city wall; and County Monument Number 26S the Roman, medieval and post-

Richard Hughes
Consultant Archaeologist
Ove Arup International
13 Fitzroy Street
London W1T 6BQ, UK

Jonathan Butler
Site Director
Pre-Construct Archaeology Ltd
Unit 54 Brockley Cross Business Centre
96 Endwell Road
London SE4 2PD, UK

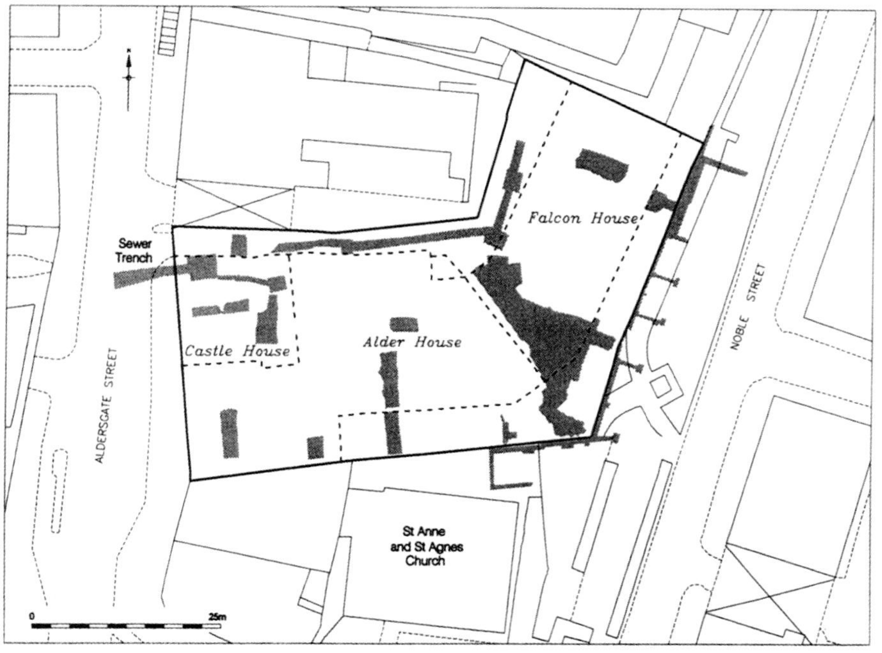

Reproduced from Ordnance Survey Superplan Data © Crown Copyright 1998

Fig 16.10 Site location

medieval city gate and city wall.

Because of the need to protect and preserve the Scheduled Ancient Monuments, the development scheme was designed by the architect and consultant archaeologist to locate foundation piling in areas of pre-existing total truncation. These locations comprised a network of pad foundations set into the natural subsoils, the level of which had been proved by 100mm coring, monitored by Pre-Construct Archaeology (PCA). The piling was undertaken by Bachy Soletanche, with their largest rig especially adapted to drill through concrete and iron railway tracks, laid in two or more levels as heavily reinforcing grillages. The rig was able to cut out precise cores of the full pile diameter, often 1500mm diameter and up to 5.5m long before descending down into London Clay! The character of every old pad was predetermined by the full depth 100mm diameter coring, as there were no old drawings of the foundations. The new basement was built within the envelope of the previous buildings. This requirement had been established as an architectural solution (often seen as a development constraint) at the start of the development and still allowed here for a very viable commercial scheme. The development scheme used a design and materials compatible with the character of the city wall.

The whole scheme involved a multi-disciplinary approach, 'partnershipping', with specific involvement by the developer and the archaeological consultant. Considerable direction was provided by Ellen Barnes, the Inspector of Ancient Monuments for London and Kathryn Stubbs, the Corporation's senior archaeological planning officer.

The strategy for the site was dual purpose. On the one hand involving an evaluation of the extent and quality of survival of the three Scheduled Ancient Monuments: the city wall, Aldersgate and bastion along the southern and eastern boundaries of the site. On the other hand, Scheduled Monument Consent was given for limited areas of archaeological excavation within the city ditch where new lift shafts were to be sunk, for a sewer heading in the north-west corner of site and for new drainage along the north-western boundary. The evaluation involved additional archaeological works, in addition to those initially required by the Corporation of London and English Heritage, resulting from discoveries. This allowed archaeological research objectives specific to the site to be formulated and adapted. There was recognition that because of the focus of the project on *in situ* preservation, there would be neither complete answers nor confident archaeological interpretations for all of the remains to be exposed. But, one objective was to see just how much archaeological data could be obtained for as little excavation as possible. Many of the areas were uncovered but not dug and in some areas the modern obstructions were removed to allow soil faces to be documented and datable artefacts recovered. The developer was supportive of this non-development related archaeological process and maintained a daily interest in the process and discoveries.

Thus, with the exception of the areas where consent for archaeological excavation was given, no medieval or Roman deposits were removed, these were left preserved *in situ*. This also allows the Corporation of London and English Heritage to consider long-term research goals and to establish heritage management objectives. After the initial monitoring of test pits, boreholes and a series of coring procedures, the excavation and watching brief were undertaken by PCA between November 1997 and March 1999 when all relevant groundworks were completed.

Area of Scheduled Ancient Monuments

In the areas of the Scheduled Ancient Monuments along the southern and eastern boundaries of the site small evaluation trenches located the city wall in three places (Fig 16.11), and the remains of a stair turret and the medieval bastion 15; however the Aldersgate itself was not found in its postulated position. A pre wall city boundary ditch was seen in these trenches following the line of, and cut by, the foundation of the later city wall. It extended 23.40m east–west across the site and survived to a width of up to 6.60m and a depth of 1.50m.

The excavation of modern disturbance allowed sections of a possible ditch associated with Cripplegate Fort and the 3rd-century AD city ditch associated with the city wall to be recorded in the south-east corner of the site. The removal of the top courses of the eastern post-medieval to 20th-century boundary wall, which was constructed on top of the Cripplegate Fort and city wall was undertaken after a detailed survey by PCA. All work was monitored by Ken Sabel, standing building specialist, who was actively involved in discussions and decisions with the archaeological and conservation consultant, Corporation of London and English Heritage.

Fig 16.11 View of the foundation of the Roman city wall

Fig 16.12 View of the medieval city ditch

Areas of limited excavation

Three areas of limited excavation were required; these were in the locations of essential new lift shafts, a sewer shaft and heading and services trenches. These works could not be designed to avoid penetrating archaeological deposits.

Six phases of medieval and post-medieval city ditch dating from the 13th century to the 17th century were revealed (Fig 16.12). The ditch was found to be at least 17.50m wide and up to 4.00m deep as excavated. Excavation and extensive sampling produced important animal bone and environmental evidence from the ditch fills. Pollen analysis allowed the plants growing in the vicinity of the ditch to be identified. Charred, waterlogged and mineralised plant remains confirmed the cessy nature of many of the ditch fills with many of the seeds having passed through human and animal bodies. The wide range of medicinal plant species identified are important in contributing to the knowledge of diet and health in the medieval period. The latest phase of ditch was dated to the mid 17th century, suggesting a reuse of the original city defences during the Civil War, in addition to the construction of an outer defensive line. This 'open area' is well illustrated on early maps.

In the sewer area an island of Roman archaeology was revealed. This survived due to the city ditch narrowing towards Aldersgate in Roman and medieval times leaving a causeway. A Roman road previously seen to the north was recorded and the possible late Roman city ditch fill was recorded in the sewer tunnel. Within service trenches along the west side of the site late Saxon/Norman city ditch fills were excavated. A gully of the same period produced large quantities of crucible fragments showing evidence of silver working.

Conclusions

This was a highly successful multidisciplinary project in which very important archaeological findings were made on the site despite the limited excavations. The site at 1–6 Aldersgate Street is of particular importance in contributing to

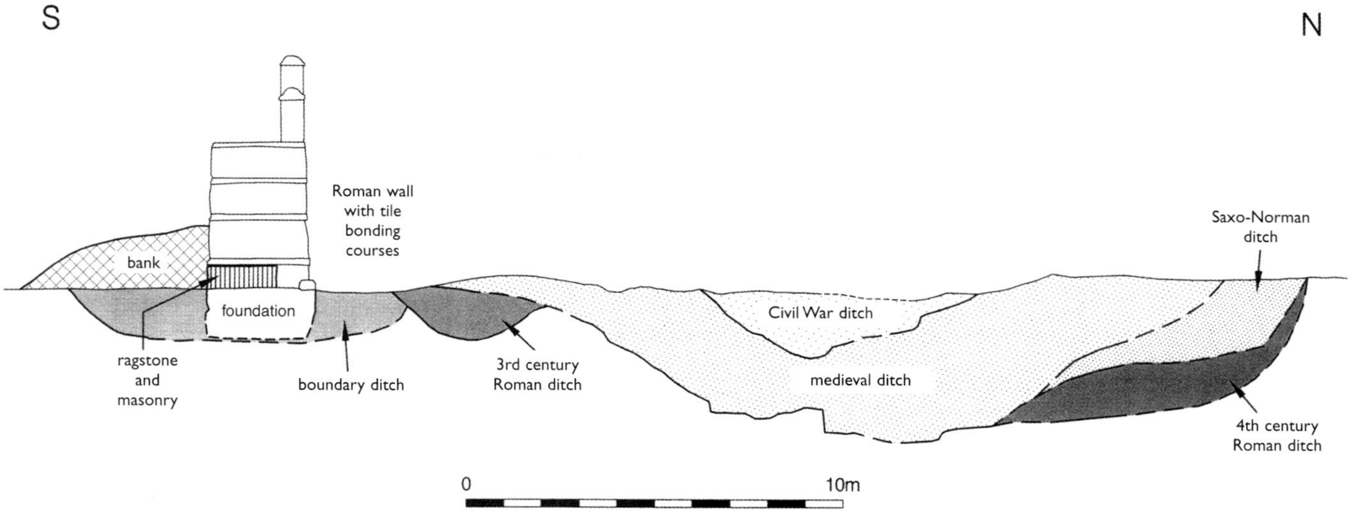

Fig 16.13 Composite section across the site

the history of London's defences from Roman times to the post-medieval period (Fig 16.13). This is the first evidence of an early Roman city boundary ditch to be found west of Cripplegate Fort. The city wall has been located with certainty, being slightly further to the north than some commentators had previously believed. A stair turret was discovered. The city gate at Aldersgate, however, was not observed. The remains of the Saxon city ditch are a rare survival, as elsewhere on the defensive circuit most evidence has been destroyed by the later medieval ditch. The presence of medieval tile in the fabric of bastion 15 substantiates its origin as medieval and contributes further proof to the suggestion that all the hollow bastions on the western circuit of the wall are of similar date. The section across the medieval city ditch was one of the best preserved in London. The several phases of recuts show that the cleaning and maintaining of the ditch was a continuous process during the medieval and early post-medieval periods. This was necessitated by the constant silting up of the water-filled moat and by the battle with the citizens of London who continued to use it to dump their rubbish. The faunal and environmental remains produced significant information about the environment of the ditch and the diet and medicinal concerns of the population. Finally, the 17th-century ditch suggests that London was protected during the Civil War by an inner ring based on its 1500 years old defences.

Many of the evaluation trenches showed tantalising glimpses of the archaeology which had to be interpreted without excavation and thus dating material. On this particular site, because of the previous excavations of the city defences, extrapolations and educated guesses could be made in interpreting the archaeological remains. On other sites this might have been much more difficult.

The development scheme allowed substantial amounts of archaeology to be preserved *in situ* for future generations. The new piling configuration and engineering records were, in a similar way to Argent's Governor's House development project, lodged with the archaeological archive allowing for the piles to be potentially reused in the next generation of building on the site, and thus further maintaining the archaeological resource for the future.

This development has provided an excellent case history where archaeology was addressed early on in the design process and this involving the developers and consideration of their commercial interests.

Bibliography

Grimes, W F, 1968 *The excavation of Roman and mediaeval London*, London

Maloney, J, 1983 Recent work on London's defences, in *Roman urban defences in the west* (eds J Maloney and B Hobley), 96–117, CBA Res Rep 51, London

RCHME, 1928 Roy Comm Hist Monuments Engl, *An inventory of the historical monuments in London: Vol 3, Roman London*, London

16d

The redevelopment of Plantation Place, City of London

RICHARD HUGHES, ANDREW LAWRENCE and DINESH PATEL

Richard Hughes
Consultant Archaeologist

Andrew Lawrence
Structural Engineer

Dinesh Patel
Geotechnical Engineer

Ove Arup International
13 Fitzroy Street
London W1T 6BQ, UK

Abstract

This paper reports on the excavation and *in situ* preservation process on the Plantation Place development site in the City of London. The impact of archaeology on the design and construction of the proposed development is discussed. Plantation Place is the largest site to have been redeveloped in the City in the past 50 years and is one of considerable archaeological value. The archaeological mitigation strategy is based on a formula of 50% excavate and 50% preservation. This satisfies the archaeological research objectives, the Corporation of London planning requirements (stemming from their unitary development plan (UDP)) and the commercial objectives of the developer. The site has yielded spectacular archaeological discoveries related to the earliest phases of Roman occupation within the City. Considerable geotechnical instrumentation data has and is being generated related to the preservation conditions.

This paper gives the background to the early studies on archaeology and ground conditions and how this led to the final archaeological mitigation strategy adopted for this site. It gives a summary of the superstructure/foundation design and describes the impact of archaeology and geology on the engineering solutions adopted, as well as the types of instruments used to monitor ground movements. The way archaeology affected the approach to the various construction activities on site and the ground movements associated with these activities and their affects on the archaeological resources are outlined. Finally some of the important archaeological findings are presented.

At the time of writing this paper the final phase of development, construction of the main piles, which support the superstructure and excavation of the new deep basement is underway.

Background

The Plantation Place development is one of the largest developments within the City of London. It occupies a prime City site, west of Fenchurch Street Station, covering a whole City block about 100m x 130m (Fig 16.14). The site lies between Fenchurch St, Mincing Lane, Great Tower Street and Rood Lane (Fig 16.14). Surrounding street levels range between +15 to 17m OD. Within this block the only two buildings that will remain are a 17th-century church (St Margaret Pattens) built by Sir Christopher Wren after the Great Fire of London in 1666 and a modern building to the south-west of the site (51 Eastcheap). The development of this site offered many challenges for designers, archaeologists and the Corporation of London, not just because of its archaeological importance but also because the form of the new building required large open floor spaces and a 15m deep basement.

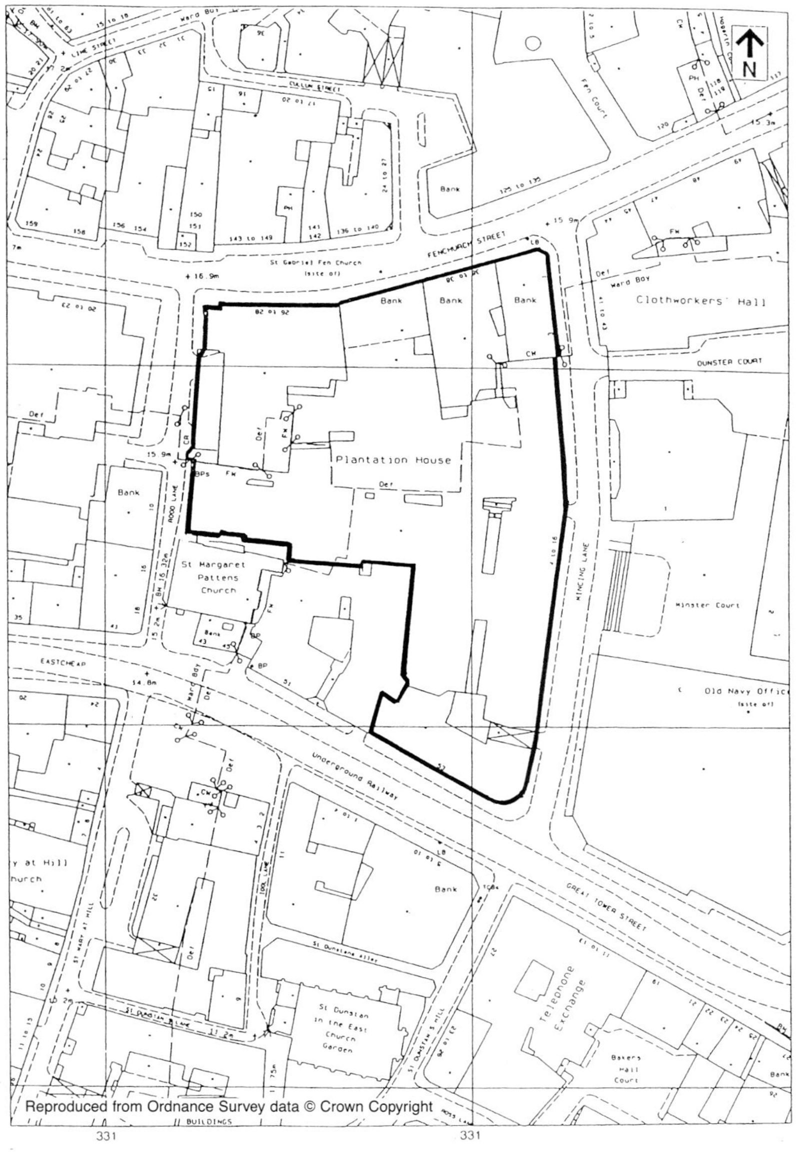

Fig 16.14 Site location

Key to the development of Plantation Place was integration of the archaeological consultant and geotechnical engineers with the Arup Associates architects, structural and building service engineers. Several years prior to the start of the site development process it had been recognised that the site had a considerable archaeological value. It has a central City location, immediately south-east of the Roman Forum and Basilica and on Fenchurch Street, a major thoroughfare from Late Saxon times onwards. Many ad hoc discoveries had been made on and surrounding the site from the mid 19th century onwards. There had been several very small-scale archaeological investigations by Cotterell in the 1930s related to the standing buildings. The sheer size of the site (some 10,000m^2) also made it of critical importance.

Early discussions with the Corporation of London focused on how the scheme would satisfy the UDP expectation of preserving the archaeological remains *in situ* – this responding to a definition of the potential resources, especially those of the Roman period, being of national value.

Arup's archaeological desk study to the developer, British Land, showed below the existing single level basement a chessboard like pattern of archaeological survival between pad footings. The present complex of buildings was found to have been formed on a network of large concrete pads ranging between 3m^2 and often 5m^2 in plan and with some areas effectively becoming continuous rafts. The engineering requirements at the time of construction of the existing buildings (1920–1950) demanded that the pads would be founded in the gravel stratum; fill soils being far too soft and variable. The old engineering drawings provided a high level of confidence about the distribution of pads and the ground conditions between them, confirming that the fills must be fully truncated.

Given the degree of truncation, the commercial objectives of the development, the archaeological research objectives and the Corporation of London requirements, the archaeological mitigation strategy was one of part excavation and part *in situ* preservation. Excavation would allow for a central new deep basement, required for the building's plant, and preservation would be focused on two transects of land. It was considered that these transects, along with preservation fronting the roads, would offer the opportunity for archaeological research sometime in the future.

The mitigation strategy approved by the Corporation allowed for full archaeological excavation of the central areas, some $4000m^2$ in extent and $12000m^3$ in volume. Given the size of this undertaking, consent was granted in mid 1998. In those areas around the new deep basement to be retained at the same level as the existing single level basement, it was accepted that the superstructure would need the support of foundations. The existing pad footings could not provide this support and a piling option was the preferred solution. The aim was to try and put the piles wherever possible through the old pad footings, a successful solution adopted at the Alder House, Castle House and Falcon House sites (see paper 16d). Local excavation at pile positions would be carried out where this was not achievable. British Land was also willing to lose space in the areas around the new deep basement by constructing the ground beams and pile caps above the archaeological resources.

The opportunity was also taken to examine some archaeological research related to understanding *in situ* preservation conditions. The new foundation designs necessitated a thorough understanding of ground conditions, mostly related to the behaviour of the Gravel, London Clay and Lambeth formations. These would all be participating in supporting the new structure. There was also a duty of care to the owners of the surrounding buildings, roads and tunnels to ensure they were not damaged by the new development. The research therefore also included reaching an understanding of how the ground behaved in response to demolition unloading, to excavation and to adding back the weight of the new building. As part of the engineering design and understanding of ground movements a detailed site investigation was carried out. The engineering and archaeological timetable for the works described in this paper is detailed in Table 16.6.

Initial studies and investigation

Previously the site was occupied by a variety of structures, which ranged between 8 and 11 storeys in height. These all had a single level basement, approximately 3.0m–4.0m deep. The perimeter walls of this basement are free standing (gravity) concrete walls and, because they support the surrounding streets, will remain a crucial part of the proposed development. Any new engineering or archaeological excavations must ensure the stability of these perimeter walls.

In addition, the District and Circle Underground railway (LUL) and Dockland Light Railway (DLR) tunnels run under Great Tower Street (Fig 16.15). These are sufficiently close to the development to warrant further consideration in the design. The LUL tunnel is a cut and cover box, while the DLR tunnels are deep twin bored tunnels in the London Clay. Dewatering within the former has also lowered the groundwater table in the underlying natural gravels at the southern end of the site.

Value of archaeological and engineering desk studies

At a very early stage in the planning process (*c* 1994) it was realised that two important studies were needed. Firstly, an archaeological desk study was required to determine the archaeological value of the site before recommendations could be made for a proper evaluation of the resources and mitigation strategy for the development. Secondly, an engineering desk study was required to determine the foundations of the previous development, the foundations for the new development and the impact of the ground movements on the surrounding buildings, roads and tunnels associated with this development.

These studies showed that the site was occupied in Roman, Late Saxon and medieval times, but that the archaeological resources were likely to be severely truncated, surviving only between the very large pad footings (up to $5m^2$) which supported the previous buildings on the site (Fig 16.16). Due to the high imposed loads from the structures, the pads were expected to be founded on the dense gravels 3.0–4.0m below existing basement levels. Engineers at the time would not have attempted to place pads on natural brickearth or archaeological deposits, as buildings would have been damaged by unacceptable settlements.

Table 16.6 Engineering and archaeological timetable for the works

Phase of works	Description	Date
Phase 1	demolish existing buildings	June 1998–April 1999
	install instruments for monitoring	July 1998–Sept 1998
Phase 2	construct secant pile wall to the new basement walls:	
	breakout old footings along line of new wall (slot excavation)	Nov 1999–Feb 2000
	install secant pile wall	Jan 2000–April 2000
Phase 3	archaeological excavation within area of new central basement	March 2000–Sep 2000
	break out and remove pad footings within area of new basement	Aug 2000–Dec 2000
Phase 4	main piling for superstructure and top-down construction (current programme)	Nov 2001–Dec 2002

These studies also showed that during World War II the far south-eastern corner of the building was heavily bomb damaged and rebuilt. It was expected that in this corner the archaeological resources would be limited as the existing basement is also deeper, typically +11 to 10m OD. The pad footings were reinforced with steel grillages below columns and that the perimeter retaining walls were concrete gravity walls with either 'L' or inverted 'T' shapes, also likely to be founded on gravels.

Initial archaeological and site investigations

Given the commercial risks of the development it was important to verify whether the above studies were correct. An archaeological and engineering evaluation of the site was therefore combined in November 1997 while the building was still occupied by tenants. The archaeological site investigation was conducted by the Museum of London Archaeological Service (MoLAS) appointed under Foundation Exploration Services. Six archaeological pits and three boreholes were carried out within the existing basement, in headroom space as low as 3.0m. The archaeological pits confirmed the predictions of the archaeological desk study and allowed assessments to be made on the accuracy of the

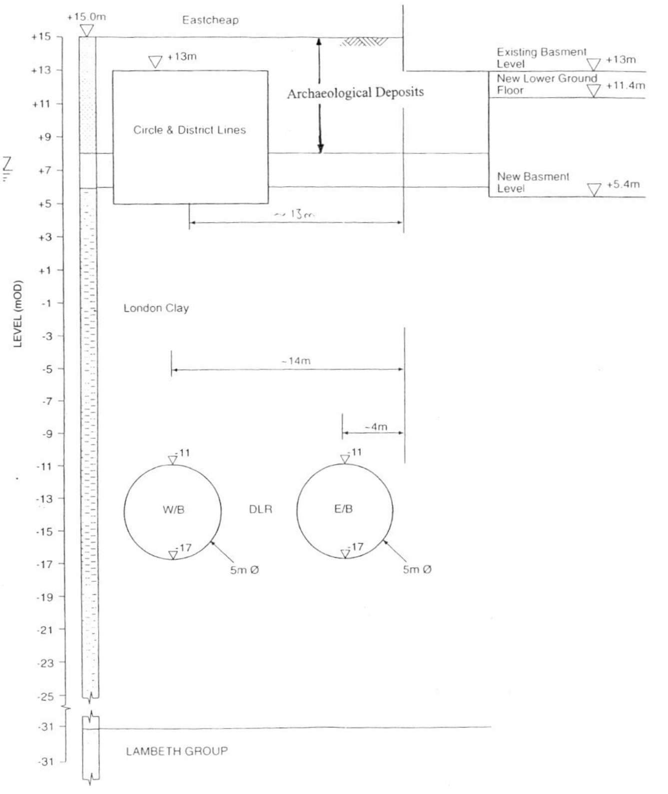

Fig 16.15 *Tunnels under Great Tower Street*

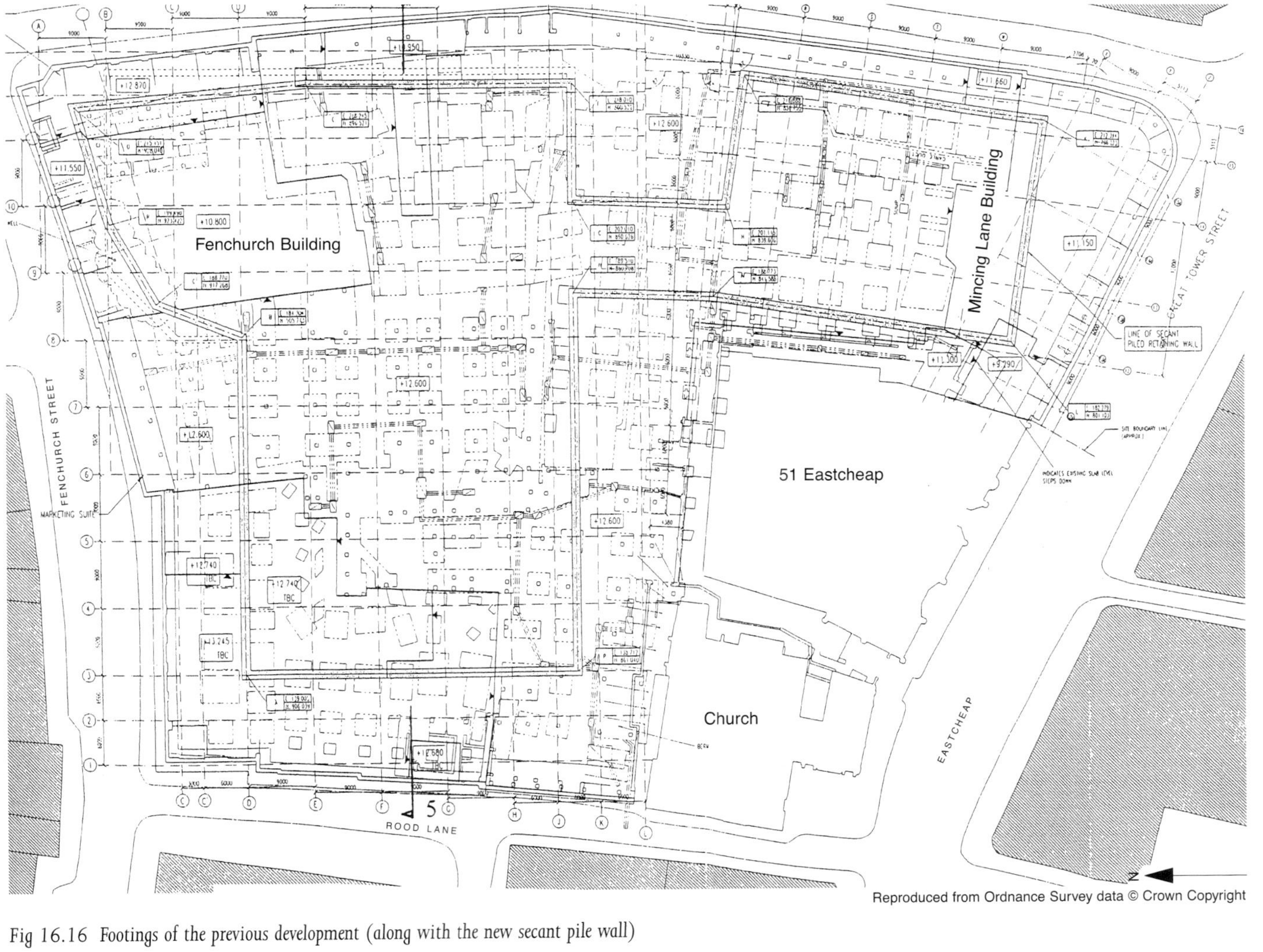

Fig 16.16 *Footings of the previous development (along with the new secant pile wall)*

pad sizes and depths shown on old foundation drawings. The pads were confirmed as being in the natural gravels. This provided greater confidence in the removal costs of pad foundations and impact on the programme for the proposed development.

Since the perimeter retaining walls were an important aspect of any proposed development, their structural integrity needed to be carefully investigated. This involved diamond drilling horizontal, vertical and inclined cores through the stem and base of the concrete retaining walls, testing the cores for strength and durability and inspecting the reinforcement in the wall from cut-outs. This allowed assessments to be made of the geometry of the walls and their stability when the propping action afforded by removal of the ground slab floor as part of the proposed demolition works to the existing basement level. The coring work also confirmed that there were no archaeological deposits behind and beneath the existing retaining walls.

Ground conditions

Table 16.7 shows the typical ground conditions as shown in the Phase 1 (pre demolition) and subsequent more detailed Phase 2 boreholes (post demolition).

The groundwater table in the gravels was at about +7.0m OD. No water was encountered in the archaeological fills. The piezometric pressure in the London Clay was found to be approximately hydrostatic to -10m OD, below which it fell to about 50% of the hydrostatic at the top of the Thanet Sand.

The undrained shear strength of the clay is given by the equation $Su = 100 + 7z$, where z is measured from the top of the clay. The bottom 5.0m of the London Clay was found to be sandy so it was recommended that piles be founded above this zone to avoid collapse of unlined bores. Suction measurements in the London Clay showed that it had an average ko = 1.5. No equivalent tests were made in the fill (archaeological soils) as it was clearly seen to be very variable.

Table 16.7 Typical ground conditions as indicated from the boreholes

Stratigraphy	Thickness (m)	Top height of strata (m OD)
Fill	up to 3.0	+13.0
Brickearth	0.5	+11.0
Terrace Gravels	5.5	+10.0
London Clay	33.5	+4.5
Lambeth Group	16.0	-29.0
Thanet Sand	13.0	-45.0
Chalk	-	-58.0

Proposed development

The proposed development is shown on Fig 16.16. In essence the new structure is split in two, the northern half being referred to as the Fenchurch Building and the south-eastern one referred to as the Mincing Lane Building. The Fenchurch Building is a 16-storey structure with a 15m deep basement; while the Mincing Lane Building is about 8 storeys and has a basement depth of about 9.0m, see cross section through both structures (Fig 16.17).

Full archaeological excavation was to be carried out within the new central basement areas of both buildings. The archaeological deposits outside this area were to be preserved, except at new pile positions where it was not possible to locate the piles through the old pad foundations (Fig 16.18). Ground movements associated with basement excavations are shown on Fig 16.19.

Top down basement construction sequence

The new basement is to be constructed using a top down method. This construction method greatly reduces the ground movements on the archaeological fills preserved outside the new central basement area. There is better control of ground movement as the basement box is gradually propped by the intermediate basement slabs. The method also leads to a fast construction programme as both the main structure and the basement excavation can be simultaneously constructed.

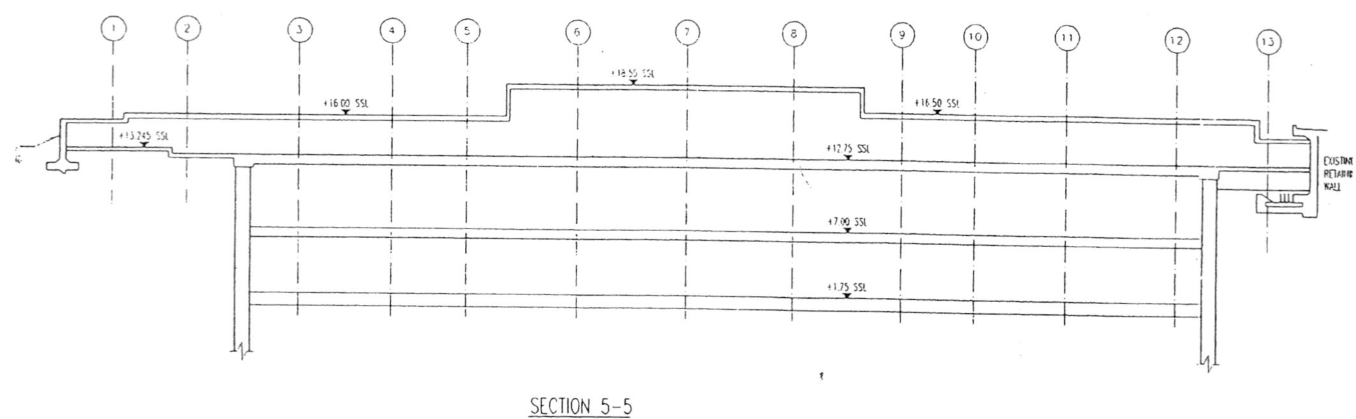

Fig 16.17 Cross section through the substructure of the proposed development (for location see Fig 16.16)

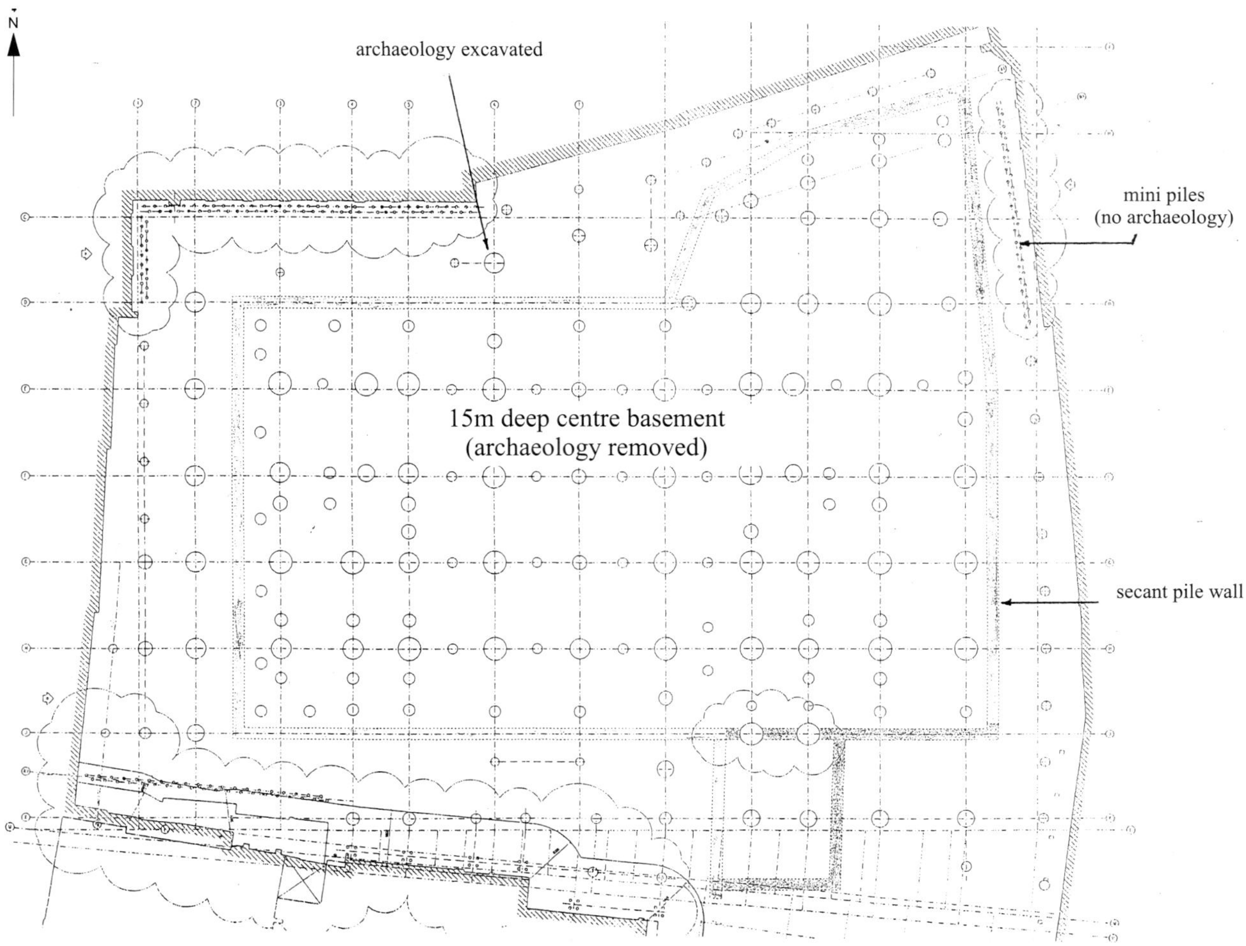

Fig 16.18 *Zones of preserved and excavated archaeological deposits along with pile positions for the new structure*

Surrounding buildings, roads and tunnels are also likely to be less affected by ground movements caused by inward movement of the basement retaining walls.

The sequence of the top down method is as follows: install new basement walls from existing basement level of +13.0m OD; excavate within central basement area to +10m OD removing the archaeology and old footings; install bored piles to support the new building; insert steel columns into the heads of the piles (15m below ground); construct new upper basement slab at +12.75m OD to prop apart basement walls; simultaneously construct structure above +12.75m OD while excavating ground below this level; build middle basement (+7.0m OD) slab whilst continuing to build structure above ground; cast lower basement raft and connect to piles (+1.75m OD), again building structure above; complete any remaining structure above ground.

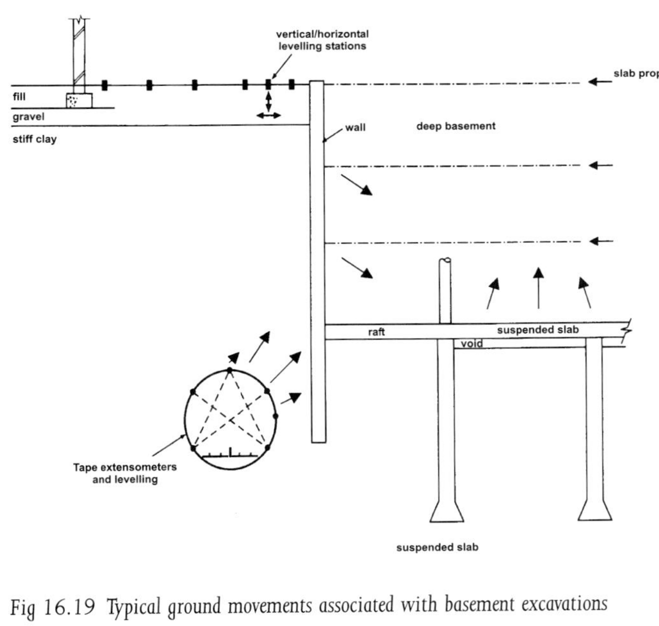

Fig 16.19 *Typical ground movements associated with basement excavations*

New basement retaining walls

The choice of wall type is very important when considering the effects of ground movements on archaeological fills. A stiff propped wall will minimise horizontal movements towards the excavation and hence will have less of an effect on archaeology while a flexible wall, such as a sheet pile wall, will deflect excessively. At Scheme Design various stiff wall options were considered for the new central basement: hard-hard secant bored pile wall (interlocking male and female piles about 1.0m diameter); diaphragm wall panels (interlocking wall panels about 4.0m long and 1.0m thick) and contiguous bored piles with minipiles behind wall (similar to secant piles but not interlocking). After carefully examining the various alternatives in respect to performance, cost, programme and impact on archaeological resources the preferred option was a secant hard-hard bored pile wall which had several advantages:

- the concrete wall was very stiff (1.0m thick) – hence any ground movement towards the excavation would be small, also minimising the impact on archaeological resources;
- pile installation is in a dry bore (ie no bentonite slurry is used) limiting contamination of archaeological fills during construction;
- the same secant rig can be used to drill through any deep concrete obstructions not removed by other means, saving construction time if a problem did occur on site;
- the wall length can be varied so that male piles can carry the vertical loads while female piles are installed short to provide stability to the excavation;
- the piles can be drilled to a high standard of verticality, typically 1:250 and sometimes better than 1:400 (using guide walls cast at the ground), ensuring good interlock to prevent water ingress over the depth of the new basement;
- the same grade of concrete is used for both male and female piles providing good long term durability.

In respect of archaeological concerns, the secant piled wall meant that the archaeological formations to be later excavated within the new central basement could be initially assessed in the guide wall 'slot' excavations. The final secant pile wall (as built) had piles of 1200mm diameter spaced at 1050mm centres at level +12.6m OD. It was about 450m in length on plan and extended to about -5.0m OD for stability, except where the walls were required to carry vertical load from the structure, the toe levels were extended to about -23.0m OD. The main steel in the wall was typically 12T40 bars.

Foundations

The proposed structure is supported on bored piles founded in the London Clay. Outside the central basement area, the column loads will be supported on individual large diameter (900–2400mm) bored piles up to 46m long but close to the existing perimeter walls they will be supported on groups of minipiles with a cap over the top. To avoid disturbing archaeological resources, wherever possible, these piles will be located over former pad footings and pre-cored through concrete before the piles are installed.

Within the new central basement area, a piled raft solution was adopted. Because of the top down method of construction the piles were installed from the existing basement level with plunged steel columns. During superstructure construction these piles would support the downward loads from the top–down method of construction. Later the raft would be cast integral with the piles.

The adoption of a piled raft foundation to support the structure over the central basement and use of long secant pile walls for the basement meant that future heave movements (caused by unloading of the ground by excavation) outside the basement would be much smaller than by conventional excavation methods. Hence, future impact on archaeological fills would also be small.

Outside the central basement, the 900mm to 2400mm diameter piles support loads of between 3.7MN to 12.8MN founded at -23.0m OD. Minipiles support 500kN loads. Within the central basement area these pile sizes support higher loads, typically 6.3 to 14.3MN, again founded in London Clay at −23.0m OD.

Demolition and excavation causes the ground to heave and move inwards towards the excavation. Careful consideration was made with regard to the engineering design of the basement and foundations of the new structure to ensure that these movements would not have a detrimental effect on preserved archaeology and the surrounding structures.

Archaeological mitigation strategy

The basis of the mitigation strategy was developed from the archaeological evaluation of the site and using the archaeological and engineering desk studies. Detailed discussions were held with the Corporation of London responding to their UDP requirements (ie part excavation and part preservation). Partial excavation satisfied the commercial objectives of British Land who required a basement to house the plant etc in order to provide a prestige development for an important tenant.

The final strategy agreed with the Corporation allowed full excavation of 50% of the potential archaeological resources within a new central basement area (reported below) and to preserve the remaining 50% along two transects of land; one from Fenchurch St to St Margaret Pattens Church and the other from Mincing Lane to Rood Lane (Fig 16.18).

Outside the central basement area, the desire for *in situ* preservation meant that the engineering design had to be fully integrated with the archaeology. This necessitated the consultant archaeologist working closely with the Arup Associates architects, structural, geotechnical and building service engineers. The engineering design, which eventually evolved from this integrated approach, tried to minimise the impact by: locating new piles through old concrete pad footings and existing perimeter retaining wall bases (ie through existing truncation in the areas of *in situ* preservation); locating main cores, lift and service pits within the central basement to avoid excavating archaeological fill in the area of *in situ* preservation; constructing ground beams and pile caps above the preserved archaeological resources, even though this meant that British Land lost some valuable space.

Since the superstructure design was still underway at the time archaeological planning consent was granted for the archaeological excavation, the actual location and size of piles through the areas of *in situ* preservation could not be fully defined. It was therefore not possible to guarantee that some piles might not go through archaeological resources.

However, the superstructure design would aim, wherever possible, to use tall concrete walls within the upper basement to transfer column loads on to piles which would go through the old pad bases. However, the height of these walls were constrained by the archaeological resources below and the ground floor slab above, and the requirements to supply air conditioning units into the new basement and so this solution could not be used everywhere. This approach was explained to the Corporation and agreed.

An additional measure was to design the new basement walls to carry vertical load. This meant deepening the walls beyond what would be required to carry horizontal earth pressures alone and would also require tall concrete walls to spread the load on to individual secant piles within the basement walls. The advantage was that columns over the wall could be supported on the wall itself rather than on two piles straddling the basement wall, one of which would go through the area of *in situ* preservation. A plan showing the final location of the pile position relating to the old pad footings is shown on Fig 16.18.

Archaeological results

The archaeological excavation has produced some significant results which will alter our understanding of the earliest phases of Roman occupation within the City. The findings of the full excavations are presented in the MoLAS report (Archaeological Interim Report) prepared for Arup Associates and British Land. Amongst the discoveries were: the foundations of a massive tower probably associated with a town house around AD 300; a major east–west Roman road through Londinium with associated mudbrick and timber buildings; evidence of the Boudican revolt of AD 60/61 (burned soil) and associated rebuilding using defensive systems consisting of double ditches and a rampart; evidence of rebuilding with masonry structures following destruction of mudbrick and timber buildings in the Hadrianic fire of AD 125–30. Roman artefacts included large amounts of pottery, many gold and other coins, a substantial corpus of painted wall plaster, a Roman fish tank and a well to the top of the London Clay. Post Roman features included Saxon/Norman pits, wells and quarries. Evidence for the Great Fire of 1666 was absent and there were few post-medieval structures apart from brick-lined wells and cesspits, which produced a few artefacts.

Instrumentation

The demolition of the buildings at Plantation Place and subsequent archaeological and basement excavations for the new development will cause unloading of the ground and will induce both vertical and horizontal ground movements (Fig 16.19). Measurements of this movement is of importance for both engineering and archaeological purposes as it gives insights as to which construction activity is responsible for the greatest settlement or heave (just as the ground settles when the weight of a building is placed upon it, so it heaves back upwards when the weight of that building is removed).

Disturbance and damage risks to archaeological deposits and artefacts is very much an unknown quantity and instrumentation allows for collecting and interpreting data, regarding ground movements in the fill. It is envisaged that at some time in the future, the results from this detailed monitoring can be collected and compared with typical 'micro' disturbances that the excavators cause by their own digging methodologies.

Potential ground movements are ground heave (due to demolition and excavation), the inward movement of the new basement walls (resulting in horizontal ground strains and settlement behind the wall) and the overturning and sliding of the existing perimeter retaining walls (following removal of existing basement slab and any propping action which it may have provided). These movements will occur both during construction (short term) and after the building has been completed (long term).

Adjacent structures/services which have been identified as at risk from damage due to ground movements are: St Margaret Patten Church, 51 Eastcheap, DLR and LUL tunnels, perimeter retaining walls, underground services and roads behind the perimeter retaining walls and possibly archaeological foundations. During the design development, various studies were made in respect to the impact of ground movements on these structures/services. It was important from an engineering aspect to make sure that these structures/services were monitored and to identify how the actual movements compared with the predictions, during all phases of construction. A programme of monitoring was therefore put in place.

Monitoring instruments

The instrumentation of the site comprised precise levelling (to measure ground heave/settlement); 3D surveying (to measure vertical and horizontal movements of buildings and perimeter walls); electrolevels (to measure tilt of the perimeter walls and the facade of the surrounding buildings); inclinometers (to measure the horizontal movement of the new basement wall during excavation); surveying points in tunnels (to measure heave); extensometers in the ground (to measure heave of the clay with depth); pneumatic piezometers (to measure changes in pore pressure in the clay caused by unloading of the ground). Many of these instruments were established before demolition of the structure had started. Later, as the basement walls were installed, inclinometers were installed ready for monitoring of ground movements caused by the basement excavation.

Selecting instruments and presentation of data

It is important to ensure that any instruments used are carefully selected and specified and that the results from them can be easily understood and related to activities on site. Fig 16.20 shows some of the processes, which were implemented in the selection and monitoring of the instruments at Plantation Place.

Without clear presentation of the data, important decisions cannot be taken when excessive movements occur requiring immediate action on site, for instance in the case of arresting the collapse of an existing perimeter wall. Some examples of the importance of presentation of the data in graphical form, used at Plantation Place during construction, are shown here: ground heave movements during archaeological excavation (Figs 16.21–16.22); typical horizontal movement of secant pile wall towards the excavation (Fig 16.23); and tilting, heave and sliding of existing perimeter wall (Fig 16.24).

Impact of method of construction and ground movement on archaeological resources

Phase I works: demolition of existing buildings

This work consisted of demolishing all the previous buildings on the site to the basement level of about +13.0m OD. The

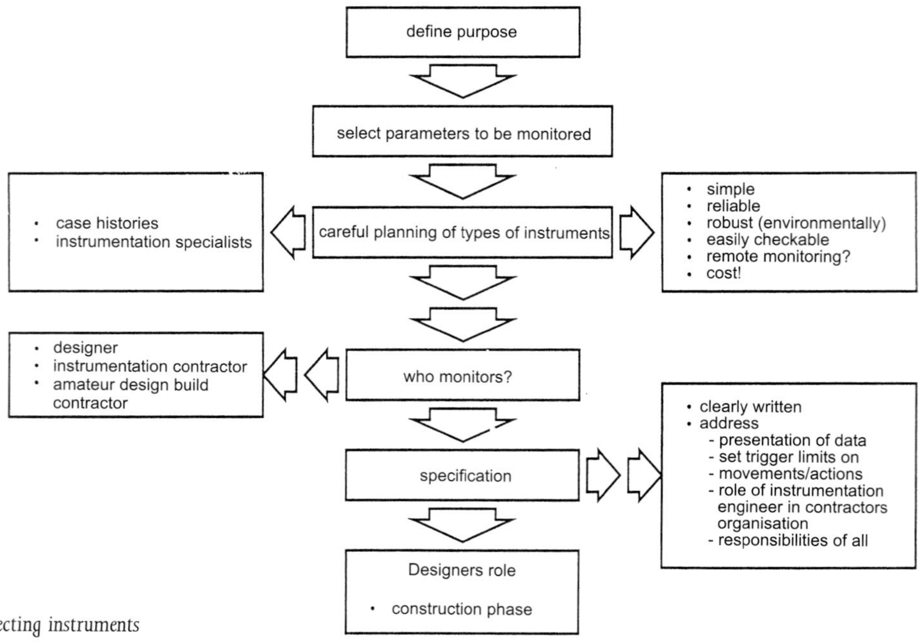

Fig 16.20 *Process for selecting instruments*

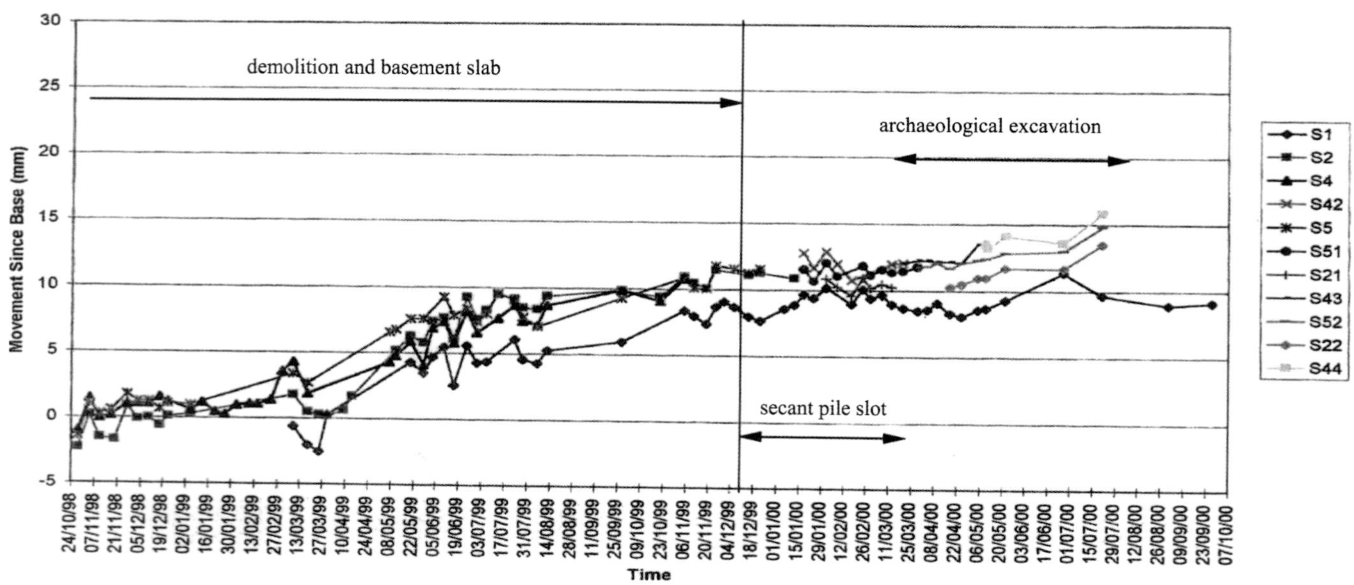

Fig 16.21 *Ground heave measured at existing basement slab level*

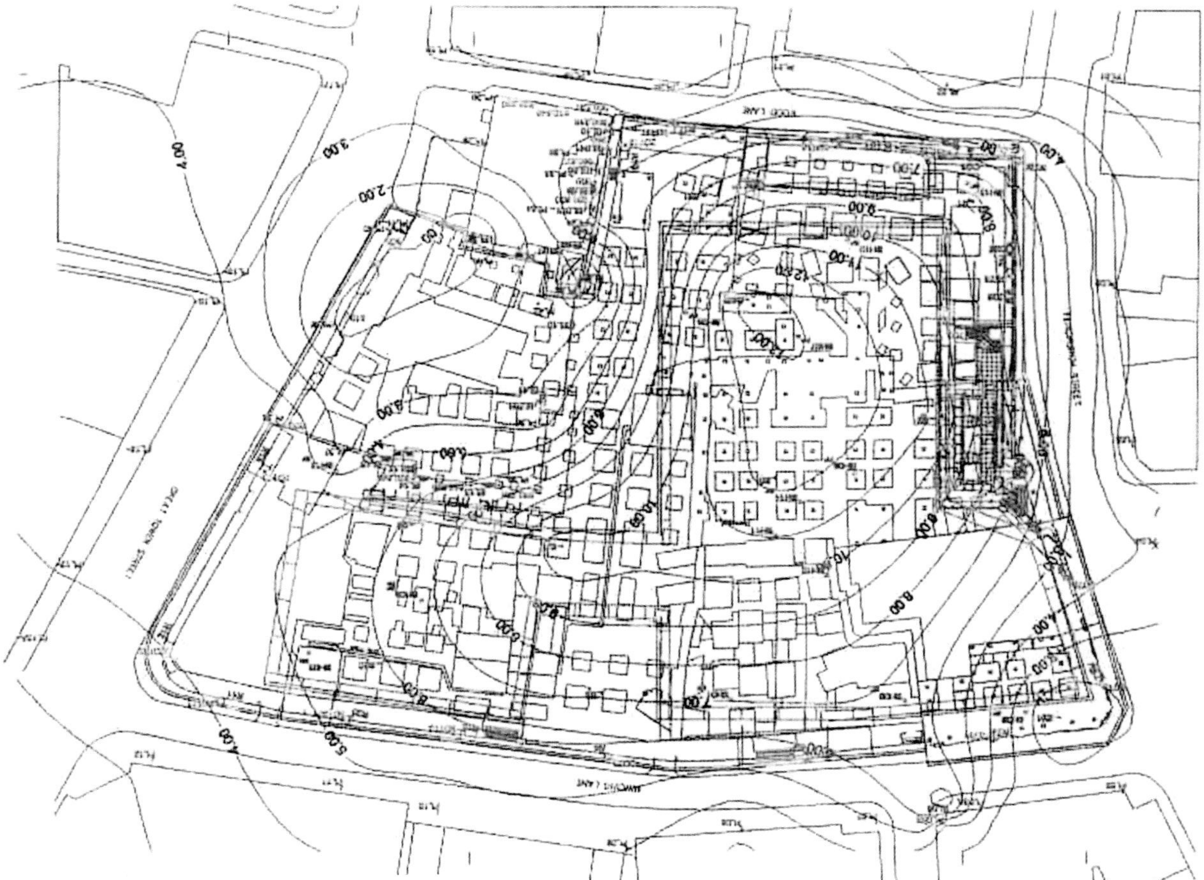

Fig 16.22 Contour map of ground heave measured at basement level

work started in June 1998 and took almost one year to complete. The demolition works was carried out by Keltbray under a subcontract with Kvaerner Trollope and Colls, who were the main contractors for this project.

The demolition involved both soft strip (hundreds of tons of Indian Granite which lined floors and walls were removed and recycled) and hard strip (removal of column/floors etc). These works had to be carefully performed given the complex nature of the structure and the sensitivity of the neighbouring occupants. For instance there were certain times of the week when demolition work was too noisy and had to be stopped while church services were conducted in peace. The demolition revealed many surprises – for instance the main floors were not solid concrete as previously suspected but hollow construction – indicating that old structural drawings cannot be entirely relied upon for accuracy.

The removal of these buildings resulted in a net unloading of the site of about 50 to 70 kN/m^2. Ground movements measured on the basement slab after demolition showed it had heaved by only 2mm at the end of April 1999, far less than the 6mm predicted at this stage when the ground floor slab was still in place. Following removal of the ground floor slab (which propped the top of the perimeter walls) and the removal of the remaining demolition rubble at basement slab level, the slab heaved to a total of 7mm over the next six months of this activity. Removal of the ground

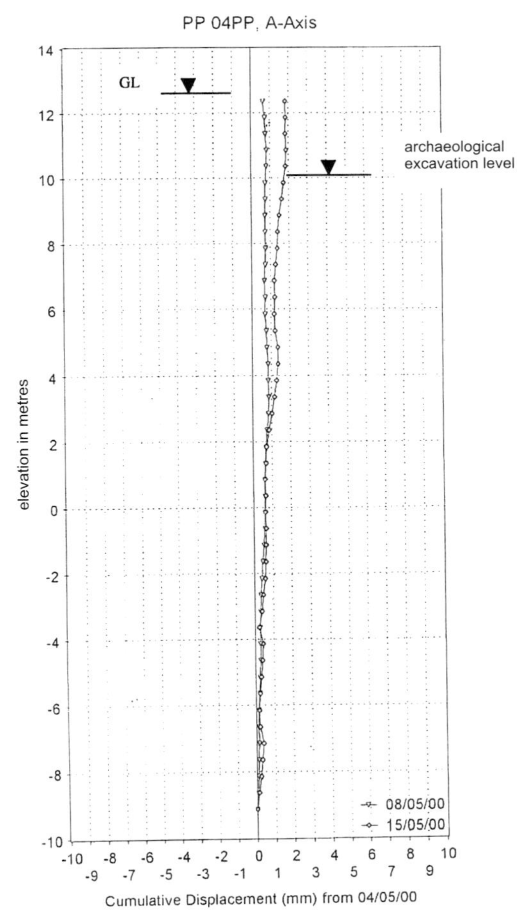

Fig 16.23 Horizontal movement of new basement wall towards archaeological excavation

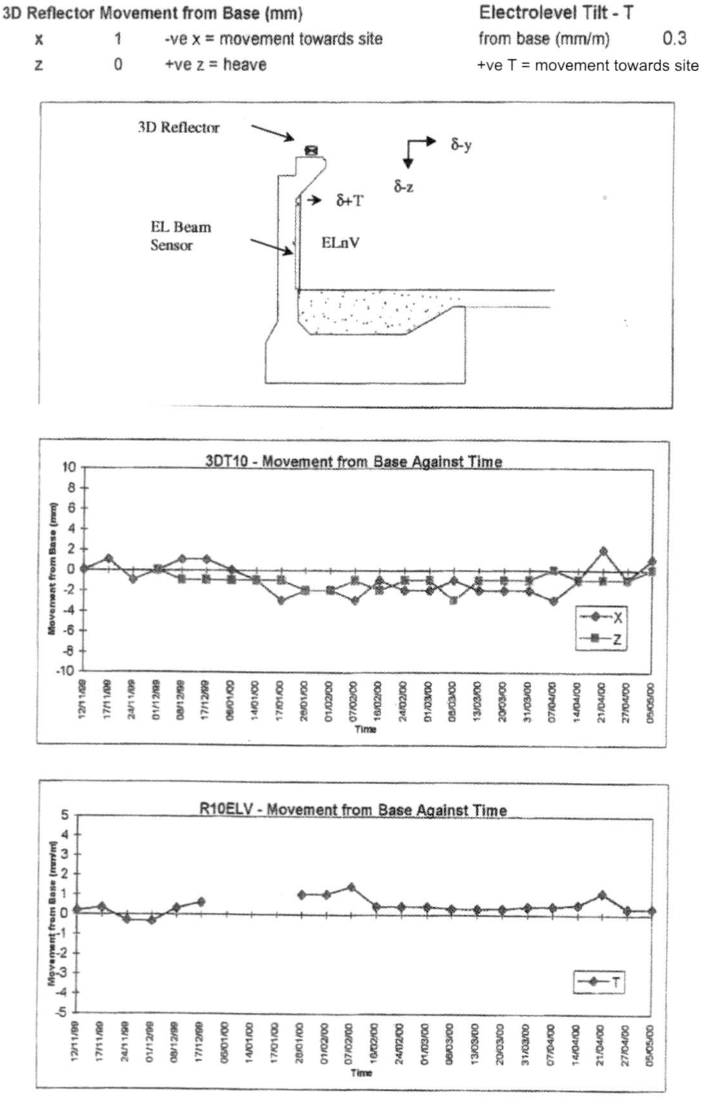

Fig 16.24 *Movement of existing retaining wall*

floor slab did not cause any appreciable movement of the perimeter walls surrounding the site.

At this time, British Land had built a glazed facade Marketing Suite in the north-west corner of the site. This would affect the next phases of work as described below.

The demolition of up to 11 storeys of buildings at Plantation Place caused only 7mm maximum heave in the centre of the site (reducing to about 2mm at the edges). Any archaeological resources within the truncated fills between the pad footings would have therefore experienced at worst this movement. This maximum movement occurred gradually over a period of about 18 months.

Phase 2 works: construction of secant pile walls to new central basement

The secant bored pile wall is 450m long and formed a box around the central basement to allow full archaeological excavations to be carried out safely within the central basement to a depth of up to 3.0m. This wall also became a permanent feature of the main development. This was preferred at Scheme Design over other forms of walls, which were of temporary construction. However, it meant that British Land had to make a key decision to install the permanent wall before the superstructure had been designed or a tenant found. This required a serious commitment to the project by British Land.

Before the new walls could be installed, it was necessary to break out the old foundations along the line of the wall. A 2.0m wide slot was dug through the old foundations, wide enough to incorporate the scalloped 'guide wall' required to guide the secant piles into place. The deep (3–4m) narrow slot could not be safely hand excavated and an archaeological watching brief took place.

Another key decision was to dig the slot as part of an advance contract to avoid any delays that might hold up the secant piling operations – this would have resulted in high downtime costs for any rigs not installing piles. Slot excavation was initially performed by saw cutting the 0.3m deep basement slab. Concrete pads, steel grillages and any other obstructions within the slot were then removed by burning steel off by hand and using large machines with pneumatic hammers and grabs. Near the Marketing Suite the slot was only 1.5m away and was carefully stitch drilled to prevent damage to the glass facade. A 0.5m deep reinforced concrete guide wall was then constructed along the line of the secant pile before piling commenced.

The slot works took about three months to perform, while the secant pile installation took another four months with all the piles being installed by end April 2000. Keltbray carried out the slot excavations and Kvaerner Cementation (in joint venture with Stent) the secant piling. The piling was executed without any major problems, although at a few pile positions very deep concrete obstructions were encountered which required coring using the secant rig.

At the end of these operations the maximum recorded heave in the centre of the site was about 10mm. This was attributed to ongoing heave movements associated with the previous demolition unloading. Part of this movement is attributed to breaking out of the old basement slab in the area of the new deep basement. The basement slab was tied to the pad footings and probably had some restraining effect on this heave movement, only released when the slab was broken out.

Breaking out the slot released the propping action of the basement slab to resist sliding of the perimeter retaining walls but this only caused walls to rotate by 1–2mm towards the excavation.

Phase 3 works: archaeological excavation

The full archaeological works within the central basement started in late March and ran to the end of September 2000, a period of about six months. Heave monitoring points, previously on the basement slab were transferred to certain pads, which were not removed until the archaeology was completed. This allowed heave movements to be measured continuously to the end of the archaeological excavations.

The archaeological dig was carried out by MoLAS and was one of the largest archaeological contracts in London, employing teams of archaeologists in three large zones on the site, phased in with other construction activities. At its height, up 70 archaeologists were employed. Keltbray supported this team with personnel and equipment for digging of non-archaeological resources, removing spoil from site, providing temporary works and making sure work was carried out safely.

The final archaeological dig was 3.0m deep (although local excavations were up to 4.0m deep). The massive $3.0m^2$ to $5.0m^2$ concrete pads within the central basement were finally removed by the end of December 2000. A small concrete crushing plant was brought on site to ensure that the pads were recycled to form aggregate for use on site. This resulted in large savings to British Land and minimised traffic problems in an area already congested during normal working hours.

Ground movements due to archaeological excavations and effects on preserved archaeological resources

During this period, the site heaved an extra 2 to 3mm (ie maximum 13mm from start of building demolition). Inclinometers in the secant wall showed that the wall and thus the archaeological fills behind the wall (*in situ* preserved resource) had only moved by about 3mm towards the excavation, much less than the predicted deflection of about 15mm. It was believed that this was as a direct result of the presence of concrete bases behind the wall. These dramatically reduced the horizontal pressures behind the wall compared with normal soil over a 3.0m height. Also, the final dig level of +10m OD was slightly higher than the founding levels of the old pad footings making sliding impossible and thus reducing the displacement of the secant wall.

Phase 4: pile installation and basement excavation

At the time of writing the main piling for the superstructure and the deep basement excavation were about to commence. Movement predictions suggest that inward movement of the basement walls due to excavation are likely to be up to 20mm at depth. Because the top of the wall is propped during construction, little horizontal movement can occur at this level. Since the archaeological fills are also near the top 3.0m they will also experience little horizontal strain. However, up to 20mm settlement is expected as ground settles behind the wall. During the course of excavation (into 2002) the actual movements will be measured using inclinometers.

In the long term the ground will also heave but it is expected that because the basement walls are deep (up to 23m below raft level), the walls will resist this heave by friction, thus significantly reducing the heave movements in the preserved archaeological fills. This is not expected to be greater than 30mm.

The net movement is therefore not expected to be very much (10mm maximum heave) in the long term after the combined effects of ground settlement and heave have been allowed for.

Conclusions

The Plantation Place development posed many challenges for both the consultant archaeologist and geotechnical engineers working with the Arup Associates design team.

The project demonstrated that integration of the consultant archaeologist within the design team and in the decision-making processes well in advance of construction is vital to the success of a project. Successful integration of redevelopment and archaeology also relied on the enthusiasm of the developer, British Land and the cooperation of the Corporation of London through all stages of the design and construction.

The success of the mitigation strategy depended on a thorough understanding of the archaeological value of the site, an understanding of the recent site history and drawings showing foundation details and how these related to the geology. The usefulness of the archaeological and engineering desk studies and the early archaeological evaluation of the site were essential in arriving at this strategy.

The paper also explains how careful consideration to the design and construction of the new structure and basement can limit ground movements induced by these works and thus, minimise impact on the preserved archaeology. Use of instruments to measure ground movements can provide excellent insights to the magnitude of movements that can occur in preserved archaeology. At Plantation Place, maximum heave/settlement and horizontal movements within the preserved archaeology are not expected to exceed about 20–30mm. These movements are small and are not expected to have an impact on the archaeology.

16e

The geotechnical instrumentation of Park Lane, Croydon, an Early Anglo-Saxon cemetery

RICHARD HUGHES,

JOHN DILLON,

JACKIE McKINLEY,

DINESH PATEL and

HILARY SHIELDS

Richard Hughes
Consultant Archaeologist

Dinesh Patel
Associate Director

Hilary Shields
Geotechnical Engineer

Ove Arup International
13 Fitzroy Street
London W1T 6BQ, UK

John Dillon
Deputy Director

Jackie McKinley
Site Director

Wessex Archaeology
Portway House, Sarum Park
Salisbury SP4 6EB, UK

Abstract

This paper presents the results from the first year of monitoring an Anglo-Saxon cemetery environment lying beneath the car park of the new development at 82–90 Park Lane, Croydon. In order to determine the long-term effects of construction on the burial environment at two locations, instruments have been installed to monitor changes in temperature, moisture content and compression of the ground. The monitoring is being supplemented by annual soil sampling to determine moisture content and various chemical parameters. Early results are presented with some interpretative comments.

Introduction

The site of a new development at 82–90 Park Lane, Croydon, contains an important Early Anglo-Saxon cemetery dated to between the 5th and 7th centuries AD. The cemetery includes both inhumation and cremation burials, many with grave goods, set in shallow graves cut into the natural gravel. The cemetery was discovered in 1897 during the construction of houses in Edridge Road, which now forms the western boundary of the site. The grave goods included spearheads and shield bosses along with the remains of 16 cremation urns. The archaeological resources deemed to still be there have been classed as being of national importance.

Prior to the new development the site was occupied by Victorian houses with gardens. At an early planning stage it was determined, at a Public Inquiry supported by English Heritage, that the more disturbed half of the site where the houses were situated could be developed with buildings. The less disturbed half of the site where the gardens were situated was to be preserved beneath a car park with instrumentation installed to monitor the *in situ* preservation conditions. Wessex Archaeology was appointed by the developer to be responsible for this work and in turn selected Arup Geotechnics as their consultant.

The main phase of excavations by Wessex Archaeology were carried out in the winter of 1999, with some further work in the summer of 2000. All or parts of 43 Early Anglo-Saxon inhumation burials were excavated, together with the remains of two cremation burials. Bone preservation was poor, with bone surviving in only 23 of the inhumation graves. Grave goods were recovered from 33 graves, predominantly weaponry (eg shield bosses, spears and swords), with some items of personal adornment (eg brooches) or dress (eg belt fittings). A single Late Romano-British inhumation burial and a probable Mid Saxon inhumation grave with a penannular ditch were also excavated. The works were generally observed by Arup to gain information about the ground conditions, particularly those encapsulating the archaeological deposits.

The work has involved the characterisation of the *in situ* deposits to understand the existing physical and key chemical properties of the ground prior to construction of the car park, the design of the instrumentation system to monitor the preservation conditions (this includes periodic extraction of samples from the ground to test the chemical properties in the laboratory) and finally the monitoring of the instrumentation over a three-year period during the development process.

Site characterisation

Any classification of soils needs to be undertaken by geotechnical engineers as an understanding of the engineering behaviour of soils is very important. This defines the type of laboratory tests to be carried out. Such characterisation of soils is an important exercise before placing instrumentation.

After demolition of the existing buildings, the engineering site investigation to characterise the *in situ* deposits was carried out at the same time as the archaeological excavation by Wessex Archaeology. The Arup work involved extracting samples for laboratory testing of moisture content, chemical parameters and compressibility characteristics, together with *in situ* testing for density. In the car park area, only limited and careful investigation of the ground above the burial environment was permitted and this was carried out close to the planned monitoring locations. The remainder of the investigation, including characterisation of the stratum in which the graves were located, was carried out in the more disturbed part of the site where Wessex Archaeology were excavating.

The stratigraphy beneath the car park is summarised in Table 16.8. At the locations beneath the car park where the monitoring instrumentation is situated no fill material is present.

Design of the instrumentation system

Throughout the design of the instrumentation system Arup Geotechnics worked closely with Soil Instruments, who have been responsible for procuring and installing the instrumentation, and maintaining the system. The selection of instrumentation on this project was dependent on reliability of the instrument under long term monitoring, accessibility of the instruments for replacement, if needed, and need for remote monitoring from anywhere within the UK with use of a modem and laptop computer (ie without visiting the site).

The final design included instruments installed at two locations in the car park to monitor changes in temperature, moisture content and compression of the ground beneath. The instrumentation was installed in June 2000. A plan of these locations, named IL1 and IL2, is shown in Fig 16.25.

In order not to disturb the burial environment situated within the top of the gravels, the monitoring instrumentation was designed to be placed in the overlying strata, ie the worked soil and the topsoil. The exceptions are the fixed extensometer rods that needed to be anchored in the gravels to provide a stable datum against which to monitor the movement of the ground above. The depths of the topsoil, worked soil and gravels beneath the finished car park slab are shown for IL1 in Fig 16.26. Between the final surface of the topsoil and the underside of the car park slab, a 100mm thick layer of sand was incorporated to spread the load. The sand is separated from the topsoil beneath by a Terram membrane and from the car park slab above by a polythene membrane.

Fig 16.25 also shows the positioning of the instruments installed at location IL1. Temperature is monitored at the two locations using four thermistors placed at different levels in the ground above the gravel in order to establish variation in temperature with depth. The depths of the instruments are such that there is one located in the worked soil, two within the topsoil and one in the overlying sand layer. Moisture content is monitored using two soil moisture sensors at each instrument location placed at different levels

Table 16.8 Stratigraphy beneath the car park

Soil type	Position of stratum	Thickness	Description
Fill material	from original ground level or below made ground	Varying	19th- or 20th-century fill, consisting of varying amounts of brick, chalk, sand, silt and clay to coarse gravel and frequent rootlets
Topsoil	from original ground level or below made ground	*c* 0.4–0.8m	moist, dark brown, very loose slightly clayey silty fine to medium sand with abundant fine to coarse gravel and frequent rootlets
Worked soil	*c* 0.8–1.0m below original ground level	*c* 0.1–0.2m	moist, orange brown, very loose very slightly clayey silty fine to medium sand with abundant fine to coarse gravel and frequent rootlets
Gravels; stratum in which the graves are situated	*c* 0.9–1.2m below original ground level	*c* 4.5–7.4m	moist, light brown, slightly silty, slightly clayey sandy fine to coarse rounded to sub-angular flint gravel; the fills of the shallow graves cutting into the gravel comprise a mixed sediment derived from the worked soil and gravels
Chalk	5.7– 8.6m below original ground level		moderately weathered very weak chalk; the chalk was not located in the investigation work of the project, but the information comes from boreholes at the site dug by Herts and Essex site investigations in 1991

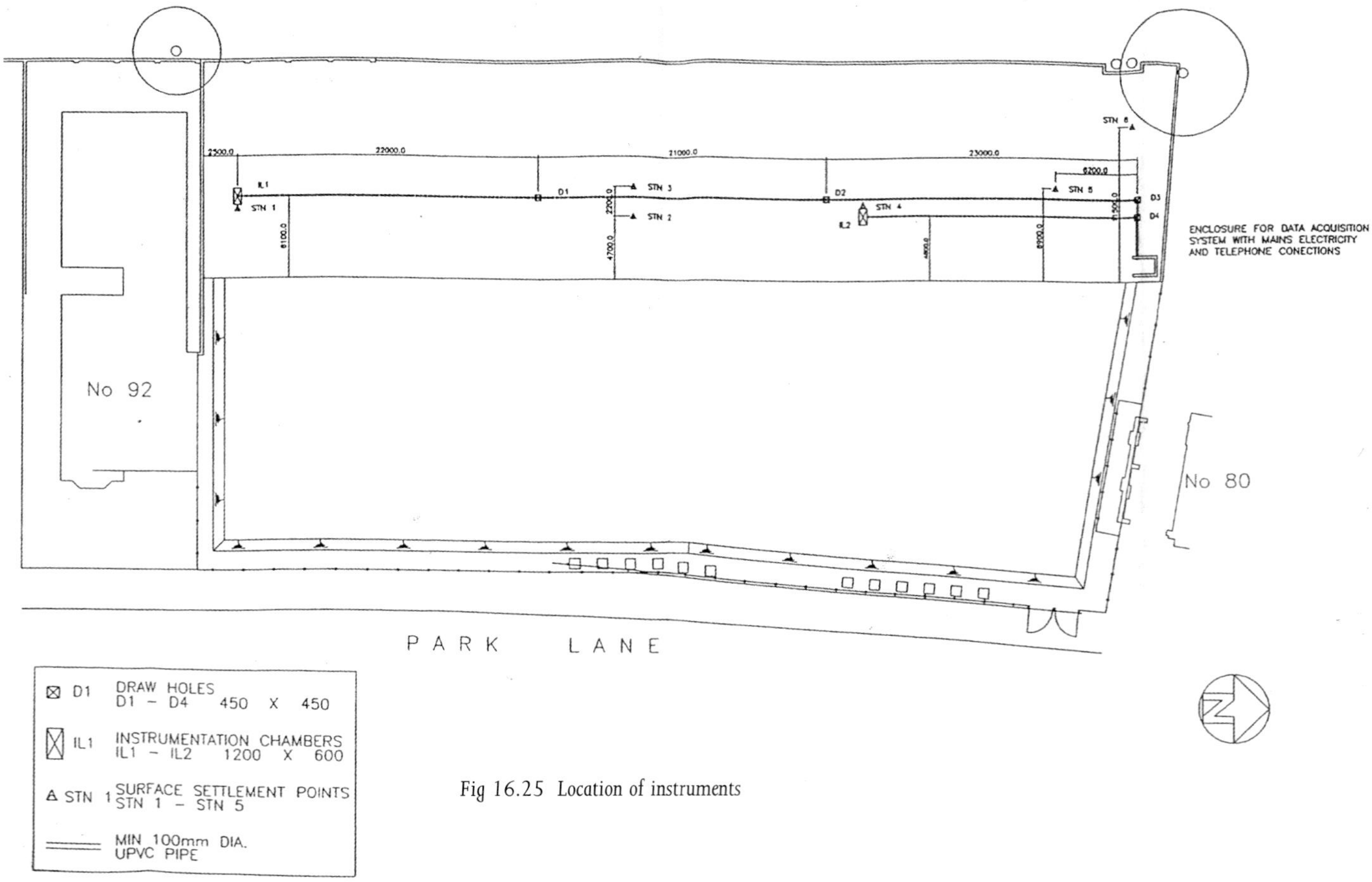

Fig 16.25 Location of instruments

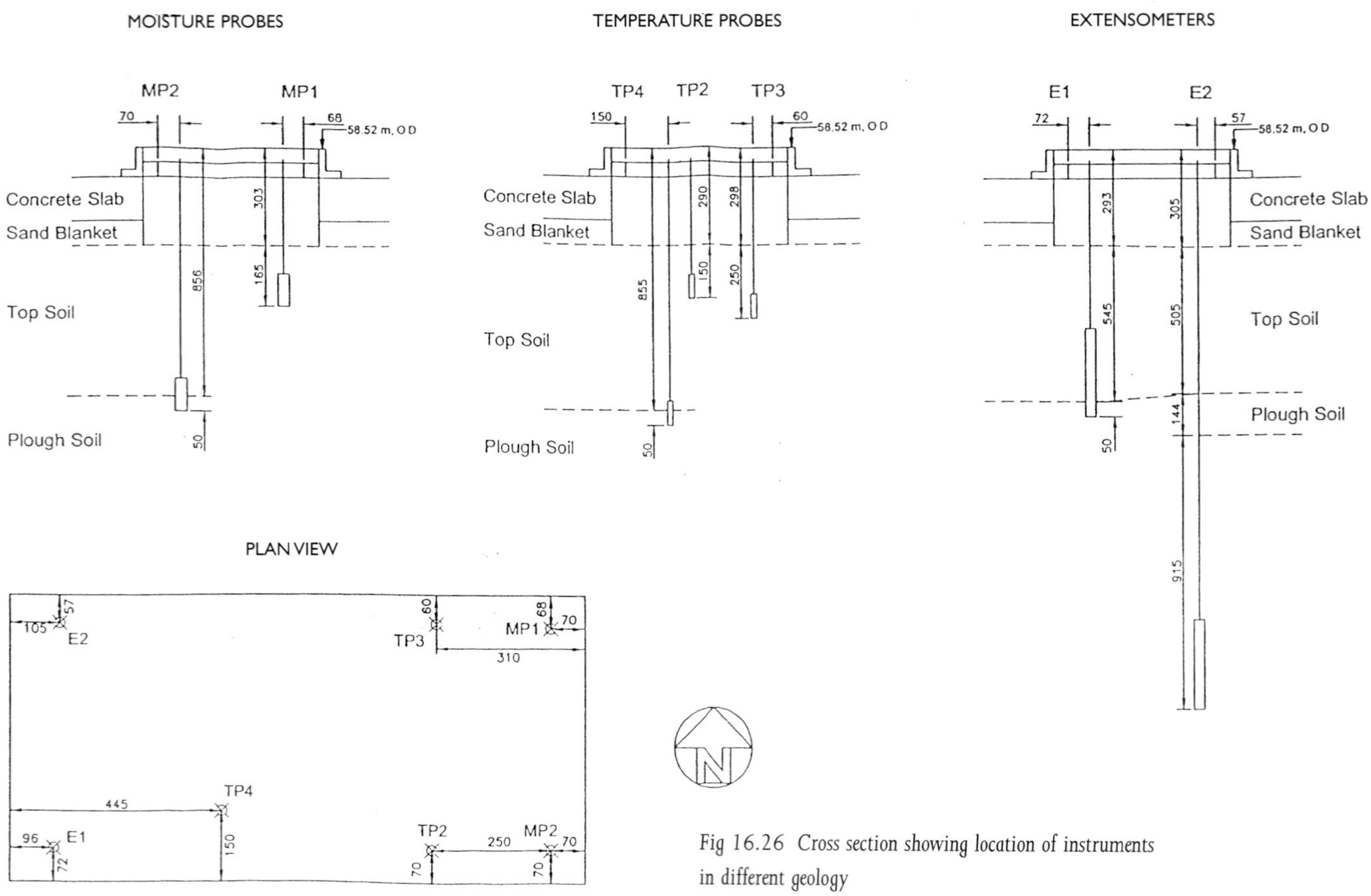

Fig 16.26 Cross section showing location of instruments in different geology

in the ground above the gravel, with one located in the worked soil and the other in the topsoil. Compression of the burial environment is monitored using a reference rod anchored in the gravels as noted earlier and another rod anchored in the worked soil. Relative movement between the two rods is measured by displacement transducers mounted from the car park slab and reading on to the top of each rod. Thus the movement of both the worked soil and the car park slab is monitored.

Monitoring results after the first year of instrumentation

The Park Lane instrumentation has been in place for one year and the following text reports on some early findings and also raises interesting results that are to be investigated.

Temperature measurements

To illustrate the relationship between the temperature measurements with depth, Fig 16.27 shows the temperature measurements at IL1 over the period from June 2000 to September 2000. The figure shows the results from all four temperature probes at the instrument locations, with probe T1 being the highest (in the sand layer), T2 and T3 being successively deeper (in the topsoil) and T4 being the lowest (in the worked soil). There was a trend for temperatures to rise throughout July to mid August, with temperatures higher closer to the surface, and to fall from mid August onwards. At the end of the period, the temperatures with depth began to exceed the surface temperatures.

Fig 16.28 shows a weekly period in June 2000 at IL1. This illustrates that there is an increasing lag in the response time with depth to a temperature variation at the surface. In addition, the temperature change corresponding to a given temperature change at the surface becomes smaller with

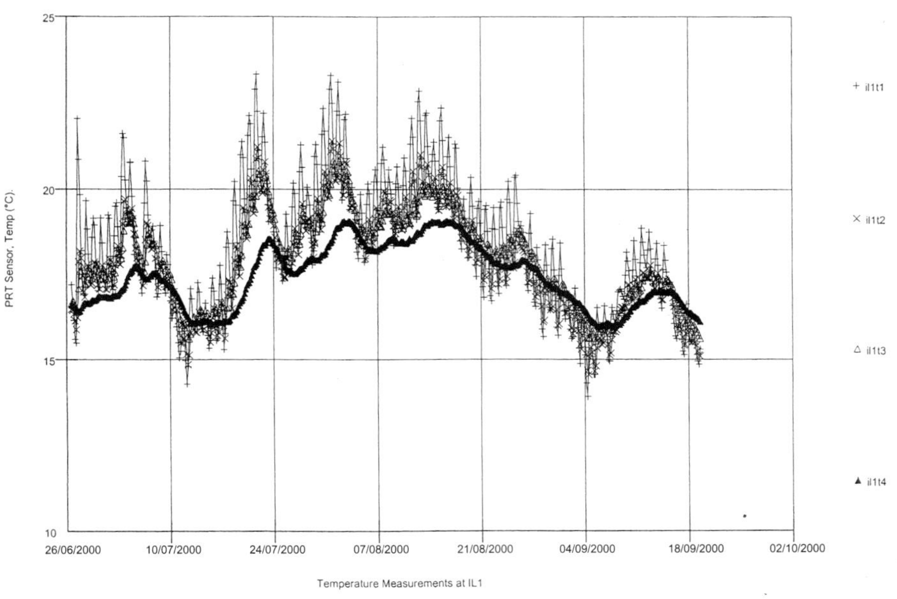

Fig 16.27 Variation in temperature with time between June to September 2000

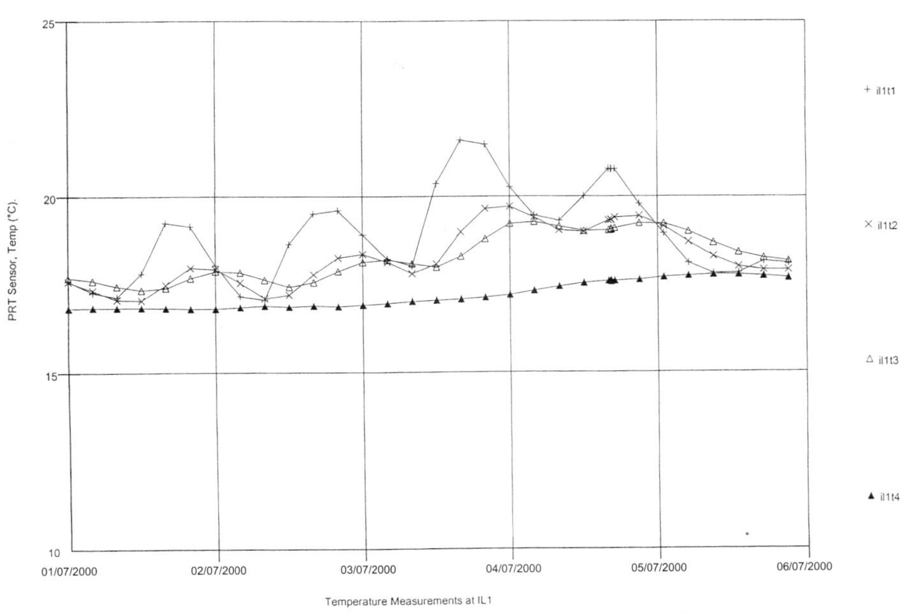

Fig 16.28 Variation in temperature with time over one week

increasing depth. This means that when the surface is heating up, temperature increases and temperatures are lower with depth. However, when the surface is cooling down, there comes a point when the temperatures at depth begin to exceed the temperature at the surface because the temperature changes are smaller with depth. In the figure, the vertical grid lines mark midnight at the start of the day. The temperatures have been recorded at four-hourly intervals. It can be seen that the temperature in the sand layer was reaching a daily maximum at about 4.00pm and a minimum about 8.00am.

Fig 16.29 shows the comparison of the temperatures in the worked soil layer (T4) between the two instrument locations IL1 and IL2 over the year since monitoring commenced. The figure illustrates the agreement between the pattern of variations in temperature between the two locations. It was interesting that initially the temperatures at IL2 were higher than at IL1 by up to about 3 degrees. It was thought that this may be because IL2 is towards the middle of the car park slab, whilst IL1 is towards the edge and is possibly partially shaded from the sun by the adjacent buildings.

In the period between September 2000 and February 2001, there was an overall trend for temperatures to fall. The temperatures at depth generally continued to exceed the surface temperatures. Although the temperatures at instrument location IL2 initially exceeded those at IL1, between September and February temperatures at IL2 fell faster than at IL1, as can be seen in Fig 16.29. At instrument location IL1 the temperature in the worked soil (T4) was then 1°C to 2°C higher than at IL2, whilst the temperature in the lower topsoil was less than 1°C higher than at IL2. The temperatures in the upper topsoil and at the surface in the sand layer became similar at the two instrument locations. The more rapid fall in temperature at IL2 than IL1 was thought to be possibly consistent with the more exposed location of IL2 in relation to IL1.

In the period between February 2001 and June 2001 there was an overall trend for temperatures to rise. As the temperatures rose, the temperature increase was greatest at the surface and becoming progressively less with depth so that by the end of the period temperatures had reverted to being generally greatest at the surface and decreasing with depth. In Fig 16.29 it can be seen that in the latter part of the year since February 2001, temperatures rose faster at IL2 than at IL1 as the ground warmed up, so that by the end of the period, temperatures at IL2 were about 2°C higher than at IL1. The temperatures at the end of the year and the difference of 2°C between the two locations are similar to what was measured at the start of the year. The more rapid rise of temperatures at IL2 than at IL1 during the latter part of the year in which warming occurred, again supports the observation that the behaviour may be consistent with the more exposed location of IL2.

The worked soil is just above the burial environment in the gravels and, therefore, the measured temperature response in the worked soil is indicative of that in the burial environment. The temperatures in the worked soil appear to be behaving as would be expected and so far are not exhibiting any unusual behaviour with respect to the other devices. In addition, the temperatures at the end of the first year were similar to those at the start.

Moisture measurements

Figs 16.30 and 16.31 show the moisture measurements at IL1 and IL2 respectively over the year since monitoring started. The figures show the results from both temperature probes at the instrument location concerned, with probe mp1 being the highest (in the topsoil) and mp2 being the lowest (in the worked soil).

The moisture measurements are given as percentage moisture, ie the volume of water per unit total volume of the moist soil, and this results from the electrical method required for data recovery. This is difficult to relate precisely to soil moisture content, which is defined as the mass of water per unit mass of solids, particularly in the case of the deposits at Park Lane where the total volume of the moist soil is likely to include some air. This is determined from oven

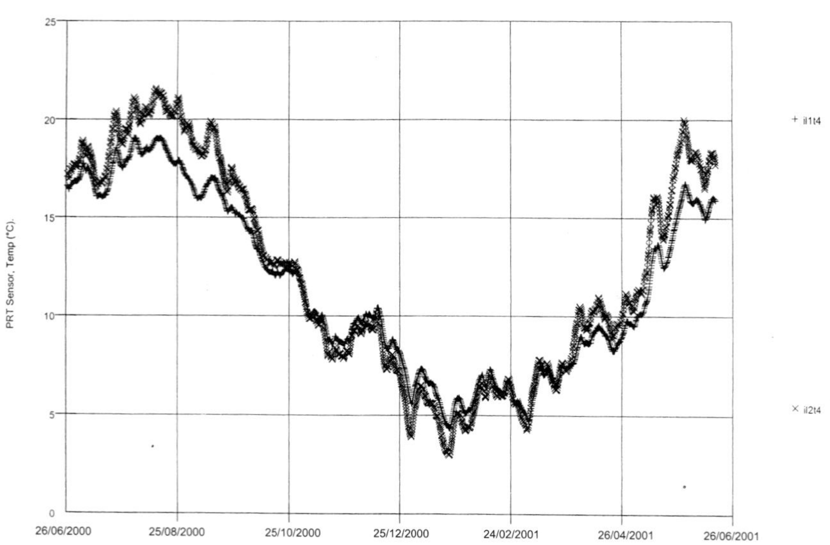

Fig 16.29 *Temperature variation with time in plough soil*

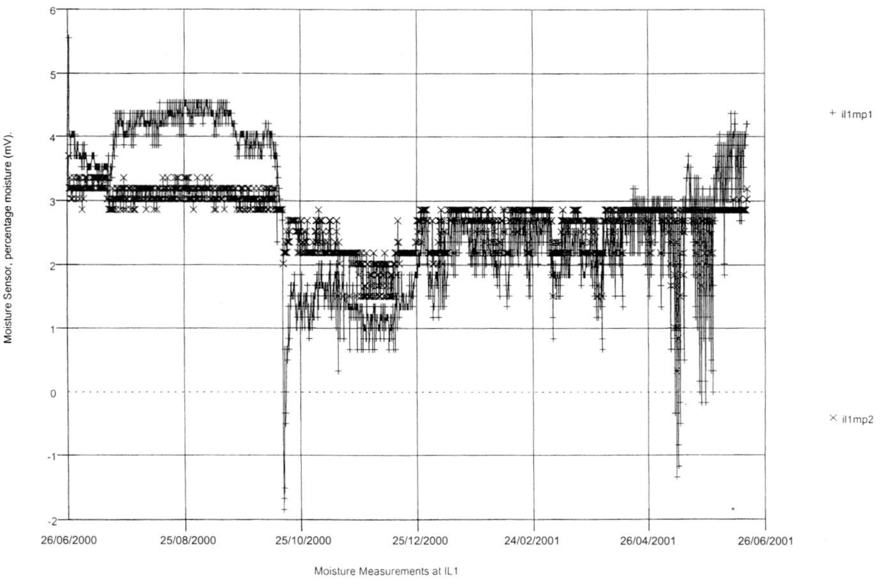

Fig 16.30 Moisture content changes over one year in IL1, above plough soil

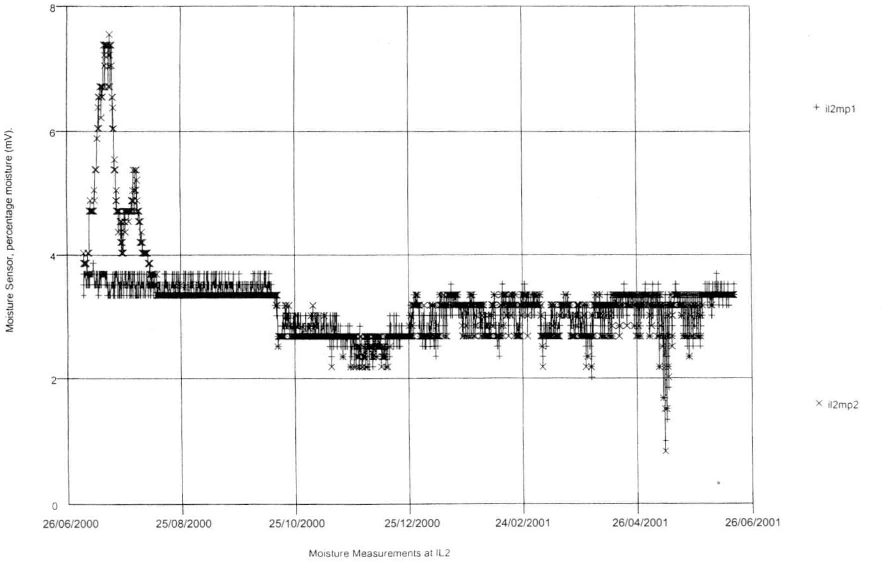

Fig 16.31 Moisture content changes over one year in IL2, above plough soil

measurement. However, it can be seen that the percentage moisture measurements are low. This corresponds to the low moisture content determined for the topsoil and worked soil deposits in the site characterisation study and sampling during installation which lay between 7 and 20%.

Fig 16.30 shows that in the first four months of the monitoring the percentage moisture in the worked soil at IL1 (given by il1mp2) remained more or less constant at about 3.1%. The percentage moisture in the topsoil (given by il1mp1) fluctuated a little over the first month, but then stabilised at about 4.3% to 4.4%. This indicated a moisture content in the topsoil which is higher than in the worked soil by about a factor of 1.4 and corresponded to the findings from the site characterisation study, where average moisture content in the worked soil was about 9% (varying from 7% to 11%) and average moisture content in the topsoil was about 13% (varying from 11% to 15%). However, the measurements from samples taken during installation at IL1 give a moisture content result in the worked soil of 18% and moisture content results in the topsoil of 19% and 20%. The reason for the higher values than in the site characterisation study is not clear. However, though they appear to indicate less of a difference between the worked soil and topsoil, the moisture content ranges in the site characterisation study did overlap. (In addition, at IL2, the measurements from samples taken during installation give a moisture content result in the worked soil of 12% and moisture content results in the topsoil of 17% and 18%, which do show a greater difference.) Overall, it was concluded that the percentage moisture measurements at IL1 fitted the expected behaviour.

Fig 16.31 shows that in the first four months the percentage moisture in the worked soil at instrument location IL2 (given by il2 mp2) fluctuated significantly before settling down to become stable by mid August. The measurements stabilised at about 3.4% which is comparable with the measurements at instrument location IL1. The reason for the initial fluctuation is not known, but it could be due to a local disturbance to the moisture regime in the vicinity of the probe when the device was inserted. The percentage moisture in the topsoil (given by il2mp1) remained constant throughout the initial four months at about 3.5%. This is lower than the measurements in the topsoil at IL1, but still within expected response based on the results of the site characterisation study and the samples taken during installation.

In the fifth month sudden reductions occurred in measured percentage moisture in both the topsoil and worked soil at both instrument locations. The reason for this is thought to be electrical. In the subsequent months the percentage moisture measurements rose so that at the end of the first year of monitoring values were similar to those at the start.

Moisture content results from the once-yearly sampling in June 2001 (see below) are given in Table 16.9. The moisture contents are similar to those measured at the time of installation.

Displacement measurements

The displacement measurements at instrument location IL1 and IL2 respectively over the year since monitoring started are shown in Figs 16.32 and 16.33 respectively. The figures show the movement of the car park slab which is labelled 'Head' because the displacement transducer readings are at the head of the rod fixed in the gravels. Also shown is the movement of the worked soil, labelled 'Anc 2', which is found from the displacement recorded at the head of the rod anchored into the worked soil relative to the displacement recorded at the head of the rod fixed in the gravels. Negative movements indicate settlement.

The car park slab at IL1 settled a little over the first three months to September 2000 (Fig 16.32). Movement was then fairly constant until settlement restarted at the end of January 2001. By the end of the first year, settlement since the start of monitoring was about 1mm. Movement of the worked soil

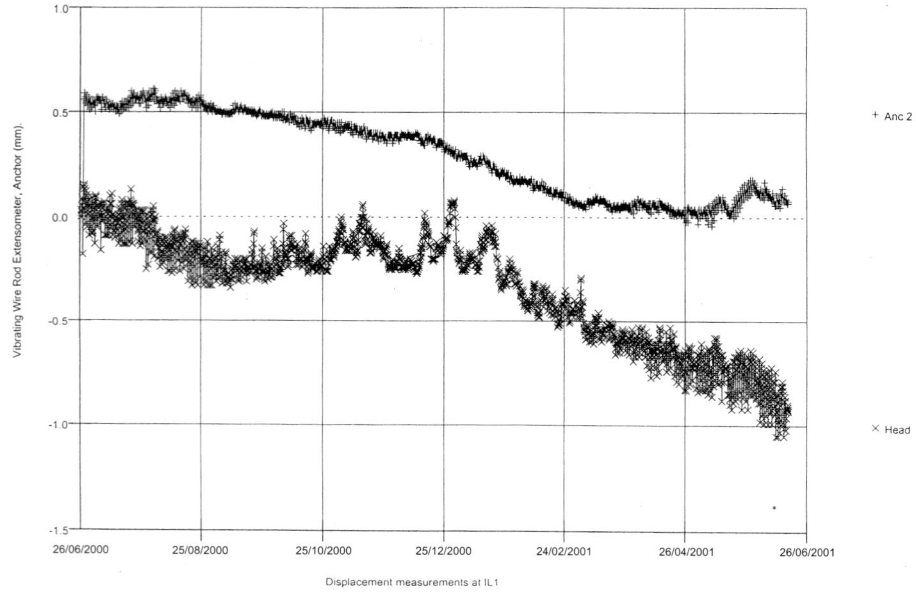

Fig 16.32 Displacement at the ground and in the plough soil over one year at IL1

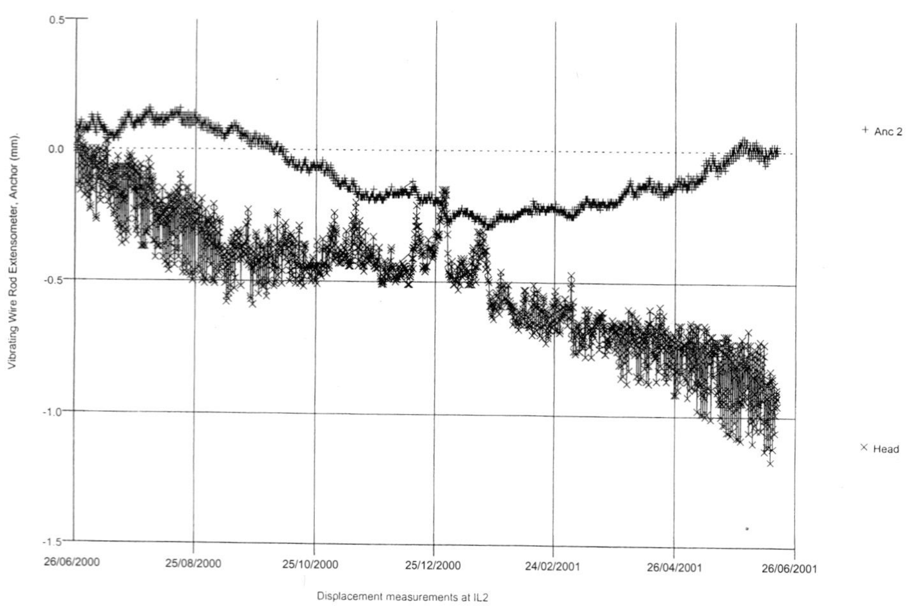

Fig 16.33 Displacement at the ground and in the plough soil over one year at IL2

became fairly stable after the end of February 2001, with an overall movement since the start of monitoring of about 0.5mm.

The car park slab at IL2 (Fig 16.33) showed a similar settlement pattern to IL1 with settlement since the start of monitoring at about 1mm after the first year. At IL2 also, settlement restarted at the end of January 2001 after being fairly constant since September 2000. After a small settlement of about 0.2mm, movement of the worked soil decreased in the last few months, bringing it back to its initial position.

It is possible that the small settlement which occurred in the worked soil during the winter months may have been due to contraction of the ground due to cooling. The lack of further settlement at IL1 and reduction in settlement at IL2 during the latter months when temperatures were rising possibly supports this. It is not clear why the car park slab settlement should have been reinitiated, though the settlement is still small.

Laboratory test results for samples taken after Year 1

In June 2001, the first set of once-yearly samples was taken for laboratory chemical testing. The test results for the sampling at the end of Year 1 are shown in Table 16.9, compared with those from the site characterisation and from sampling at the time of installation. It can be seen that the values obtained for the different parameters are similar to the previous values, except for the phosphate test results. The phosphate levels in both the topsoil and the worked soil are at least twice the values previously measured, the increase being greatest in the latter. The test method was the same and it is not thought to be a difficult test, therefore it is possible that this represents a real change. Investigation will be carried out into whether phosphate levels alone could change whilst all the other parameters remain unaffected.

Summary

The selection of instrumentation on this project was dependent on reliability of the instrument under long-term monitoring, accessibility of the instruments for replacement, if needed, and the need for remote monitoring from anywhere from within the UK with use of a modem and lap tap computer (ie without visiting the site). This objective was achieved. This requires close working between archaeological, geotechnical and instrumentation contractors to develop a strategy and implementation of the scheme.

The readings for the first year show a consistency illustrating that the instrumentation is working and is providing useful data. The data is now being collected by a datalogger and linked to the office by a telecom connection. Responding to the findings it will be possible to modify as necessary the frequency of instrument readings.

The worked soil is immediately above the burial environment in the gravels and therefore the measured temperature response in the worked soil is indicative of that in the burial environment. The temperatures in the worked soil appear to be behaving as expected and so far are not exhibiting any unusual behaviour with respect to the other devices. The temperatures at the end of the first year of monitoring and the difference of 2°C between the two locations are similar to what was measured at the start.

Despite sudden reductions which occurred in measured percentage moisture in both the worked soil and topsoil in the fifth month, at the end of the first year the moisture

Table 16.9 Test results for the sampling at the end of Year 1, compared with those from the site characterisation study and from sampling at the time of installation

	Topsoil; site characterisation	Worked soil; site characterisation	Topsoil; installation	Worked soil; installation	Topsoil (year 1)	Worked soil (year 1)
Moisture content %	11–15	7–11	17–20	12–18	15–18	14–15
Ph	7.5–7.6	7.9–8.0	7.3–7.6		8.0–8.3	8.4
Total SO4 %	<0.01–0.03	0.02–0.03	<0.01		<0.02	<0.02
Water soluble SO4 g/l	<0.01–0.02	0.01			0.0013–0.0066	0.0011
Organic matter %	1.6–3.2	0.5–1.8	1.4–2.3		2.7–3.4	4.7
Carbonate %	<0.5–2.9	1.7–2.1	6.5–7.2		2.4–2.8	2.8
Chloride %	<0.01	<0.01			<0.01	<0.01
Water soluble chloride mg/kg	<10	<10	<10–11		5.8–11	4.7
Phosphate mg/kg	<10–15	<5–11	<10–13		29–31	41

measurements were similar to those at the commencement of the monitoring. Additionally, the moisture contents from the annual sampling are similar to those measured at the time of installation.

Settlement of the car park slab since the start of monitoring was about 1mm. Examination of behaviour over the last year shows that at the end of January 2001 settlement recommenced after being fairly constant since September 2000. It is not clear why the settlement should have been reinitiated, though the movement is small. In the worked soil at IL1 there has been an overall small settlement of about 0.5mm since monitoring began. At IL2, movement of the worked soil decreased by about 0.2mm in the latter months, bringing it back to its initial position.

In June 2001, the first set of annual samples was taken for laboratory chemical testing. The values obtained for the different parameters are similar to those from the site characterisation and from sampling at the time of installation, except for the phosphate test results. The phosphate levels in both the topsoil and the worked soil are at least twice the values previously measured, the increase being greatest in the latter. It is possible that this is a real change. Investigation will be carried out into whether phosphate levels alone could change whilst all the other parameters remain unaffected.

Acknowledgements

Wessex Archaeology would like to thank Frogmore Developments who commissioned the archaeological works, and in particular Barry Kitcherside, Robert Nicholls and Simon Casey. Arup Geotechnics, Soil Instruments and Price and Myers all played an important role in the design and implementation of the preservation and monitoring system.

17

Waterlogged archaeological remains, environmental conditions and preservation *in situ*: a case study from the Tower of London

GRAHAM KEEVILL, DAVID HOGAN, MAT DAVIS and DAVID HOWELL

Graham Keevill
Archaeological Consultant
85 Kynaston Road, Didcot
Oxfordshire OX11 8HA, UK

David Hogan
Wetland Soil Scientist
Royal Holloway Institute of Environmental Research (RHIER)
Huntersdale, Callow Hill
Virginia Water, Surrey GU25 4LN, UK

Mat Davis
Land Quality Coordinator
Environment Agency
Kingfisher House, Goldhay Way
Orton Glodhay, Peterborough PE2 5ZR, UK

David Howell
Conservation Scientist
Historic Royal Palaces
Apartment 21, Hampton Court Palace
Surrey KT8 9AU, UK

Abstract

Excavations in the moat at the Tower of London during 1995–1997 revealed a series of timber structures in excellent condition. Some of the remains were of incomparable significance for understanding the castle's history. The presumption was that the structures should be preserved *in situ*, not least because they were often only partially revealed in the excavations. It was important to ascertain whether the reburial conditions in backfilled trenches would provide the requisite conditions for continued preservation. Therefore Historic Royal Palaces, in partnership with the Royal Holloway Institute of Environmental Research, began to monitor both reburied and unexcavated environments within the moat. This paper describes the reasons for the project, the range of the monitoring regime, and the results of data gathered since 1997. Finally it suggests that the results are of wider significance for those working towards the *in situ* preservation of organic-rich archaeological deposits.

Introduction

The Tower of London is one of the world's most important historic sites. It is protected in English law as a Scheduled Ancient Monument, and its universal significance has been recognised in its designation as a World Heritage Site. It is a complex place, but it cannot be a static entity to be preserved, unchanged and unchangeable into the future. It is home to a living community with its own needs, and it is one of the most visited historic sites in the world. This inevitably creates daily and longer-term pressures on the Tower's fabric. Historic Royal Palaces have responsibility for managing, interpreting and presenting the site. In 1996 they initiated a project (the Tower Environs Scheme) to enhance and improve the setting of the castle. The Scheme included a proposal to re-flood the moat, a defensive feature surrounding the outer curtain wall. The moat had been filled with water from the time it was dug out under Edward I in 1275–1281 until it was drained and filled in on the orders of the Duke of Wellington in 1843. The concept of reintroducing water into the moat offered distinct presentational and interpretative benefits for the visitor, but it also presented a potential threat to archaeological remains buried within the fills of the feature. This threat needed to be quantified as part of the feasibility studies into the concept, and an archaeological evaluation was therefore carried out by the Oxford Archaeological Unit on behalf of Historic Royal Palaces during 1995–1997.

This was probably the most extensive and intensive programme of archaeological work ever carried out at the Tower. The results were excellent, both providing much new information on the development of the defences and showing that the re-flooding was feasible so long as the necessary excavations were restricted to the upper 1.5–2.0m of the moat fills. These levels were of 18th- to 20th-century date. The lower levels, however, were waterlogged and the anaerobic conditions meant that organic materials were exceptionally well preserved. Finds ranged from small artefacts such as leather shoes and a wickerwork fish trap to large timber structures such as piled foundations, part of a historically documented mill, and the base of a timber bridge (Keevill 1997; Impey 1997). The small artefacts were removed during the excavations and subsequently conserved for long-term storage and display. The structures, however, had to be retained *in situ*, not least because several of them were fully integral with the foundations for the outer curtain wall and related features.

The most important discoveries belonged to the middle and later decades of the 13th century. A previously unknown stone building with an attached timber structure (interpreted as the base of a bridge) was found in the centre of the west moat. The timbers were mostly of oak, but unfortunately they were too fast-grown for dendrochronological dating. Fortunately beech piles and poles were also found in immediate association with the masonry, and these did date very successfully. Indeed most of the beeches had only been subjected to minimal conversion and thus still retained the bark. This meant that the dendrochronology dates were accurate to the felling season, and all the timbers were dated to the winter of 1240–1241. This coincides with a well-known reference in the chronicle of Matthew Paris to the catastrophic collapse of walls and a great gatehouse at the Tower in 1240 (Impey 1997). The excavated masonry probably formed a forebuilding on the approach to the great gate. Later 13th-century foundation rafts consisting of beech piles were found at both the south-east and south-west corners of the moat. Again they retained bark, and again excellent dendrochronology determinations were achieved, ranging across the years 1275–1277. These dates can be matched with medieval documents specifying the supply of beech timbers for foundation works under dams and mills in exactly these areas of the Tower and its moat.

How could we be sure that the reburied remains would actually be preserved? This was an obvious question to ask given that the exposure of the organic materials by the excavations had inevitably altered the original environment which had surrounded and preserved them. The soil surrounding the timbers had been carefully stockpiled separately during the excavations, so that it could be put back around the woodwork during backfilling. It was hoped that this would ensure long-term preservation of the organic materials, and all available advice supported this. It did not seem to be quite enough to leave it at that, however, and so Historic Royal Palaces initiated a programme of environmental monitoring. This was carried out by the authors following extensive discussion with English Heritage and others. The intention was to track a series of appropriate indicators over a period of time in order to check the reburied environment against 'control' measurements in unexcavated parts of the moat. The monitoring equipment was installed in the west moat, where the most important archaeological discoveries had been made.

The initial focus was very much on the historically important timber structures and determining whether the reburied environment was working for their preservation. Further considerations came to the fore as the project developed. For instance, it was clear that some of the timber features in the moat had suffered considerable deterioration in the past. Some pile and beam foundations under the southern arm of the outer curtain wall had rotted away, while others had survived in reasonably good condition. Elsewhere, there appeared to be a quite distinct level above which timber would not be preserved – in effect a decay horizon. The monitoring was not able to assess these problems directly because the sample points were not in the relevant location, but they could address potentially related issues such as water level and tidal influence from the River Thames. Another matter which has grown considerably in relevance was the potential for large-scale neighbouring redevelopment projects to influence the environment in the Tower moat. Such a threat was no more than a potential one when the monitoring started, but the subsequent start of a major redevelopment on the adjacent site at Tower Place made it very relevant. Finally some heed had to be paid to the prospect of a rising water table in London and how this might affect the moat. All these issues have been addressed to a greater or lesser extent during the monitoring programme.

Methods

The monitoring involved investigating the burial environment of the Tower moat where the established conditions are conducive to preservation of organic materials, and determining whether the conditions will continue. While the primary aim of the study has been to determine present environmental conditions in order to prescribe management recommendations for the site, the results contribute to the wider interests of archaeologists and others involved in long-term site preservation strategies. These include the development of a protocol for monitoring sites containing buried preserved organic remains, which would have both national and international significance for the study and preservation of archaeological remains. The instruments were installed jointly by all the authors. Field measurements were largely collected by David Howell, and the resulting data have been interpreted by David Hogan and Mat Davis.

Experience gained in monitoring environmental parameters in a range of materials and ecosystem types was used to select an appropriate monitoring regime with minimal disturbance to the site. The target priorities for investigation were to determine hydrological properties, in particular soil/sediment water regime (depth, duration and pattern of waterlogging) and associated soil moisture profiles, together with the related oxidation–reduction balance. The main characteristics of the ground and surface water chemistry were also determined to give an indication of the presence and pattern of any pollution or nutrient inputs likely to affect rates of organic matter decomposition.

Water level measurements

Eleven boreholes were sunk using a combination of hand auger, power auger and 'split spoon' power auger. Field descriptions together with sampling and analysis of the borehole material provided important information on the soil and substrate to a depth of 4.0m below ground surface. Dipwells were subsequently installed in six of the boreholes. Each dipwell comprised a narrow diameter plastic tube, slotted at the base and fitted with an end cap. A geotextile membrane with a clean washed and lime free gravel pack were used to provide a filter around the slotted section of tube. The borehole was backfilled above this with recovered spoil to near ground level, where a bentonite clay seal was inserted to prevent infiltration of surface water.

Soil moisture measurements

An estimate of the volumetric moisture content of the deposits within the moat was made using a neutron probe. Access tubes for use with the neutron probe were carefully inserted by hand into the remaining five boreholes to ensure the necessary close contact between the access tube and surrounding deposits. Each tube was constructed from aluminium with a sealed nose cone at the base and removable rubber bung at ground level.

The probe data recorded three separate readings (over a 16 second count rate) at 100mm intervals down each access tube. The mean of these readings was converted into an estimated volumetric moisture content using a standard calibration line. Such a calibration line can be determined by theoretical, drum, or field calibration (Anon 1970). Where possible a site specific calibration line is produced for the actual deposits present at a study site. Unfortunately the lack of data on the character of urban archaeological deposits and the requirement of the project not to excessively disturb the deposits to be monitored prevented the determination of a site-specific calibration line. A standard calibration line supplied by the Institute of Hydrology for a loam soil type had to be selected to give an approximation of the moisture content of the moat deposits. Readings were collected on three occasions during 1998 and 1999.

Reduction and oxidation conditions: redox potential

Redox potential gives an indication of the intensity of oxidation or reduction occurring in the soil system (Gambrell and Patrick 1978) and can be measured at specific points using permanently installed platinum electrodes and a portable voltmeter. Soil micro-organisms oxidise organic materials during the process of respiration, using oxygen as an electron acceptor under aerobic conditions. When oxygen becomes depleted under waterlogged conditions, ions such as NO_3^-, Mn^{4+} and Fe^{3+} become utilised in turn for electron transfer, the electron availability being measured by redox potential. Indicative categories of oxidation and reduction in terms of redox potential are indicated in Table 17.1.

The measurement of redox potential was carried out using a series of platinum electrodes, made to uniform design at RHIER. Each probe comprised a tip of 0.5mm gauge platinum, fused to a copper terminal wire with the connection waterproofed by means of a sheath of shrink-fit sleeving. A short length of copper wire core at the opposite end of the probe was exposed by paring back the sheath. An additional band of shrink-fit sleeving was also applied here to prevent water penetrating the inside of the wiring sheath. Probes were rinsed with dilute hydrochloric acid and distilled water and calibrated in the laboratory with quinhydrone in pH buffer 4.0 (+ 218mV) before being installed in the field.

All probes were installed in triplicate to cater for variability in substrates which influence redox potential. A mean value of the three readings was used for the interpretation of results. Probe tips were inserted horizontally into the excavated face in two trenches sufficiently to encounter undisturbed material (at least 200mm), and in waterlogged material in the pit floor. The trenches were subsequently backfilled, leaving a short length of wire exposed for connection to a portable meter measuring redox potential. Elsewhere probes were inserted by making a small hole to the required depth using a screw auger. The probes were inserted down the access holes and pushed in at the bottom until firm contact was made between the platinum tip and material at the bottom. The probes were then left in place for the duration of the project. Most were in place for the winter of 1997, with some later supplementation in February 1999.

Following initial collection of readings to check probe performance and stability due to settling in, a datalogger was used to obtain a series of continuous records over monthly periods. Interpretation of the redox data requires a correction for the calomel half-cell by the addition of 244mV (a subtraction of the Eh value of the reference cell, in this case -244mV) to the field measurement. A further correction to pH 7 is conventionally made to allow for comparison of results with other studies, though in this case no correction was necessary since soil reaction is close to neutral.

Table 17.1 Categories of redox potential

Redox potential (mV)	Category
>+400	oxidised
+400 to +100	slightly reduced
+100 to −100	moderately reduced
-100 to −400	highly reduced

Determination of water quality

Water quality was an additional consideration, as this was likely to influence processes of decomposition taking place within the materials containing the remains of interest. Water quality studies can also give indications of the likely flow paths of hydrological support for waterlogged materials, thereby indicating likely sources of actual pollution or potential risk. Sampling to provide information on groundwater quality was carried out in six piezometers, and also from the brick culvert which runs around the moat and discharges surface water into the Thames. Laboratory analyses can indicate the location and magnitude of pollution likely to reduce the effectiveness of preservation within the burial environment. This might include saline intrusion by tidal influence from the river.

Water samples were collected in 1998 and 1999. Piezometer tubes were purged of water using a hand pump and allowed to refill in order to obtain a fresh representative sample, which was pumped into a sample bottle. A separate sample of water from each location was taken in an acid-washed bottle for phosphorus determination. Other samples were also taken from the locations of three manhole inspection covers along the line of the culvert and from the river. Samples were dispatched immediately to the laboratories at RHIER, where analyses were either carried out immediately or following short-term freezer storage. The following laboratory determinations were carried out to facilitate characterisation of hydrological inputs to the system and the identification of possible contaminants: pH, electrical conductivity (EC), nitrate (NO_3^-), ammonium (NH_4^+), phosphate (PO_4^-), sodium (Na^+), potassium (K^+), calcium (Ca^{2+}), magnesium (Mg^+), chloride (Cl^-) and sulphate (SO_4^{2-}). In addition, some direct measurement of water quality parameters using portable meters and probes was made during a site visit in February 1999 to collect hydrological data.

Results

Water levels

The depth in metres of the water table below the ground surface in the six dipwells (D1–6) and along the culvert (C1–3) are shown in Table 17.2. In general, the level rose and fell consistently across the readings, declining from March 1998 to the lowest level in April and returning to autumn/winter highs. The water table fluctuation among wells which are closely spaced, however, suggests that lateral variation in the character of deposits has an important local influence on hydrological properties such as horizontal hydraulic conductivity. Comparison of recorded water levels with individual profile properties indicates that the water table was usually found in an olive or greyish brown silty clay beneath the made ground, occasionally rising into the latter. There was no indication of the water table falling within either the perched aquifer or any of the highly reduced organic-rich layers. Neither was there any evidence for significant tidal or rainwater influence on the water table.

Table 17.2 Water levels in dipwells and along the culvert

Date	D1	D2	D3	D4	D5	D6	C1	C2	C3
10/03/98	1.20	1.19	1.90	1.22	1.31	1.13	1.77	1.94	-
30/03/98	1.32	1.32	1.34	1.26	1.38	1.30	2.76	2.82	-
16/04/98	0.91	1.01	1.06	1.03	1.03	1.05	1.15	1.34	-
17/09/98	1.25	1.62	1.90	1.64	1.98	1.36	-	1.97	-
11/02/99	1.29	1.33	1.78	1.28	.136	1.20	1.88	2.05	2.18

Water levels recorded within the culvert were extremely variable, indicating (probably rapid?) response to discharge (rainfall), and little relationship with water table depths. Much of the water carried in the culvert derives from run off within the area of the Tower, carried directly to the culvert via a system of drains. Though little is known about the possible contribution to the water table in the moat by water discharging (and sometimes backing up) in the culvert, the discontinuous nature of the moat deposits suggests that direct and rapid effects are minimal.

Soil moisture

Neutron probe measurements for March and September 1998 and February 1999 showed a very similar pattern and level of moisture content among the three monitoring runs throughout most of the profile depth. The only exception occurred in the upper 400mm, where rainwater effects on water balance can be expected. The February and March figures, in the 15–40% range, represent wet conditions of the winter period while in contrast September, towards the end of the dry season period, gave figures of 10–20%.

A large drop in moisture content occurred immediately below the topsoil. This may be an effect of neutron loss within 200mm of the surface giving an underestimation of soil moisture. This is not evident from the backfilled area, making it more likely that the effect is due to the presence of rubbley made ground below the topsoil. This material has a lower moisture retention capacity than the overlying topsoil and turf. Wet weather immediately prior to the installation of the access tubes had also provided an input of moisture to the surface layer. Moisture content falls to a relatively uniform value from the middle into the lower levels of the moat. Data from access tubes 1 and 2 indicate an elevated moisture zone between 2.3m and 3.0m depth. It is unclear precisely what this represents but it probably reflects a particular characteristic of the moat deposits at this location, such as the presence of a perched water table over the underlying impermeable London Clay, or the input of water from the revetment wall. Data from the backfilled trench presented a more variable soil moisture profile, not consistent with the patterns found in the unexcavated ground.

Redox potential

Redox potential in the backfilled excavation was measured in a six-week run commencing on 21 August 1997, a little over three months after backfilling had been completed. The probes had been located close to the bottom of the excavations immediately below the important mid 13th-century oak timbers near the centre of the trench, immediately above the same timbers, and within the upper made ground. The plots indicate that the most stable conditions occurred below the remains, where a potential of about +50mV (moderately reduced) was maintained. The most reduced conditions of between -100 and -150mV (highly reduced) were recorded immediately above the remains, indicating the presence of constant anoxic conditions of the burial environment at that location. By contrast, the topsoil was much less reduced, generally in the range +300 to +400 (slightly reduced) and less stable than deeper layers. A diurnal pattern is apparent, possibly relating to daily fluctuations in temperature. These results are encouraging because they suggest that burial conditions had returned to a state which should ensure the continued preservation of the organic remains remarkably quickly after the backfilling had taken place.

Water chemistry and quality

The results of water analyses from 1998 are shown in Tables 17.3–17.4. These preliminary results can only be indicative and interpreted in general terms, though they do provide indications of exceptional concentrations and ranges in important chemical characteristics, highlighting where more detailed investigations might be warranted. Results also allow some general comparison among individual sites and between the perched aquifer of the moat fill and the surface run off via the culvert drainage system. Seasonal effects can also be seen in comparing the same location at different dates. Further sampling was carried out in 1999–2000 but has not yet been fully analysed.

Most water pH measurements were in excess of 7, indicating varying degrees of alkalinity. Measurements of conductivity (an indication of the concentration of dissolved salts) from the culvert varied more than those from piezometers, probably due to the rapid response of the former to local rainfall events and a dilution effect on discharges concentrated in the culvert. Rainwater inputs to the perched water table in the moat deposits will be limited to direct addition through surface infiltration, some of which will be intercepted by evapotranspiration by the grass. The highest conductivity measured in the piezometers was within the backfilled area (P6), perhaps because additional concentrations of solutes are likely to have become dissolved within the groundwater. Well P1 in an unexcavated area displayed relatively large amounts of sodium and chloride ions. These could reflect contamination from brackish tidal waters, though no evidence of hydrological pathways is available to substantiate this idea. Nitrate-N concentrations were generally found to be low, the highest being in the

Table 17.3 Results of water chemistry analysis (water sampled May 1998)

Location	pH (uS)	EC ppm	NO_3^- ppm	NH_4^+ ppm	$PO4^{3-}$ ppm	Cl^- ppm	SO_4^{2-} ppm	Na^+ ppm	K^- ppm	Ca^{2+} ppm	Mg^{2+} ppm
Culvert	7.42	2600	2.03	0.15	285.4	207.0	116.4	58.0	10.2	253.3	26.92
P 1	7.54	2950	0.09	1.96	1821.4	185.8	132.4	64.5	26.8	347.3	39.96
P 2	7.39	1740	0.01	1.36	98.3	77.3	37.7	31.2	13.3	244.0	17.94
P 3	7.50	1160	0.01	1.58	42.4	21.9	11.3	8.7	4.8	177.6	15.98
P 4	7.51	1740	0.05	1.09	28.5	31.3	31.5	12.1	6.9	243.8	13.77
P 5	7.58	1490	0.07	0.77	68.7	67.8	41.2	29.9	11.7	214.7	14.23
P 6	6.98	3920	0.05	0.49	24.5	109.3	434.4	48.4	16.5	1289.0	57.12

Table 17.4 Water quality data (water quality data recorded 17th September 1998)

Monitoring point	Temperature (°C)	Electrical conductivity (µs/cm)	PH	Dissolved oxygen (mg/l)	Dissolved oxygen (%sat)	Redox potential (mV)
P1	15.4	1800	7.0	3.1	31	+43
P2	15.2	1500	7.0	3.3	33	-129
P3	15.1	1000	7.0	3.7	37	-93
P4	15.2	1400	6.9	2.6	26	-106
P5	insufficient sample to analysis					
P6	17.0	2500	6.8	3.8	39	-106
Air temperature	22.0	Warm dry weather				

culvert, but with very low or undetectable concentrations in the piezometers. In most piezometer water analyses, most of the N occurred in the form of ammonium rather than nitrate. This is typical of the anoxic conditions found in wetlands. The more oxygenated conditions of flowing water in the culvert showed the reverse, with nitrate-N dominant over ammonia.

The value of the project

The monitoring has been of obvious value for Historic Royal Palaces in their management of the Tower's archaeological resource. The work is still ongoing, but the results to date appear to show consistently that the reburied environment around the timbers found in the west moat has returned to conditions which should ensure the continued preservation of those remains. The water level has remained well above the timbers, so that they continue to be waterlogged. The redox measurements show that the environment has returned to the anoxic conditions which are critical for long-term preservation. Water quality appears to be more variable in the backfilled trench than in unexcavated ground, but this is probably not surprising and does not appear to be an immediate cause for concern. Historic Royal Palaces, therefore, can be satisfied with the results so far, though there is no cause for complacency.

The project also has wider relevance to developing strategies and protocols for site management, especially in wetland environments. Moats are a classic type-site for waterlogging in many ways, even in a drained and backfilled example as at the Tower. Arguably they have been under-represented in past consideration of wetlands, which have tended to focus (understandably enough) on large-scale landscapes such as the Somerset Levels or the fenlands of central eastern England. The Tower moat is not now a visibly wet feature in the way that it was planned, intended and created. Nevertheless it still provides a waterlogged environment in its lower levels where the most important archaeological remains have been found and can be anticipated, and apparently in its middle levels as well. The correlation of water level and chemical information with height values (Ordnance Datum levels) from the moat excavations is important in determining preservation conditions in the past, now and (at least potentially) for the future. Such information should be of wider relevance for studies of wetland management and especially, perhaps, for the analysis of moated sites and the behaviour of preservation conditions within them.

At the same time the data can allow us to consider the physical horizons where decay becomes apparent, and the related conditions which cause this. Interestingly this horizon does not necessarily reflect current water levels. A complex and substantial late 18th-century timber structure was found at the south-western corner of the moat, immediately to the south of the area where monitoring has been in progress. The timber structure lay approximately 1.0m below the current water table suggested by the monitoring data, but the uppermost timbers had clearly reached a quite advanced state of decay in the past. Presumably this relates to the draining and backfilling of the moat in the 1840s, which must have caused severe dewatering for some period of time. When excavated the timbers appeared to be stable in their current condition, and the relevant trench was backfilled very soon after the timbers had been exposed. Presumably, though, they could now be exponentially vulnerable to any major fluctuations in the buried environment. Such a realisation is important for the Tower authorities, but the question of decay horizons and past mechanisms for decay again appears to be one of wider relevance.

Conclusions

The monitoring equipment is still in place, and there is the potential for extension of the monitoring regime in the future. The existing study has focused on the west moat because of the great significance of the archaeological remains discovered here. Important deposits, features and structures survive in many other parts of the moat, however, and it would undoubtedly be valuable to increase the physical extent of the study. The south moat, in particular, has been identified as an area where the condition of buried timberwork is a matter for concern. Excavations here in 1996 showed that some piles and beams in the foundations, under the Byward Postern and the outer curtain wall adjacent to it, had partially or wholly rotted away at some time in the past. This would appear to be one area where environmental monitoring would be valuable. Meanwhile the data collected so far provide a valuable contribution to the ongoing debate on the protection and preservation of our wetland habitats in their many and varied forms.

Bibliography

Anon, 1970 *Neutron moisture gauges: a guidebook on theory and practice*, Technical rep ser 112, International Atomic Energy Agency, Vienna

Gambrell, R P, and Patrick Jr, W H, 1978 Chemical and microbiological properties of anaerobic soils and sediments, in *Plant life in anaerobic environments* (eds D D Hook and M M Crawford), 1 edn, Vol 1, 375–423, Michigan

Impey, E I, 1997 The western entrance to the Tower of London, 1240–1241, *Trans London Middlesex Archaeol Soc* 48, 59–75

Keevill, G D, 1997 The Tower of London, *Current Archaeol* 154, 384–7

18

Archaeology and urban regeneration: lessons from Blackfriars multiplex, Gloucester

JOHN PUGH-SMITH, RICHARD SERMON and ALBERT WILLIAMSON-TAYLOR

Abstract

The Government's rallying call to the public and private sectors to unite and achieve the regeneration of our inner urban areas is bound to lead to tensions. One of these is how new development should fit with its surroundings particularly in this country's historic towns and cities. However sensitively designed, the commercial demands to achieve a viable and lettable scheme now require modern retail and leisure development to achieve maximum floorspace and deep and complex foundations to carry the necessary loadings. Since archaeology is a 'finite and non-renewable resource' (DoE 1990, paragraph 6) where physical preservation should, where possible, be always achieved (DoE 1990, paragraph 8) how can these tensions be resolved? How much destruction is acceptable? This paper suggests that solutions can be found if matters are approached on a team basis, and, if a bespoke solution is evolved working as much upwards from the archaeology as from the building downwards.

The Gloucester case study

In the early 1990s Gloucester City Council, in association with Arrowcroft plc, a development company, began to progress a scheme for the redevelopment of central Gloucester to provide a new retail development with associated facilities to be known as 'the Blackfriars scheme'. The area for this regeneration initiative comprises the site of a former electricity generating station and a series of cleared areas used for car parking. It is also of significant archaeological and historical importance, at both national and international levels, lying in the south-west quadrant of the old Roman city (Fig 18.1). It was inhabited during the Anglo-Saxon and

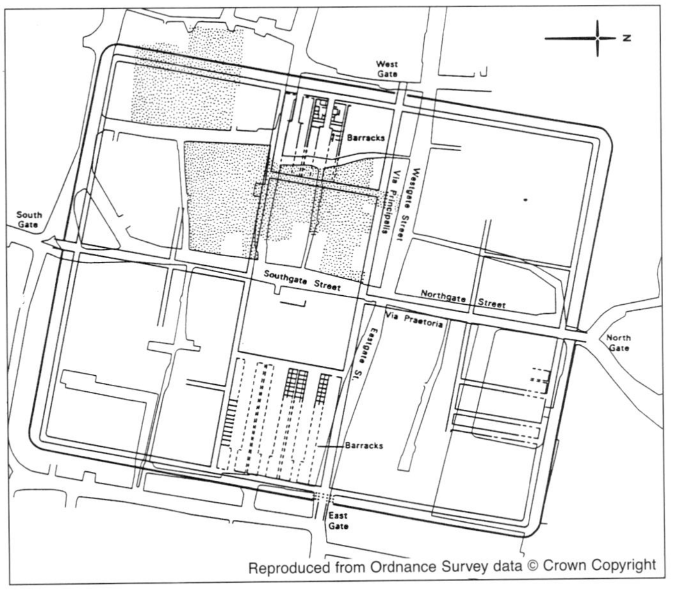

Fig 18.1 *Roman Gloucester*

John Pugh-Smith
Barrister
39 Essex Street Chambers
London WC2R, UK

Richard Sermon
Principal Archaeologist
Gloucester City Council
Gloucester Archaeology Unit
The Old Fire Station, Barbican Road
Gloucester GL1 2JF, UK

Albert Williamson-Taylor
Structural Engineer
Adams Kara Taylor
Chapter House, 2 Farringdon Road
London EC1M 3HN, UK

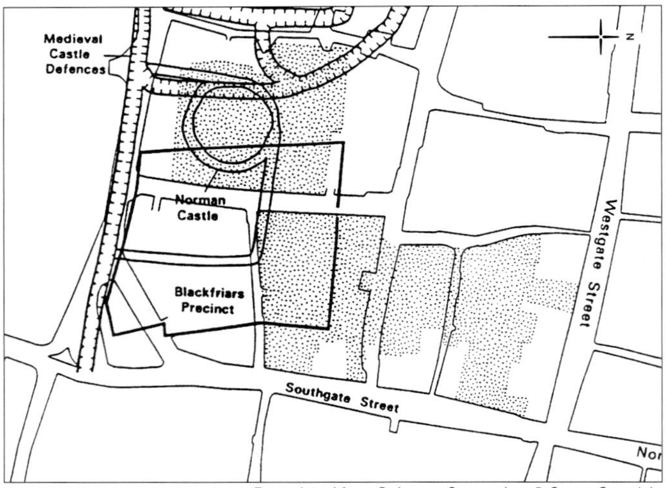

Reproduced from Ordnance Survey data © Crown Copyright

Fig 18.2 Medieval Gloucester

medieval periods. Its continued importance is reflected in its associations with the Norman and later medieval castle and the establishment of the Blackfriars Priory (Fig 18.2).

There are several areas of Scheduled Ancient Monument located within and adjacent to the proposed development. These are Blackfriars Priory and Glevum Roman Colonia. These designations indicate the national importance placed on the archaeological remains in this part of the city. The proposed development also lay entirely within the City Centre Area of Principal Archaeological Interest (Gloucester City Council 1996).

The first archaeological assessment of the land was by the Gloucester Archaeology Unit (GAU) in 1988 (Atkin 1988). This work was followed in 1990 by field evaluation of the land to the west of Ladybellegate Street, which was then planned to be the site of a new magistrates court (Atkin 1990), and in 1993 by evaluation of the surface car park to the east of Ladybellegate Street (Atkin 1993). In 1994, the County Archaeology Service carried out further assessment and field evaluation of the proposed magistrates court site (Parry 1994; Reilly1994), which in 1995 lead to the archaeological excavation of some of the proposed pile locations (Parry and Reilly 1996). Then in 1997, Arrowcroft plc's consultant archaeologist, Mike McCarthy of Carlisle Archaeology Unit, carried out a new assessment of the area (McCarthy and Zant 1997), which in consultation with the City Archaeologist, Richard Sermon, and English Heritage lead to the production of an evaluation brief (Sermon 1997). By this time the magistrates court scheme had been withdrawn and the area was incorporated into the current Blackfriars development scheme.

In early 1999 Arrowcroft submitted an application (Ref 99/29/FUL) to build a multiplex cinema with below ground, multi-storey car parking on land to the west of Ladybellegate Street (Figs 18.3–18.4), making the best use in the natural fall of the land to accommodate the bulk of the proposed building. The architect for the scheme was Stefanie Fischer of Burrell Foley Fischer and the engineer was Albert Williamson-Taylor of Adams Kara Taylor.

Following the receipt of the application, Richard Sermon advised the planning authority that the below ground car parking would have a significant impact on the known archaeological deposits in this part of the city. As a response

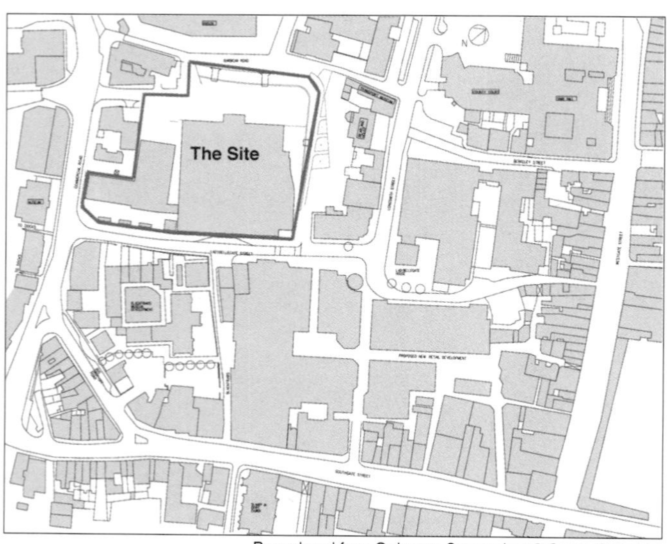

Reproduced from Ordnance Survey data © Crown Copyright

Fig 18.3 Blackfriars multiplex location

Fig 18.4 Proposed development

the City Archaeologist and English Heritage produced a mitigation brief (Sermon 1999), with specific guidance for a number of different mitigation zones (Fig 18.5) identified across the site:

A – with an intact archaeological sequence where preservation *in situ* of all significant deposits is the recommended mitigation strategy;

B – truncated down to 13.00m OD by the former Midlands Electricity plc building, where the recommended mitigation strategy is to locate piles in areas of previously exposed natural geology (avoiding archaeological structures and deposits), and areas previously excavated for the magistrates court foundations;

C – with large deep-cut Roman and medieval defensive ditches, where a dispersed pile foundation would not significantly disrupt the archaeological sequence or its future interpretation;

D – largely comprise a Scheduled Ancient Monuments (Glevum Roman Colonia). In the evaluation brief, it was agreed that these areas would be assessed in light of the results from adjacent excavations and evaluations

In response, Lisa Donel of GAU was instructed by Arrowcroft to work with their architects and engineers on the production of the mitigation strategy (Donel and Taylor 2000). This work began with the adoption of three broad engineering principles: to minimise the extent of excavation required for the construction of the car park; to minimise the number of vertical penetrations through the ground and to minimise the extent of excavation required to construct the foundations.

With these principles established the design team set about addressing the archaeological issues, the first of which was the below ground car parking. To avoid disturbing the archaeological deposits which follow the natural ground slope from east to west across the site, the lowest level of car parking (12.00m OD) was reduced substantially to a narrow band running parallel to Barbican Road, whereas in the submitted drawings it had covered the whole area (Figs 18.6–18.7). This amendment has placed formation levels for the Level 1 (12.00m OD) and Level 2 (15m OD) car parking well above the known level of significant archaeological deposits (Fig 18.8).

In response to the need to reduce the number of vertical penetrations of the ground, the engineers recommended large diameter piles (over 2.0m) combined with Y-shaped transfer columns (Fig 18.9) to provide a 19m piling grid with only 54 pile locations (Fig 18.10), ie larger but fewer piles. By way of comparison, in 1995 the proposed magistrates court was to have occupied the central part of the current site. A comparison of the number of piles and the presence of piles, pile caps and ground beams with those in the same area from the present scheme, showed the following: the magistrates court with 131 piles with caps and ground beams compared to the multiplex cinema with 20 piles with no caps or ground beams.

In addition, there would be two further advantages with

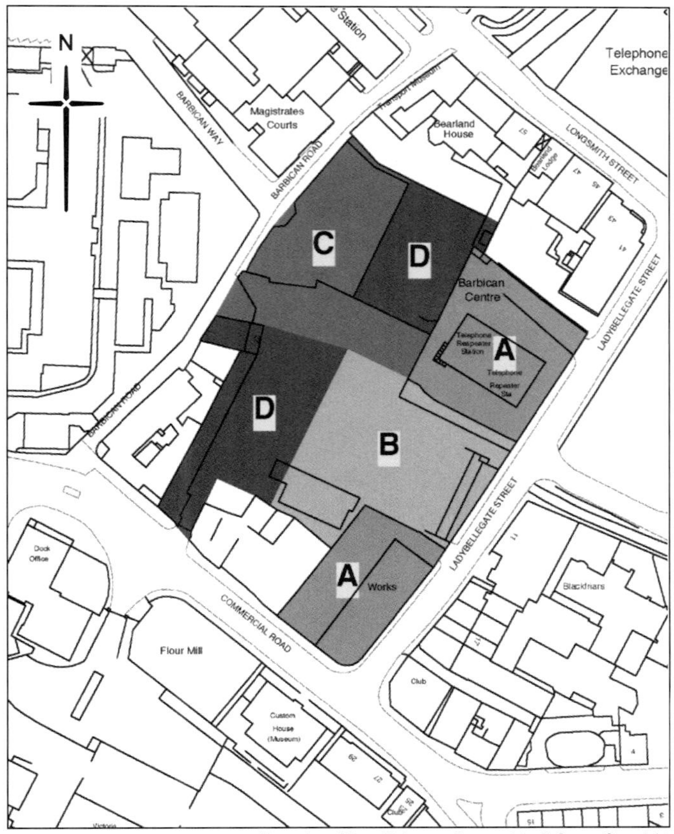

Reproduced from Ordnance Survey data © Crown Copyright

Fig 18.5 Archaeological mitigation zones

these large diameter piles. Firstly, the piles would be of a sufficient diameter and depth that they could be reused for future development, thus providing sustainable protection to the archaeological deposits on the site. Secondly, each of the pile locations could be archaeologically excavated and recorded (preservation by record) within large diameter concrete pre-cast shaft lining units (Figs 18.11–18.12), an approach which was explained in the written scheme of investigation (Donel and Taylor 2000). While the loss of any archaeological deposits would be regrettable, the view was taken that all significant deposits affected by the development would be subject to controlled archaeological excavation and preservation by record. Furthermore, the projected deposit loss of less than 3% would still be well below the 5% loss suggested to be acceptable for development in York (Ove Arup 1991). This paper arose from the first detailed study to set down a level of acceptable deposit loss with regard to development in an urban historic environment. The figure of 5% acceptable loss quoted in the study, referred to deposits that would be completely destroyed, normally by piling, and <u>not</u> subject to archaeological excavation or recording.

By excavating each of the large diameter pile locations, all of the archaeological deposit that would be threatened by the development would then be subject to controlled archaeological excavation and recording (preservation by record). This would then allow over 97% of the archaeological deposits to be preserved *in situ*, the innovation being that with even with this type of piled foundation, none

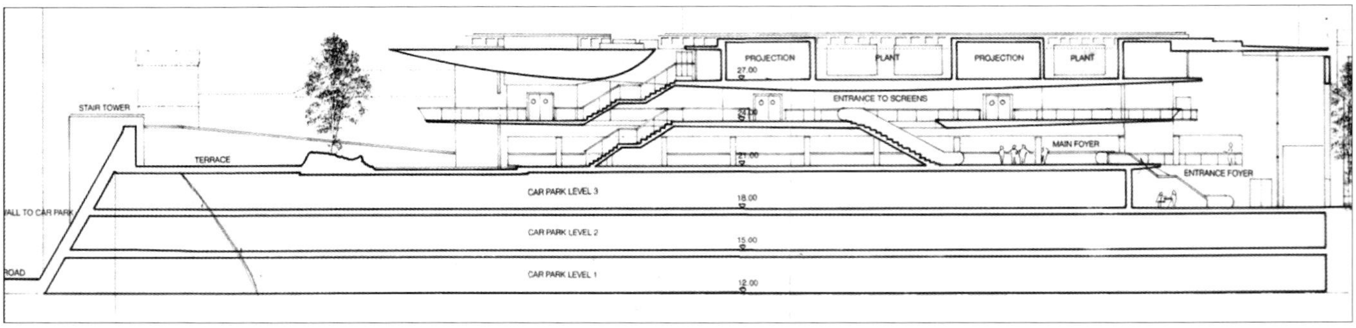

Fig 18.6 *Cross section across proposed multiplex development prior to archaeological mitigation*

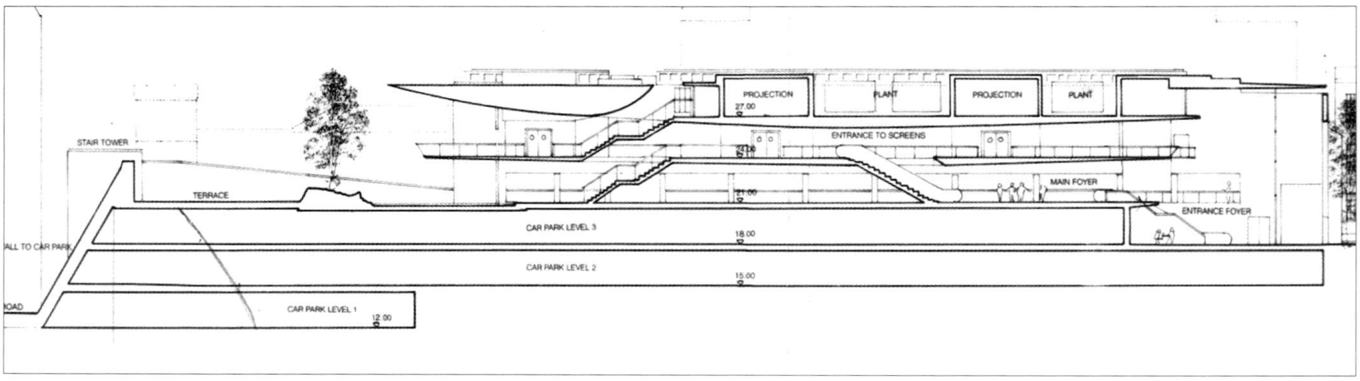

Fig 18.7 *Cross section across proposed multiplex development after archaeological mitigation*

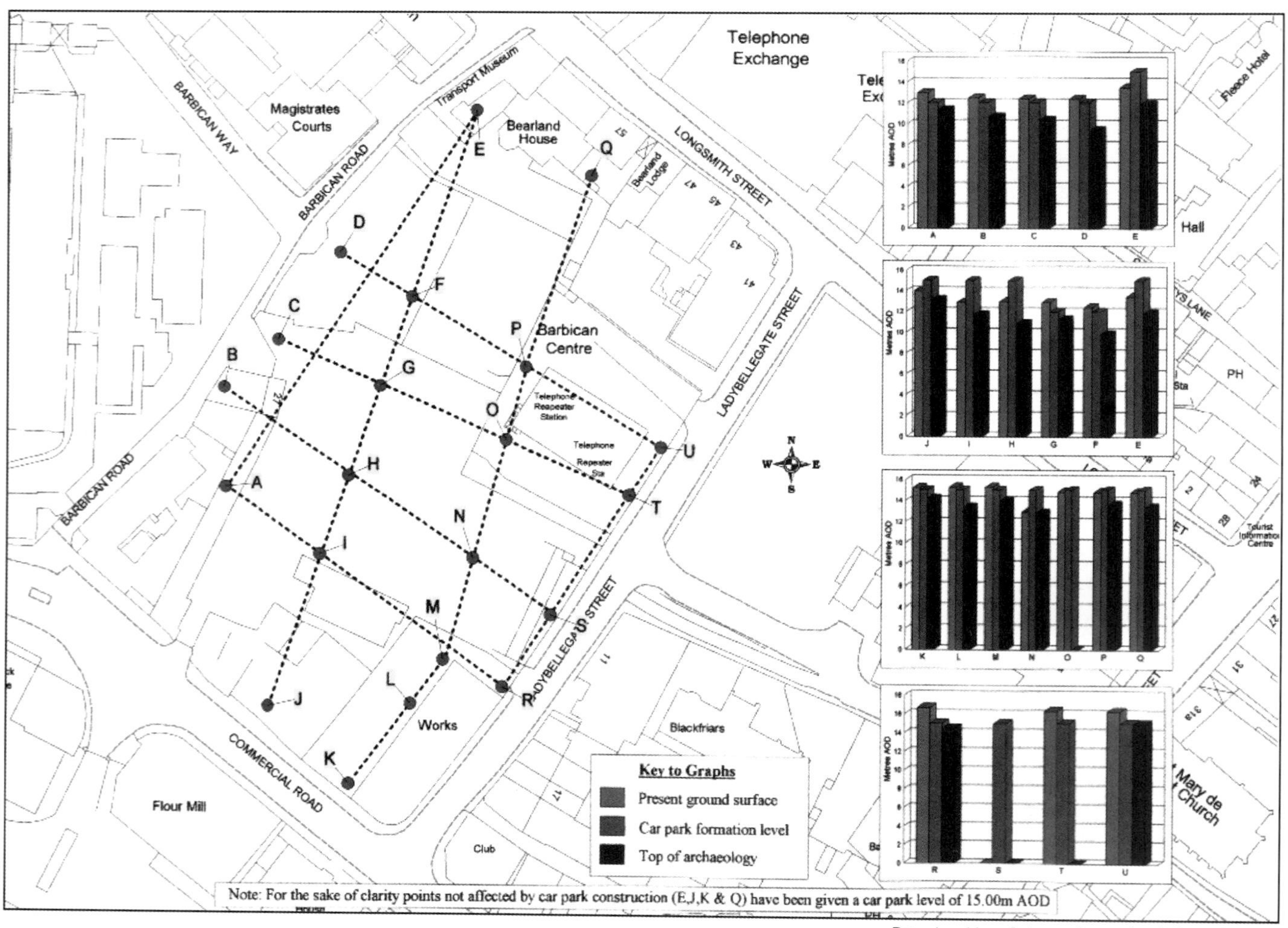

Fig 18.8 *Deposit profiles across the site*

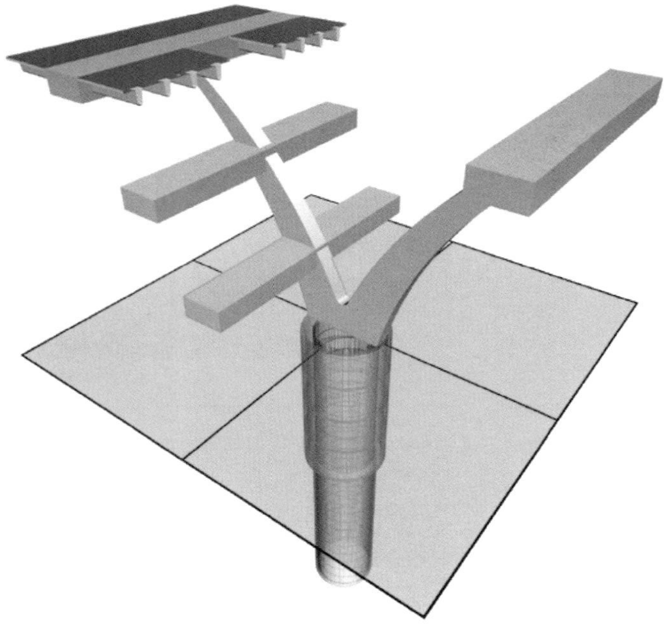

Fig 18.9 Piles and transfer columns

of the threatened archaeological deposits would be 'lost' (destroyed without record) in the sense of the York study.

In response to the need to minimise the extent of excavation required for the foundations (see above), the design team recommended using upstand ground beams at the perimeter of the site, rather than the traditional approach of employing downstand ground beams. They also located all lift shafts, stairwells and service ducts within the car parking levels, and, therefore, above the level of any significant archaeological deposits. As a means of validating the work of the design team, the Museum of London Archaeology Service (MoLAS) was asked to review the overall approach. Its full statement was appended to the mitigation strategy (Donel and Taylor 2000), from which the following closing comment was drawn: 'Overall, the design team present a thorough, considered and innovative mitigation strategy, fully supported by a range of drawings. It appears to successfully address the objectives of the planning brief and provides a strong framework for the preparation of detailed structural and archaeological designs' (George Dennis and Paul Falcini, MOLAS).

Following the production of the Blackfriars multiplex archaeological mitigation strategy and the written scheme of investigation (Donel and Taylor 2000) and their submission in support of the amended planning application, some concerns were raised as to the actual levels and nature of archaeology in those parts of the site that had not previously been evaluated. Following discussions with the City Archaeologist and English Heritage, 12 test pits were excavated in order to supplement existing knowledge as to the depth of modern deposit over the archaeology. The test pit results revealed that the proposed basement floor heights for the new development would lie well above the observed archaeology (Fig 18.8), and that the overall approach of the design team was sound.

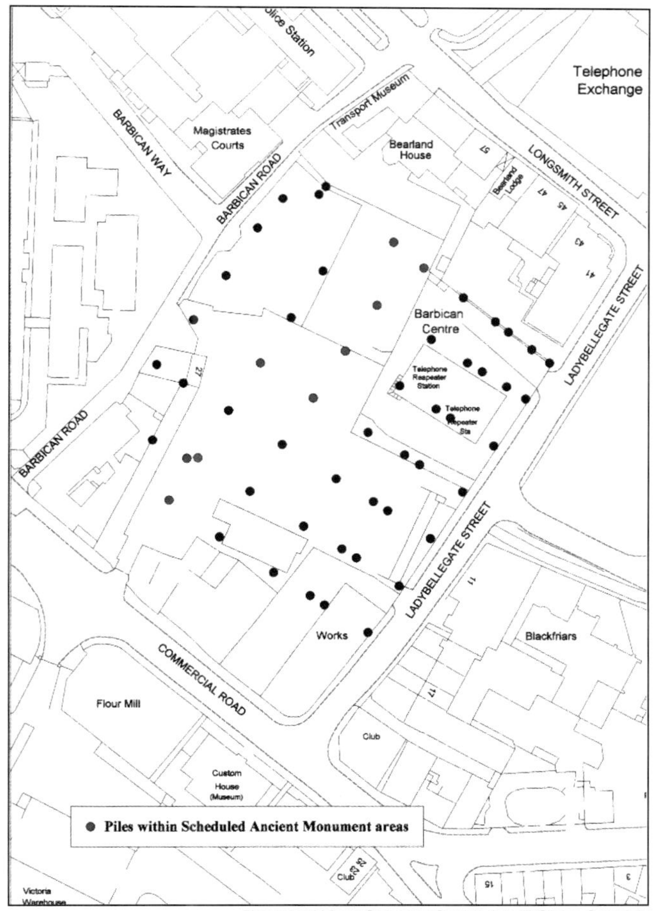

Reproduced from Ordnance Survey data © Crown Copyright

Fig 18.10 Proposed pile locations

Scheduled Monument Consent was granted by the Department of Culture, Media and Sport on 10 June 2002. However, by this stage, the original scheme had been abandoned due to a successful legal challenge by one of the objectors during the course of the concurrent inquiry into the related Compulsory Purchase Orders (R (on the application of British Telecommunications plc) v Gloucester City Council [2001] EWHC Admin 1001; [2002] JPL).

Some lessons

Engineering and archaeology are both professions where not only an understanding but also the provision of solutions are paramount. Assumptions are made, investigations and research are undertaken. For the archaeologist, understanding and preserving the past interprets how future developments on a site should progress. For the engineer, creating future spaces and development is influenced by understanding elements of structure from the past.

The engineer can provide a solution as to how development should be built. To provide substructural solutions requires an understanding of the archaeology and the reasons for its preservation. An examination of the

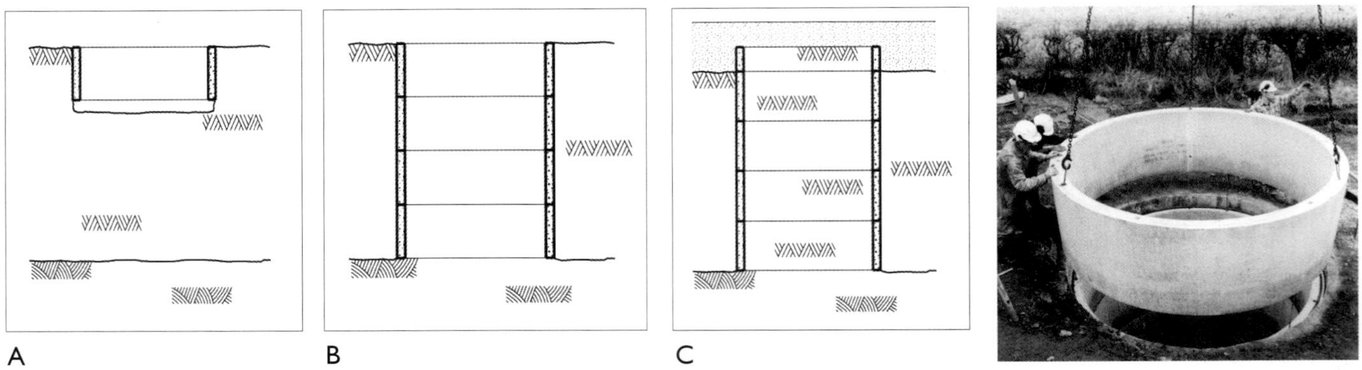

Fig 18.11 *Method of pile excavation*

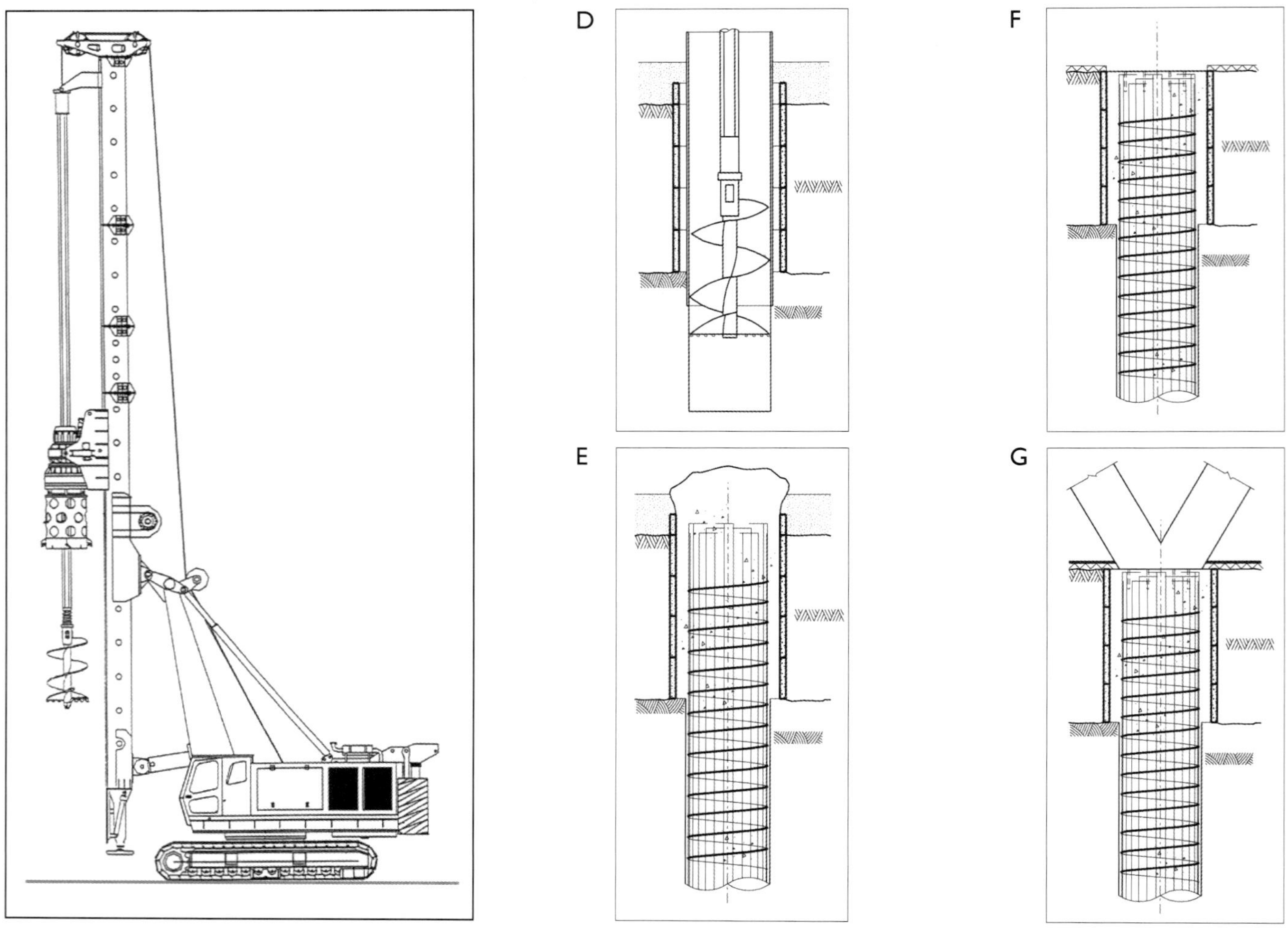

Fig 18.12 *Method of pile construction*

engineering history of the foundations of a former building can be as important as the building itself. By way of a further example, whilst the foundations of a former mill may have led to destruction in those areas during the time of their construction, the loadings they have carried may well have led to substantial preservation of archaeological deposits within the voids between them. A raft solution, anchored solely in these disturbed areas, can then be utilised. It can be designed so as not to impose any greater loads on the site than had previously existed, thus minimising the risk of further compression of the archaeological deposits, and, at the same time give the architect sufficient, clear-span floor areas for new retail units.

By understanding the importance of preservation coupled with issues raised by developments, appropriate solutions can be developed. These solutions should reflect the top-down approach of the architecture while accommodating the bottom-up requirements of the archaeology. The interplay between the disciplines can then result in bespoke, tailor-made solutions.

Putting this in the context of urban regeneration, the aspirations to revitalise our urban centres, make better use of land through denser occupation will mean that fresh solutions will need to be found to overcome the problems of developing on sites of archaeological interest and where other heritage issues arise.

Accordingly, it is in the interests of all parties that a holistic approach to preservation is adopted. Modern development and the rejuvenation of urban spaces are not mutually exclusive from the preservation of archaeology.

Conclusions

From experiences both in Gloucester and subsequently, in London, the following points emerge: there is a need for bespoke solutions to be found if regeneration of historic city centres is to proceed, successfully, hand in hand with archaeological preservation; the concept of forensic engineering needs to be developed as an essential requirement from the early stages of a scheme's development; the design solution should be developed as much from the archaeology upwards as from the building downwards; the concept of sustainable mitigation, by which the same foundation design can be reused again, also needs to influence the design solution; there needs to be a far greater team approach taken by all involved from the public as well as the private sectors. Only in these ways can potential conflict be removed from a difficult, complex and yet rewarding process.

Bibliography

Atkin, M, 1988 Gloucester, the south-west quadrant of the ancient city; the archaeological implications of redevelopment, unpub rep

Atkin, M, 1990 Excavations at the Magistrates Court site, Gloucester, unpub rep

Atkin, M, 1993 Ladybellegate Street car park, Gloucester; an archaeological evaluation, unpub rep

DoE, 1990 Department of the Environment, *Planning policy guidance note 16: archaeology and planning*, London

Donel, L, and Taylor, A, 2000 Blackfriars Multiplex archaeological mitigation strategy, unpub rep

Gloucester City Council, 1996 *Interim adoption City of Gloucester local plan*, Gloucester

McCarthy, M R, and Zant, J, 1997 Blackfriars, Gloucester, an archaeological impact study, unpub rep

Parry, C, 1994 An archaeological assessment, unpub rep

Parry, C, and Reilly, S, 1996 Summary report, *Glevensis* 29, 27-8

Reilly, S, 1994 An archaeological evaluation, unpub rep

Ove Arup, 1991 York development and archaeology study, unpub rep

Sermon, R, 1997 Archaeological evaluation brief, unpub Gloucester City Council rep

Sermon, R, 1999 Archaeological mitigation brief for the Blackfriars Multiplex, Gloucestershire, unpub Gloucester City Council rep

19

Naturally reversing the effects of time with the calcite *in situ* precipitation system

EDWARD KUCHARSKI

and FIONA CHOW

Edward Kucharski
Lithic Technology Pty Ltd
23 Winchester Way
WA 6149, Australia

Fiona Chow
Geotechnical Consulting Group
1A Queensberry Place
London SW7 2DL, UK

Introduction

A new technique has been developed that allows the preservation of archaeological remains *in situ*. The calcite *in situ* precipitation system, or CIPS, allows the precipitation of natural calcite cement in porous materials, producing increases in strength without significant reductions in porosity. The method allows the effects of time to be reversed in a fast and environmentally friendly manner.

Calcite *in situ* precipitation system

CIPS is a low viscosity, water-based, non-particulate solution allowing easy penetration into fine-grained materials. Natural calcite cement (otherwise known as calcium carbonate or lime) is precipitated from the solution, coating the particles and bridging their contacts, thus producing an increase in mechanical strength and stiffness. The amounts and rates of calcite delivery can be controlled, with full cementation occurring between eight hours or three days after application. The pore space between the particles are left open allowing multiple CIPS injections and large increases in cementation strengths. Silica sands have been cemented into sandstones with similar strength to concrete (40 MPa UCS). The solution is not harmful to the environment and has neutral pH.

Restoration of the Whaler's Tunnel, Fremantle, Western Australia

The Whaler's Tunnel was carved out of a soft limestone outcrop in 1837 by the Fremantle Whaling Company and left unlined. The tunnel lies beneath the Round House, the first prison in Western Australia and a local landmark (Fig 19.1).

Fig 19.1 The Round House, Western Australia

Over 160 years of water seepage the calcite cement has been washed out of the limestone leaving the tunnel walls weak and friable in places. Concerns over safety led to its closure in 1996. In the restoration scheme, designed by Ove Arup and Partners, CIPS was used to strengthen the limestone walls, to replace the lost calcite and provide increased resistance to wind and water erosion. CIPS solution was sprayed directly onto the walls and absorbed through capillary suction (Fig 19.2). The work was completed in one week showing a marked increase in the strength of the walls, without any visible changes in appearance of the limestone.

Fig 19.2 Spraying CIPS solution on to the walls of the *Whaler's* Tunnel

Other applications

The CIPS technique has numerous potential applications in the protection of stone and rock, preservation of archaeological remains, strengthening of castle and church walls, reducing the exfoliation of masonry buildings, strengthening historic lime mortar.

20

Seahenge (Norfolk): from the sea to the beach to ... where?

MIKE CORFIELD

Mike Corfield
(formerly Chief Scientist, English Heritage)
Heritage Science and Conservation Consultant
Northwood, Stafford Road
Gnosall, Staffordshire ST20 0ES, UK

Abstract

In 1998 a circle of timber posts was seen to be eroding from the beach at Holme-next-the-Sea on the north coast of Norfolk. The circle surrounded an upturned tree stump that had been exposed on the beach at low tides for some time previously. The identification of the timber as a Bronze Age ritual structure generated great interest and considerable debate about its future. This paper explores the issues that have arisen in the last four years and considers how the decisions made about the timbers' future might impact on the development of policies for the conservation of structures in the intertidal zone.

Introduction

The coastal zone is a highly dynamic region, the full utilisation of which was not appreciated until comparatively recently. It had been relatively neglected by archaeologists until some fifteen years ago. In 1986, in response to a proposal to build a barrage across the Severn Estuary from near Penarth on the Welsh side to Brean Down near Weston-super-Mare on the English side, the Severn Estuary Levels Research Committee was set up. Under the aegis of this committee extensive survey, particularly of the Welsh side, of the estuary took place. The results of this work have been considerable (Bell 1994; Bell et al 2000). Significant discoveries include Mesolithic footprints (Aldhouse Green et al 1992), trackways, timber piled house platforms and boats. The Wootton Quarr Intertidal Survey on the north coast of the Isle of Wight was funded by English Heritage; its objectives were to investigate the formation of the submerged landscapes of the Solent and its intertidal wetlands, and to develop survey and recording techniques, threat assessment methodologies and management options for the archaeology of the intertidal zone. A spin off from this project is the Wootton Quarr Reburial Project which is investigating the type of environments suitable for reburying timbers from coastal excavations, and the physical, chemical and hydrological parameters to be assessed. Other coastal surveys include the Hullbridge Survey in Essex (Wilkinson and Murphy 1995), and the Lindsey Survey (Brookes et al 1990).

More recently, a survey of coastal zone archaeology has been undertaken and published as *England's coastal heritage* (Fulford et al 1997). The report provides a comprehensive review of coastal processes and the (then) potential impact of sea level rises; it reviews the extent of the known archaeology of the coastal zone, but observes that there had been limited systematic surveys. Despite this, the report demonstrates the rich survival of archaeological evidence in the coastal zone ranging from sites that have exploited the resources of the sea to those formerly on dry land but which have been inundated by the advance of the sea. Temporally, the sites range from the

Palaeolithic to post-medieval fortifications.

However, much of the archaeology of the coastal zone is buried under the accumulations of marine silts and only becomes exposed when the tides and prevailing winds promote erosion. In this context, a major contribution to *England's coastal heritage* is the chapter by Long and Roberts focusing on sea level changes and coastal processes. Allen puts this into the archaeological context. The complexities of managing in the dynamic environment of the coastal zone archaeology is covered by Champion, Firth and O'Regan. These issues are also discussed in this volume by Murphy and Trow (see paper 28).

More recently, concerns about the impacts of changing climate have added to the attention being paid to the management of the coastal zone. Studies of climate change impacts are being carried out in the English regions: one of the first published covers south-east England, and identifies sea level increases as one of the major challenges to be addressed (Surrey County Council 2000). While rises in the level of the sea may be presumed to increase the security of coastal archaeology, it is likely too that there will be an increase in coastal surge tides in the summer and this will increase site erosion. Coastal low-lying lands may have to be sacrificed as indefensible. Shoreline Management Plans are being prepared for all coastal areas. The first of these plans is for discreet lengths of the coast and identifies where there are threats to coastal assets, including important habitats, but not archaeological sites. The second-generation plans will recognise that relatively short lengths of the coast cannot be managed without reference to the processes of erosion and accretion that might take place over much longer distances. To address the wider context of coastal change, regional coastal groups have been formed: the Standing Conference on Problems Associated with the Coastline (SCOPAC) covers 200km of the south coast of England from Lyme Regis to Brighton and includes 29 local authorities and agencies. Bray et al (1997) have summarised the issues that need to be considered in the development of sea level rise predictions. These include late Holocene sea level indicators derived from geological and archaeological evidence feeding into the data, recent global and local changes, adding in global warming, climatic and oceanographic modelling and information about local wave climates and storm surges to produce a regional sea level scenario.

There are few more dynamic coasts than the beach at Holme-next-the-Sea which is located at the point where the coast turns south-west into the Wash. A study by the Department of Marine Studies and Coastal Management at the University of Newcastle for the Environment Agency (Anon nd) has shown that the beach evolution is influenced by three important factors: the strong north-east/south-west tidal systems interacting with the tides from the Wash; the offshore sandbanks; and the ebb delta at Thornham harbour which ties up large volumes of sand removed from the beach. The effect of these is that the beach profile is steepened between successive winters, the surface is lowered and during storms the dunes are sacrificed. It is the only beach in the area that is behaving in this way, and it is not thought possible to prevent it by stabilisation methods such as beach recharge because the underlying causes cannot be addressed.

Analysis of sediment cores taken after the discovery of the timbers showed that the beach had been subject to major environmental changes in the past. Analysis of pollen and plant remains indicate that at the time the monument was constructed the land surface was a salt marsh. Soon after, dunes developed on the seaward side of the monument and marine influences were excluded, and this led to the development of peat under reed-swamp and alder carr. The development of the peat must have been rapid because there is no evidence on the wood of its having been attacked by old terrestrial or marine boring insects or crustaceans. Later still the dunes were breached by the sea and the peat was buried under the build up of sediments which protected the monument until they in their turn were eroded and the timbers exposed. (Murphy, pers comm).

When the circle of timbers first appeared there was some uncertainty about its date and nature. Some thought that the structure might have been a medieval or later fish trap. The first step was to confirm its date by dendrochronology. A segment was cut from the central trunk and this was examined at the University of Sheffield. A good sequence of rings was present that suggested a date at the end of the second millennium BC or the beginning of the first. To achieve greater precision the sample was sent to the Belfast radiocarbon dating laboratory for wiggle matching; this, and the subsequent use of Bayesian analysis, gave a precise date of 2050 BC for the trunk and 2049 BC for the surrounding timbers (Bayliss et al 1999). Further study of the timbers revealed a previously unseen ring which brought the dates of felling for both elements of the monument to the same year.

Meanwhile, the story had become front-page news in the national press, and an enthusiastic journalist from the Eastern Daily Press had christened the monument 'Seahenge'. Although it had none of the attributes of a henge monument the name stuck and the site became a place of pilgrimage for people of various pagan persuasions and new age cults.

With the significance of the monument confirmed by its date, attention was given to its fate. I was asked to examine the options for the monument and identified three: preservation *in situ*; excavation, analysis, conservation, and deposit in a museum; excavation, analysis, and reburial after investigation.

Preservation *in situ*

Preservation *in situ* implies that the monument would be left undisturbed, with some form of protection from the ongoing processes of erosion and a covering of material that would keep it waterlogged and in an anaerobic environment. This would have been a considerable challenge. It would have been necessary to enclose the monument within a ring of close piling which was sufficiently deep to ensure that it would have a reasonably long life on the eroding beach. The timbers would have been covered with a dense moisture retaining soil that would exclude air, and they would have to be fully saturated. To achieve this, the height of the structure would have to be below the high water maxima so that the rising tide would maintain the wetness of the covering material. It would therefore be necessary to cover the soil with a geotextile to prevent it being washed away and have this held in place with heavy rocks. The weight of the rocks would have to be carefully calculated to ensure it was sufficient to withstand the tide and hold the covering material in place, but not so heavy that its weight would risk damaging the fragile wood.

The structure would have been a formidable intrusion into the seascape and would doubtless have attracted large numbers of visitors to the beach. Its construction would have involved pile driving and other heavy mechanical plant on the beach. This would have been unacceptable because the beach is an internationally important habitat for wading birds; it is a Ramsar site, which is an international designation which promotes the wise use of wetlands. Its focus has been on ecological issues but currently attention is being given to adding cultural heritage management to its remit (Miles 2001). The beach is also a candidate Special Area of Conservation (SAC), and a UK Site of Special Scientific Interest (SSSI). The Norfolk Wildlife Trust which is responsible for managing the beach was already concerned that the publicity surrounding the monument was attracting large numbers of people to the beach and the bird population was being disturbed at a time when they were nesting in preparation for breeding. If a defensive structure was to be built it would have had to have been at a time between nesting when the noise and activity would not have impinged on the birds.

The crucial issue that finished the preservation *in situ* option was the recognition, as Dix et al explain (see paper 4), that a fixed structure like this on the beach would actually increase the rate of erosion due to the scouring action of the tides flowing around it. A second timber circle close by which is being observed and recorded *in situ* is showing evidence of this effect, with a pronounced erosion trench developing.

Excavation

We therefore brought forward plans for excavation and the Norfolk Archaeological Unit began work in February 1999. The start of the excavation resulted in another option for the monument being thrust upon us. There was a strong feeling in the pagan groups that the monument was a sacred site and that it should not be disturbed. Their view was that it should be allowed to die naturally by being swept out to sea by the tides. The site was occupied and David Miles, English Heritage, Chief Archaeologist, took part in a meeting with neo-pagans and Druids at which he was able to persuade most of those present of the potential of the timbers to reveal important information that would add to the significance of the site. However, one of the people present and a small number of others were not persuaded and the stand-off continued. Eventually a High Court injunction had to be obtained so that the excavation could begin.

The conditions for the excavation were far from ideal. The work had to be carried out between the tides, and when the site became exposed by the ebbing tide, the water had to be pumped out before work could recommence. It was also important that the work was done quickly to avoid disturbing the birds. Although there were no signs of the original land surface it could be seen that the timbers had not simply been driven into the ground, but had been carefully placed into a prepared trench. The timbers had been carefully prepared; the bottom of each one was carefully squared off, and on a couple of them, side branches were left in place and the ends of these were cut square.

It was of course necessary to get the timbers into suitable storage as soon as possible after they were excavated. The nearest place with suitable facilities for storing waterlogged wood was the Flag Fen Archaeological Centre, near Peterborough. Here they could be put into existing tanks; the stump, which was too large to be immersed in the available tanks, had to be sprayed to keep it wet. Fortuitously the wood analysis was to be done by Masie Taylor who was based at Flag Fen. Some of the local people of north Norfolk were outraged that the monument had been removed from the village and protested loudly. The then chairman of English Heritage promised that the timbers would be returned to Norfolk when the analysis had been completed, and that if a Museum would be prepared to accept them, English Heritage would pay for their conservation.

Maisie Taylor's investigation of the timbers has been outstandingly successful. She has been able to identify at least 40 different axes used to shape the timbers. Using the information from the dendrochronology it has been possible to demonstrate a complex arrangement of them around the trunk. Together this suggests a much greater use of copper alloy axes than was expected at this early stage of the Bronze Age and a high degree of social organisation. It also raises questions about the ritual significance of the monument. The full details of the analysis will be presented in the official report in due course, but meanwhile Francis Pryor has written a lively report in his book Seahenge (Pryor 2001).

With the timbers safely housed at Flag Fen we had to give consideration to the two remaining options: conservation or reburial. As the first step the condition of the wood was analysed and a conservation strategy was developed. The state of the wood was such that it could be conserved either by pre-treating it with polyethylene glycol and freeze drying it, or by fully replacing the wood with polyethylene glycol. There was nothing about the condition of the wood to suggest that reburial would not satisfactorily preserve the wood for future excavation and study.

Reburial

Because of the greatly increased scale of archaeological activity in the coastal zone and the likelihood that considerable quantities of worked timber will be recovered, English Heritage is currently supporting a research project that is investigating reburial with the aim of developing protocols for identifying suitable sites and optimum burial environments. Reviewing the visual evidence of timbers which have been recovered during excavation, reburied, and subsequently re-excavated, suggests that little change seems to occur. It is difficult to be categorical about this because until very recently there have not been any instances where wood has been subjected to full scientific evaluation of its condition before being reburied. The evidence therefore is subjective and based on the original excavators' memory of its appearance. This was the case when the Neolithic Sweet Track, which is preserved within the Shapwick National Nature Reserve in the Somerset Levels, was examined to determine whether it had suffered as a result of seasonal falls in the water levels. Although the wood of the trackway *appeared* to be in good condition, scientific examination showed that almost all the wood substance had been lost; however it was impossible to say whether there had been further decay since the original excavation because of the lack of baseline data (Brunning et al 2000).

In determining a suitable site for reburial of the Holme timbers we used the following parameters: sufficient depth of anaerobic sediments to ensure that the timbers would be safe from erosion for the foreseeable future; ease of access for the plant needed to dig the burial pit; minimal disturbance of the wildlife; ease of access for long term post burial monitoring.

The cores taken from the beach and from locations nearby showed that there were some 5.0m of muds above the basal gravels, and these would be more than sufficient to provide a secure burial environment for the wood (Wells and Canti nd). The burial site would be higher up the beach than the original location of the monument – immediately seaward of the dunes. We felt that here the timbers would be safe from erosion for many years to come. It was intended that there would be intensive monitoring of the proposed site over six months before the reburial took place. The monitoring would provide a continuous record of the water table, the pH, redox potential and conductivity of the water, and an assessment of the activity of cellulose degrading bacteria. After burial we would continue the monitoring with the addition of probes in the wood to measure its degree of saturation and salinity. We felt that this would give a high degree of confidence that safe reburial has been achieved.

While the plans for reburial were developing, the accumulation of evidence had greatly increased the significance of the monument and of the timbers and a strong lobby developed against the reburial option. It was forcefully put forward that too little was known about the effects of reburial to be sure that the wood would not deteriorate in the ground. Particular concern was expressed that because the saline water in the wood had been replaced by fresh water, it may be at risk from osmotic effects if it were returned to a saline environment (Marchant 2001). Theoretically the fresh water in the interior would be drawn through the cell walls of the wood, which would act as semi-permeable membranes, in an effort to equalise the salinity on either side of the membrane; with an inexhaustible supply of saline water the interior of the wood could lose all its water in this way. This was not considered a serious risk because the timbers had already been through a cycle of change from fresh water to salt water in prehistory, and laboratory experiments on small wood samples showed no evidence of osmosis and distortion of the wood. Nonetheless, in the final analysis English Heritage's Advisory Committee and Commissioners took the view that the timbers were of international importance and that the owners should be advised that they should be conserved so that they would be readily available for further study, or display if this became possible.

Conclusions

The saga of Seahenge exposes many of the issues that have to be considered when dealing with structural remains in the intertidal zone. Preservation *in situ* will rarely be an option on coasts that are eroding, though studies of coastal sediment movement may show that the erosion process may be reversed or stabilised, and this should be a consideration in the decision process. Attempts to prevent erosion by use of hard defences will almost always be doomed to failure because of the increased scouring effects that occur around fixed structures, furthermore, the scouring may also place adjacent structures in jeopardy. Therefore, for structures like Seahenge which are deemed to be internationally important excavation and conservation may be the only solution. For structures that are considered to be significant, to have potential for future investigation, but no immediate plans for further study, reburial will be the preferred option. Finally, the majority of structures exposed on shorelines will simply be recorded, possibly sampled, and left to their fate as the sea encroaches.

Acknowledgements

The analysis of the information from the Seahenge project is still in the progress, therefore much of the information contained in this paper has been provided by the specialists involved in the work in advance of the publication. I am grateful to Bill Boisemeir of the Norfolk Archaeological Unit for information about the excavation and to Peter Murphy, English Heritage Regional Adviser for Archaeological Science for the East of England for his interpretation of the evolution of the coastline at Holme-next-the-Sea. Maisie Taylor has been generous in sharing the results of her study of the wood, and many other matters relating to the monument. Brian Ayers, the Head of Archaeology in Norfolk Museums Service has led the negotiations about the future of the timbers. Ian Panter and Mark Jones have carried out the condition surveys, and many colleagues have commented on the issues through the 'sea-sites' and 'conservation-research' mail lists. Finally, I must thank my colleague David Miles, English Heritage's Chief Archaeologist, and Philip Walker the Inspector of Ancient Monuments covering Norfolk for their support and encouragement throughout the Seahenge saga.

Bibliography

Anon, nd Holme Dunes: coastal processes and geomorphology study, unpub Department of Marine Sciences and Coastal Management, Univ Newcastle rep

Aldhouse Green, S H R, Whittle, A W R, Allen, J R L, Caseldine, A E, Culver, S J, Day, M H, Lundquist, J, and Upton, D, 1992 Prehistoric human footprints from the Severn Estuary at Uskmouth and Magor Pill, Gwent, Wales, *Archaeologia Cambrensis* 141, 14–55

Bayliss, A, Groves, C, McCormac, F G, Baillie, M, Brown, D, and Brennand, M, 1999 Precise dating of the Norfolk timber circle, *Nature* 402, 479

Bell, M G (ed), 1994 *Archaeology of the Severn Estuary, Severn Estuary Levels*, Res Comm Ann Rep, Univ Wales

Bell, M, Caseldine, A, and Neumann, H, 2000 *Prehistoric intertidal archaeology in the Welsh Severn Estuary*, CBA Res Rep 120, York

Bray, M, Hooke, J, and Carter, D, 1997 Planning for sea level rise on the south coast of England: advising the decision-makers, *Trans Inst British Geographers*, ns 22, 13–30, London

Brooks, I P, Hunt, C O, Dorning, K J, Harding, J L, and Jones, A M, 1990 *The Lindsey coastal survey* 1989–90, Lincoln

Brunning, R, Hogan, D, Jones, J, Jones, M, Maltby, E, Robinson, M, and Straker, V, 2000 Saving the Sweet Track: the *in situ* preservation of a Neolithic wooden trackway, Somerset, UK, *Conservation and Management of Archaeological Sites* 4, 3–20, London

Fulford, M, Champion T, and Long, A, 1997 *England's coastal heritage: a survey for English Heritage and RCHME*, Engl Heritage Archaeol Rep 15, London

Marchant, J, 2001 Banished to a watery grave, *New Scientist* 24 March 2001, 7

Miles, D, 2001, Ramsar designation and the case of Seahenge, in *The heritage management of wetlands in Europe* (B Coles and A Olivier), 157–64, Europae Archaeologiae Consilium and WARP, Brussels and Exeter

Pryor, F, 2001 *Seahenge: new discoveries in prehistoric Britain*, London

Surrey County Council, 2000, *Rising to the challenge, impacts of climate change in the south east in the 21st century*, Kingston upon Thames

Tomalin, D J, Loader, R D, and Scaife, R G, (eds) in prep *Coastal archaeology in a dynamic environment: a Solent case study*, English Heritage monogr

Wells, J, and Canti, M, nd Palaeoenvironmental investigations associated with an early Bronze Age sub-circular structure in the present inter-tidal zone near Holme next the Sea, Norfolk, English Heritage, unpub Centre for Archaeology rep

Wilkinson, T J, and Murphy, P, 1995 *The archaeology of the Essex coast, volume 1: the Hullbridge basin survey*, East Anglian Archaeol Rep 71, Chelmsford

21

Zeugma archaeological project 2000: the conservation management programme

ROBERTO NARDI and KRISTIAN SCHNEIDER

Abstract

In 2000 a third of the ancient city of Zeugma has been inundated by the reservoir of the Birecik hydroelectric dam. The Packard Humanities Institute (PHI) funded and organised an international rescue operation based on emergency archaeology and conservation. The Centro di Conservazione Archeologica, Rome (CCA), carried out the conservation programme based on *in situ* conservation, risk mitigation and reburial of the archaeological remains. After the inundation a shore maintenance programme was implemented to monitor and, if necessary, adjust the new equilibrium established along the lakeside. The ethical considerations of the operative choices, the decision-making process and the project strategy are outlined. The conservation programme is described in detail giving information of the technical operations and on the use of traditional techniques and materials (lime based mortars) for consolidation and protection of archaeological remains left *in situ*.

Introduction

Zeugma is the name given to the site of two towns located on opposite banks of the Euphrates, in the south-eastern part of present-day Turkey. One of the successors of Alexander the Great, Seleucus I Nicator (312–281 BC), founded the twin towns, naming them Seleucia and Apamea after himself and his wife. In the 1st century AD, the towns passed under Roman rule. A legionary garrison, the Legio IV Scythica, was established there to protect what was the only permanent bridge across the Euphrates for several hundred kilometres. For two hundred years, the towns constituted an important trade link between the Roman and Parthian empires. At its peak, the city may have had a population of 50,000–75,000. On the west bank, the city extended over approximately 2000ha. With the dwindling power of the Roman Empire in the 3rd century AD, the city was sacked and burned in AD 252 by the Sassanids. Zeugma is mentioned for the last time in 1048 just before the arrival of the crusaders, after which historical sources fall silent.

In the 1970s, the German archaeologist Jörg Wagner identified the site of the ancient city. Research excavation, principally by David Kennedy from the University of Western Australia, Catherine Abadie-Reynal, Professor at the University of La Rochelle, and the Gaziantep Museum, has taken place over the past 15 years, furthering our knowledge of the international significance of the site (Kennedy 1998).

In 1989, the valley that hosts the site was included in the south Anatolian Development Project as the location of the Birecik hydroelectric dam. This meant that approximately a third of the city area would be inundated.

Extraordinary archaeological remains were discovered during the 1999–2000 excavations, drawing public attention

Roberto Nardi
Director

Kristian Schneider
Project Manager

The Zeugma Conservation Management Programme
CCA Centro di Conservazione Archeologica
Rome, Italy

to the problems of the selection of this site for inundation. In this context the Packard Humanities Institute (PHI), in cooperation with the Turkish Authorities, funded and organised an international programme for emergency archaeological rescue operations, based on excavation and on-site conservation: the Zeugma Archaeological Project (ZAP 2000).

The archaeological excavations were conducted under the supervision of the Ministry of Culture of Turkey under the umbrella of the Oxford Archaeological Unit (OAU), which coordinated the other archaeological missions from Turkey and France, while the Southeast Regional Development Administration (GAP) provided the logistics for the project. The CCA was asked to manage and implement a conservation management programme based on *in situ* conservation of the archaeological remains. During 2000 a full team of professional conservators implemented the conservation plan, including field assistance to the archaeologists; first aid during excavation; preventive conservation; protection; reburial; and risk mitigation of the area before flooding.

Fig 21.1 *A view of one of the 20 trenches of the site*

Fig 21.2 *The conservation team at work cleaning a mosaic*

Premise

Before the PHI-sponsored rescue excavation, the zone to be inundated had been excavated on an extremely low budget, with little manpower. This forced the excavation team to work to stringent schedules entailing a number of sacrifices. Wall paintings and mosaics were the first to fall victim to these conditions, although some of them were lifted by unspecialised workers, but without sufficient documentation. After one month's excavation 700m^2 of mosaics and about 100m^2 of wall paintings had been recovered by a multitude of different methods and piled up, exposed to the elements, in a museum garden.

Under the new regime such sacrifices were not to take place; our aim as far as possible was to leave the site and its structures *in situ*. This would limit archaeological inquiry to a minimum number of trenches. This was only partially accepted by the other project groups, and much of the area destined for inundation has been submitted to archaeological investigation. To convey the magnitude and complexity of the project, the upshot was an excavation carried out by 100 archaeologists, 300 workmen and heavy mechanical gear in temperatures above 50°C, with only a couple of weeks of warning (Fig 21.1).

In response we organised a team of 20 professionals for three months on site and set priorities according to the level of water, the speed of the rise in water level, the projected timescales of the archaeological excavation and the fragility of the archaeological structures (Fig 21.2).

The result was an action plan to design measures for preventive conservation during excavation in agreement with the archaeologists, in order to mitigate the physical impact on structures and finds and limit the amount of conservation work in future. Limited (minimum) timescales were set for excavation according to the needs of *in situ* conservation. A discussion forum (for archaeologists and conservators) was set up to plan an integrated approach for those cases where the rising water level did not allow further intervention. All artefacts with conservation requirements were documented and information was gathered for assessing priorities. Finally the conservation treatments were carried out and the reburial programme was set up and implemented.

Reburial of archaeological remains: was this a choice?

Protection by reburial with carefully placed fill and full conservation on the materials is one of a number of options available for reducing the impact of the deterioration processes that start with archaeological excavation and are caused by natural and human factors. Its principal advantage is that it leaves the remains untouched beneath a protective layer of fill so that they will be available to future generations for study. The principal disadvantage is that not all the remains will be accessible (Shilston and Fletcher nd).

Today conservation *in situ* is an obligation for the scientific community, which has to assume the responsibility not only for discovery, study and complete documentation of the material remains of the past, but also for their conservation as a unitary context for the future. Monuments or remains, which pertain to a unique archaeological context, cannot be treated like individual artefacts; otherwise, one pays the price of the dispersal of homogenous contexts or, worse still, the destruction of potential information. Only examining all the stratigraphic components of an archaeological site, together with all the elements that make up the external context (landscape) as an integral whole, now and in the future, can provide complete and comprehensible historic information. A basic condition then is that these contexts should be preserved for the future so that interpretations can be confirmed or contested and changed.

In the context of this project the dam is one event in the lifetime of the site. The dam is considered economically viable for the next 50–80 years. After that time, the progressive obliteration of the lake by the sediment transported downstream by the Euphrates will render the dam's hydroelectric power plant inoperative. This most probably will not coincide with the end of the lake itself, as it will be used for other purposes for some time, but certainly the life span of the lake has to be measured in centuries at the most, rather than in thousands of years. Thus the dam does not represent complete destruction, but just one of the many events in the 2300 years of the site's lifetime.

At the same time, the act of removing single elements from their original context (as occurred in Zeugma for the mosaics and frescos of area A) without protecting what is left *in situ* risks being a culturally limited and scientifically dangerous operation, which can be the origin of erroneous historic interpretations and irreparable damage. Such an operation is seen to justify the concept that one thing must be saved and another might be left to destruction according to a subjective judgment, which is debatable. This leaves to chance the survival of anything that remains *in situ* and destroys information and original materials in a single operation with no return.

Therefore our decision was to document as much as possible, recover what would suffer damage if left on site and prepare the rest for a long sleep, awaiting future generations of archaeologists. In practice, this meant that during the excavations of July to October 2000, 240m^2 of mosaics and 40m^2 of wall paintings were consolidated and protected on site, as opposed to 60m^2 of mosaics that were lifted because they would have been destroyed by the flooding. These figures contrast with the previous spring when 700m^2 mosaics and more than 100m^2 of wall paintings were lifted in only a couple of months.

The strategy applied at Zeugma is based essentially on the following principles: cleaning and integral documentation of all the archaeological structures; removal of the small finds and those structures whose condition did not guarantee their survival after inundation; conservation *in situ*, without disassembly or destruction of complete contexts, of all the structures able to receive the treatment of protection and safeguard before reburial and inundation; protection of the original surfaces by the application of lime whitewashes and hydraulic, lime-based mortars, copying ancient techniques that have, on various occasions, proved their effectiveness in terms of protection; reburial of the protected and consolidated structures; putting in place a barrier to protect the reburial against wave action and implementation of a shore maintenance programme. Such a strategy should preserve the archaeological context as a whole. In addition, once under water the environmental conditions should be stable in terms of thermo-hygrometric and chemico-physical conditions and there will be no risk of unauthorised excavation. Finally, and most importantly, no destructive excavations or restoration will have been undertaken.

The project

The archaeological project was a full-scale, massive excavation campaign featuring, at its peak, 100 archaeologists, about 300 workers, 3 hydro-mechanical excavators and various tractors and trolleys. An area of 8700m^2 was excavated in 20 different trenches. This took about 6000 archaeologist's working days. The entire conservation process took 1000 working days for the operations on site during excavation, 1800 working days in the laboratory on finds and mosaics lifted from the site and 600 working days for maintenance of the shoreline.

This obviously has a significant impact on the conservation strategy in the field and afterwards. To understand the choices made on site we should briefly analyse the various stages of possible deterioration and the main risk factors involved during these stages. There were four different stages of ambient conditions for the archaeological remains during the project: excavation ie at the moment of excavation/uncovering; open air exposure ie after being excavated; inundation ie during the flooding; submersion ie after the flooding.

When an archaeological site is buried, it is generally preserved for centuries without suffering substantial alterations (de Guichen 1984). This is because without air and light, the stable conditions slow down the natural ageing process and the tendency of all materials to return to their primitive chemical state. This is why the majority of archaeological excavations reveal objects, structures and decorated surfaces still in a reasonable state of preservation and legibility. All materials have a favourable environment for preservation and will react in a different manner to external stress. In general, waterlogged or desiccated organic materials such as leather and wood are most at risk and liable to disintegrate immediately after excavation, but inorganic materials such as metal and stone can also suffer irreparable damage. In fact, all excavated objects, whatever the material, are fragile and potentially liable to physical damage. In general, the greatest danger for all types of material occurs during excavation, when there is a change in the equilibrium of the environmental and conservation parameters. It is now recognised that stable environmental conditions are more important than the environmental parameters themselves. In fact the same objects can survive in very different ambient conditions such as desert tombs, permafrost tundra burials, muddy riverbeds or salt-water environments; so long as they reach a stable condition, they may last for millennia.

Once exposed, however, the materials suddenly encounter changes in atmospheric conditions (temperature, humidity and light) which are drastically different from their previous state. A process of interaction with the surrounding environment then begins so the material can reach a new equilibrium, which in the majority of cases means a change in the physical and chemical structure of the material itself. Once unearthed, apart from the processes of chemical and physical deterioration, the materials might be subject to other kinds of damage: mechanical damage during excavation or handling, neglect, theft and vandalism, poor or erroneous restoration, wind erosion or water damage, rain, earthquakes and floods, biological deterioration caused by plants, insects or animals, etc.

To preserve archaeological structures and objects, conservation treatments must be carried out during and immediately following excavation. A considerable investment of resources and energy is required; from the moment archaeological excavation begins and the recovery of structures and finds starts, the ratio in terms of energy required for digging and then for preserving is 1:20 (Nardi 1986). From a conservation perspective the less we excavate the more we preserve for the future.

In the first stage, excavation, when the archaeological remains are unearthed, they pass from a stable microclimate to a new, continuously changing outdoor environment. This means a dramatic temperature change and a rapid loss of moisture content with risks of loss of surface layers, especially if decorated, formation of tenacious insoluble surface deposits, cracks and deformation in fragile organic materials, cracks and rapid, massive corrosion through oxidation and hygrometric variations in metal objects, etc. Furthermore there were the 'usual' risks of loss of information through the fragility of many ancient artefacts and material evidence (crumbling of wall plasters and wall paintings caused by detachment from their original support, loss of mosaic tesserae in areas with lacunae and/or disruption, crumbling of fragile objects during lifting, etc).

In the second stage, open air exposure, there are still microclimatic risks due to exposure to an outdoor environment, but this was a relatively brief period (the project started in July and the excavation area was completely flooded by 15 October) with relatively stable conditions, and so was not the main risk. A much higher risk was the anthropogenic damage to the structures – more through neglect than through vandalism. It is human nature to find the easiest and/or straightest route from one point to another. This route might or might not coincide with the path laid out – a phenomenon we all have observed in public parks, near subway stations or other points of public interest. We normally accept that a certain amount of damage will be sustained by establishing such a natural route. An archaeological excavation is no exception to this rule. It is a rather common phenomenon that the level of awareness drops significantly once the excavation work is finished and the drawings have been made. The sense of abandonment is tangible and especially fragile structures on or near 'useful' pathways are at the utmost risk of damage in a very short space of time.

The third stage, inundation, carries a very high risk of complete destruction of the archaeological remains, as wave action can undermine the structures themselves, leading to complete caving in of walls and floors, washing out of earth or stone walls, the detachment of plasters and frescos, disruption of tesserae and tumbling pillars, hygrometric shock caused by rapid soaking of materials with water, etc.

The fourth stage, submersion, is possibly of least concern from a conservation point of view, as the materials will be once more in a stable underwater environment, which as we know has a much higher ability to create a suitable conservation environment than any outdoor situation. Given water's high density, the temperature variations will be minimal and there will be no hygrometric variations. The water movement can be considered non-existent as the flow speed of 500m^3 per second will be distributed over the whole section of the reservoir – approximately 40,000m^2 resulting in a nominal speed of 1.25cm/s – and most probably concentrated in the central area, both in longitude and altitude of the section. There will be an area 2.0m high beneath the maximum level that is affected by continuous fluctuation through the normal exercise of the dam. Anything in this area obviously runs a very high risk of complete destruction.

Excavation

The rapid evaporation of the moisture content of the unearthed materials had to be avoided at all costs. Crystallisation of soluble salts should not occur on the original surfaces, but on a sacrificial layer. Endangered structures or decorative surfaces should be consolidated in parallel with the excavation. Small finds should not be exposed to direct sunlight, and pass slowly to their new moisture content in a controlled environment. All highly perishable artefacts should receive treatment as soon as possible. Incorrect or inexperienced handling of delicate artefacts must be avoided.

Open air exposure

Access to the excavated areas had to be limited. Delicate or perishable structures had to be secured and pathways near them blocked. Small finds of all kinds had to be moved to the finds store on a daily basis.

Inundation

The archaeological investigations had to be completed in time to allow the completion of the *in situ* conservation treatments and the reburial programme. The conservation treatments had to be finished on time, so that the materials used had the time to set and harden. All fragile structures had to have a protective layer in order to mitigate the transition phase from humid to water-saturated status inside the materials, and a sufficient adherence between the substrate and the artefact had to be secured. All excavated areas had to be backfilled and protected efficiently from erosion by wave action.

Submersion

The protective measures had to be controlled and where necessary additional work done in order to stabilise the shoreline in the proximity of important archaeological remains just above the water line or in the fluctuation level.

The strategy

Based on the principle of minimum intervention, the methods used were as quick and simple as possible, giving the maximum outcome in the least time. In the first stage, the main activities were concentrated on risk mitigation for the structures and objects, and secondly on securing the already treated artefacts and areas and preparing them for inundation. Obviously these stages occurred together as the lower parts of the site started to be reached by the water, while in the higher parts the excavation still proceeded.

All priorities were discussed with the archaeologists and a division of responsibilities was agreed. The basic measures of find protection were carried out by members of the archaeological team, while the conservation professionals looked after the more complex, technical operations and assisted with advice. Handouts, written information material, lectures, personal contact and collaboration gave an overall awareness of basic preventive conservation measures. Trench supervisors acted as the key contact for discussing programmes and priorities.

The conservation programme

For the sake of clarity the activities on the site will be presented by category rather than in chronological order. We would like to stress that careful planning and timing of each individual operation was crucial to the whole operation, as in the end the rising water would not pause for any reason. So the first step was to set up an interactive timetable for all the trenches, where every single artefact in need of treatment was included. Needless to say, these timetables needed continuous updating as the excavation evolved. Beyond the times for treatment of individual artefacts, the timetables contained a deadline for the archaeological investigations in order to accommodate time for backfilling prior to the flooding.

The conservation programme was carried for the entire area of excavation and was divided into seven operations: first aid on finds and conservation treatment (safe excavation, stabilisation and recovery of fragile small finds); lifting mosaics and architectural elements; documentation using moulds for graffiti and inscriptions; *in situ* conservation and protection of mosaics, wall paintings and wall plaster; temporary preventive protection measures; planning and management of the reburial programme of the excavated areas; post-excavation maintenance of the shoreline.

First aid on finds and conservation treatment

During the Zeugma 2000 excavations, thousands of small finds, mostly ceramic, bronze, iron, glass and some organic materials (ivory, wood, bone, etc), were retrieved. As is widely recognised, most of these items suffer from dramatic microclimatic changes when passing from buried damp conditions to hot, dry open-air exposure. Furthermore, due to the need for archaeological drawing, photography and documentation they often have to stay in place for a prolonged period of time. This means exposure to direct sunlight with all the negative effects that entails for damp objects, including loss of surface due to mechanical stress caused by the rapid volume increase of the water contained and deformation and/or cracks in organic materials such as ivory or bone.

The main goal of first aid for small finds on site was to minimise the amount of damage suffered by the objects between their initial exposure and their later recovery in the conservation laboratory, and to save the maximum amount of information from those objects. In general, there were two categories of objects: those that were sufficiently stable to be picked up and stored without additional support and secondly those that were too heavy or too fragile to be lifted without prior stabilisation and/or additional support.

For the first category some very basic – simple, but efficient – techniques were adopted to mitigate the impact of changing climatic conditions. Direct exposure to bright sunlight was avoided by storing the recovered objects in ready-made geo-textile bags. Iron and bronze objects were put in zip-lock polyethylene plastic bags with holes punched in them to permit the microclimatic conditions to migrate slowly from the previous state to the present one. The holes were necessary to avoid the formation of condensation inside the rather small volume of the bags. These punched plastic bags were then also put into geotextile bags in order to avoid them being temporarily stored in the shadow of some wall, only to finish in bright sunlight a couple of hours later. These operations were carried out by the archaeologists.

In the case of large or fragile objects, a block lift was carried out by the conservation team. The methods and materials used varied according to the kind of artefact recovered. For every block lift, a recovery record sheet was made to keep track of all materials used. This was necessary to identify the materials that might have come into direct contact with the artefacts and therefore might interfere with any future archaeological analysis. Furthermore, a sketch was made to record the exact position of all lifted fragments. All artefacts were moved to the finds store on a daily basis for data processing and further conservation treatment.

Throughout it has been our policy not to leave an overcrowded finds store to an uncertain future at the end of the excavations. Therefore, during and after the excavation most of the excavated metal objects were treated. Stabilisation of the objects had priority over any other considerations. Bronze objects were cleaned and then washed in demineralised water to remove all active chlorides. Iron objects were cleaned, micro-sandblasted to remove loose corrosion layers and then treated with a professional rust converter based on tannic acid.

The finds store was organised so that all fragile artefacts were preserved in polyethylene boxes and the microenvironment was stabilised by using silica gel with a colour index indicator. The state of the silica gel and the objects is still monitored on a weekly basis. A total of 1200 metal objects and 200 worked stones were treated. This took eight months amounting to a total of 1200 conservation working days.

Lifting of threatened mosaics and structures

Some structures and fixed artefacts which would not have survived the flooding were removed during the excavation. Three figured mosaics and a geometrical one were removed from site (totalling 75m^2), as well as a stone fountain and some stone architectural pieces. The latter were protected with geotextile before being moved by heavy mechanical gear under the supervision of a conservator who advised on the positioning of the ropes (and not chains).

The lifting of the mosaics was a much more complex operation. The first step was accurate cleaning of the whole surface. Special care was taken to remove the deposits of soluble salts on the surface, as these hamper the adherence of the canvas covering needed to secure the tesserae. Then the cutting lines were determined, applying as general rule respect for the design of the mosaic. Cutting of figured scenes was avoided. The documentation was carried out using transparent paper. This proved to be more accurate than acetate or similar synthetic materials, as these shrivelled under direct sunlight and temperatures above 45°C. Once finished, the paper rolls were scanned and transferred to computer aided design software (CAD).

While this operation was underway, the surface of the mosaics was glued over with a double canvas layer, applied crosswise. The first layer was slightly less tightly woven than the second one to improve adherence to the individual tesserae. Both layers were washed and trimmed before being applied in order to reduce differential shrinkage when the glue was setting. A polyvinyl glue was used, it was brushed on the surface and the canvas was laid on top. The canvas was then beaten with sturdy hand brushes to squeeze it into the joints between the tesserae. While the first layer was being put in place, the cutting lines were drawn on the canvas. After the second canvas layer had been applied in the same way, these lines were copied to the second layer as a guide for cutting.

The cutting was carried out with a modified angular cutter. Instead of the normal abrasive cutting disks, a diamond-studded glass-cutting blade was used. The resulting cuts were approximately 1mm wide.

The lifting process began at a free side, normally the entrance and long iron chisels were driven underneath the mosaic. The dividing point was between the preparatory layer and the final mortar layer that holds the tesserae. Once completely detached, the fragment was turned upside down onto a ready-made wooden support covered with geotextile. Every fragment was immediately identified by its number and orientation. At the end of the day, all the material was moved to the laboratory for restoration treatment.

As with the small finds, all four of the lifted mosaics were cleaned, reset and reassembled, after the excavation. At the time of writing, one of them is already on display at the Gaziantep museum; the others are in storage ready for display.

Documentation using moulds for graffiti and inscriptions

Latex moulds were taken of inscriptions and graffiti that were to remain *in situ* or were too heavy or fragile to be easily handled. The first step in this process was careful cleaning of the original, whether it was an inscription, a stone carving or graffiti on wall plaster. The purpose was to remove any surface dirt so that all three-dimensional features would be clearly recognisable; this involved removing soil, sand, small stones, etc from within letters and incisions. The next step was to check for any fragile parts in need of consolidation, as they could be damaged by the process itself. These were consolidated before the mould was made. Then the latex was spread with a soft brush over the original surface, with special care taken to fill all letters and/or incisions. Latex sets rapidly, forming a thin layer. Depending on the dimensions and kind of inscription, further layers were applied to reinforce the mould and to guarantee an optimum rendering of the three-dimensional features of the original. Once completely set, the mould was carefully peeled off. The mould was a perfect negative of the original, and was then scanned. The latex moulds proved highly useful for inscriptions, because they can be read on a light table – even the right way round – and they are readily accessible for further consultation by specialists.

In situ conservation and protection of mosaics, wall paintings and wall plasters

During excavation, conservation advice was provided to the archaeologists to prevent damage to surfaces. In particular, when mosaics and wall paintings were found, the excavators were asked to leave a 100mm layer of soil on the surface in order to be able to mitigate the effects of rapid exposure to atmospheric conditions and to avoid the formation of layers of insoluble salts and soil clusters on the surface due to crystallising salts (Costanzi Cobau 1986; Costanzi Cobau 1990). This layer was then removed by the conservators during the cleaning operations.

Any decorated surfaces were treated and cleaned by the conservation team in order to avoid damage and to prevent the formation of tenacious surface deposits. The proper, rapid treatment of the surfaces immediately after uncovering them also meant that the original colours showed up without being blurred by dust or deposits. Inscriptions, graffiti and any other details were clearly recognisable. This permitted more accurate documentation and interpretation.

It took 1000 working days control the microclimatic during cleaning: remove soil layers on the surface and clean; document; edge, consolidate detached areas and remove roots; protect the surface prior to reburial.

Microclimatic control during cleaning

During cleaning, the surfaces under treatment were kept shaded in order to slow down the evaporation process. This was necessary because of the climatic conditions, which were unsuitable for uncovering such surfaces. The highest risk was the evaporation of the humidity contained in the newly unearthed layers, with the consequent crystallisation of soluble salts and all its devastating effects.

Removal of soil layer on the surface and cleaning

In order to avoid the complete exposure of the original surfaces for more than 30 minutes, the removal of the soil layer on the surface and the final cleaning of the parts just uncovered were carried out at the same time, proceeding in layers of about 200mm. This technique permitted the control of the times of exposure to the air and the ambient heat, producing a controlled drying-out process of the original structures. In this way, formation of insoluble surface deposits was avoided prior to their complete removal and enabled the partial removal of the soluble salts.

To soften the earth, which dried out rapidly through exposure to the air and the high ambient temperatures, prior to removal, vaporised water was applied with hand-held sprayers. This was done in order not to damage the original layers, which often were softer than the material being removed. The removal of the soil was accomplished with the help of scalpels, up to the last layer directly in contact with the original surfaces. The subsequent cleaning was carried out with brushes, toothbrushes, cotton buffers and a slightly biocidal solution of water and 2% Neodesogen.

After cleaning, as a precaution in the case of the wall paintings, drying the surface was controlled by the application of tissue paper moistened with water, where the remaining soluble salts could crystallise.

Documentation

Complete graphic and photographic documentation was done after cleaning. A same size drawing was produced on sheets of polyethylene by tracing all the lines of the figures and the geometric patterns. All the drawings and photographs were processed using CAD to produce a detailed illustration for documentation of the conservation process. All the data about the general condition of the floors, the areas of lacunae or degradation were traced on the base plans.

Edging, consolidation of detached areas and removal of roots

This operation was carried out in order to re-establish cohesion between individual layers and between the plaster and the wall or between loose tesserae. It was accomplished by applying a consolidating material, which, by penetrating, improved the physical characteristics of the plasters. In the case of frescos and plaster in imminent danger of crumbling, the excavation was carried out by the conservators in parallel with consolidation. Where necessary in the case of wall paintings, temporary props were built. Areas at risk were secured with cotton gauze linings applied with high concentration acrylic resin (15% acryloid in acetone), while exposed areas were consolidated.

The normal consolidation procedure was mechanical removal of soil deposits and roots along the borders or in lacunae using scalpels and a vacuum cleaner; rinsing of detached areas with a 1:1 solution of water and ethyl alcohol with complete removal of soil deposits between the different layers and between loose tesserae; infiltration with the help of hypodermic needles of a mixture of hydraulic, lime-based consolidant (hydraulic lime/Lafarge and sieved brick powder in a 1:1 mixture); mending cracks along the borders with a lime-based mortar (hydraulic lime/Lafarge, lime, sieved brick powder, calcareous stone dust in a ratio 0.5:0.5:1:1).

Surface protection prior to reburial

A protective layer of hydraulic lime-based mortar was applied on the consolidated and cleaned surface. This new surface, 20mm thick on wall paintings and 50mm thick on mosaics, will protect the original from the mechanical stress caused by the water and will serve as a sacrificial layer. Moreover, the hydraulic properties of the mortar will slow down the saturation process and prevent an immediate soaking of the underlying materials. At the same time, the hydraulic properties of the mortar will lead to a continuous hardening once submerged and therefore to a progressive improvement of the resistance to mechanical stress over time. In this way any stress will be mitigated by the new layer protecting the original surface. The choice of the protective materials was determined by their characteristics of high resistance, outstanding mechanical and hydraulic qualities, long-term durability and complete chemical and physical compatibility with the original surfaces.

The application of the protective layer comprised a lime whitewash brushed on in direct contact with the surface, followed by the protective mortar layer. The lime whitewash separates the original from the protective mortar layer. In the future this will allow easier removal of the latter (Figs 21.3–21.4). Once set up, the protective layer, made of hydraulic lime/Lafarge, lime, brick powder and stone dust in a 0.5:0.5:1.5:0.5 mixture, was applied. This layer was extensively worked in to achieve a good adherence to the underlying surface and to improve the setting process of the aerial components of the mixture.

Temporary preventive protection measures

Together with the trench supervisors, the safest routes – for both people and archaeological remains – to various areas inside the trench were determined. Where necessary, additional paths, especially for wheelbarrows, were established outside the excavation area. Some endangered areas were declared off-limits until the conservation treatments were finished. In order to protect the exposed structures during conservation treatments, the working areas were cordoned off with ropes and wooden poles. Other areas were permanently sealed off and only authorised personnel were admitted. 'Natural' passages next to delicate artefacts were blocked, and alternative passages were staked out and reinforced with wooden planks. When, in certain areas, the only access to an area had to pass by or over a delicate artefact, special care was taken to secure and protect the artefact in question. This meant edging of unprotected or damaged borders of mosaics, wall paintings and plasters, closing lacunae in mosaics that were subject to continuous passage, protecting borders or passages with wooden planks or covering already cleaned mosaics in order to protect them from soil and dust.

Fig 21.3 *Applying the lime whitewash to a mosaic*

Fig 21.4 *Covering a mosaic with hydraulic, lime based mortar*

Planning and management of the reburial programme for the excavated areas

At the end of the archaeological investigation, all the trenches were backfilled following a strategy designed with regard to the constraints dictated by the timetable (general calendar, availability of machinery and materials) and site logistics (accessibility, topography, location of trenches). The techniques and materials used during the reburial were chosen according to their availability locally, the conservation needs of the archaeological remains left *in situ* and the topographic features where the trenches were sited.

As outlined above, it was clear right from the beginning that all trenches had to be backfilled to ensure mitigation of environmental changes during inundation of the wall plasters, wall paintings and mosaics and to prevent the collapse of the archaeological structures unearthed.

The building techniques encountered during the excavations at Zeugma are basically of four types: massive cut stone wall buildings; constructions made of massive stone pillars alternating with loose stone/mud infill; mudbrick walls with a high percentage of gravel and loose stone constructions using a poor lime mortar as a binding material. The foundations are often poor and sometimes completely lacking; fortunately most of the more complex structures are sitting directly on the bedrock. The masonry is made of the local bedrock which is a very soft mudlime stone.

Due to the nature of the materials and the construction techniques used, all of these structures would be in extreme danger if exposed to direct wave action. The water would rapidly wash out the mud used as building material or undermine the structures not built on bedrock and all standing structures would collapse. Even the stone structures would be subject to a considerable amount of erosion, given the softness of the local stone.

The Zeugma site is sited in a wide basin, just a few hundred metres away from the Birecik Dam, which produces particularly intense wave action on windy days. This leads to considerable erosion of the soil covering the archaeological remains which could reach the bedrock especially on the steep slopes facing the prevailing wind (north-west) and in the valleys. Several surveys have demonstrated that archaeological remains in these locations are particularly scarce. The bedrock was normally covered with less than 1.0m of soil. Most of the existing terrace walls already seemed to be severely damaged; it remains unknown whether this was due to agricultural activity on the site, erosion or natural disasters. The archaeological remains discovered in these locations are mainly in small natural terraces or rock-cut rooms.

Comparison of these observations with the preliminary studies led to the following considerations: the critical moment for the archaeological structures is during and immediately after the inundation; soil coverage alone would not be sufficient in the excavation areas which were exposed to wave action; coverage of the excavated areas with geotextile could be functionally effective but would most probably attract looters in the case of exceptional low water conditions; the structures needed a 'neat packaging' to keep the mud used as building material in their walls from being dissolved when they passed from humid to water-saturated conditions. Therefore, it was decided to build up different layers of fill, from fine and soft materials being used in contact with the archaeological remains (soil from 0–5mm with some major inclusions) to gradually bigger and heavier material (sand 0.5–5mm, gravel 10–30mm, stones 50–200mm) which would be exposed to wave action while the water was rising. The exact sequence depended on the topographical features of the individual 'trench environments' (Fig 21.5)

While in the gently descending valleys, a simple soil/sand filling was sufficient; more exposed or steeper areas needed heavier sealing materials to withstand the wave action during inundation. Soil and sand were still used to pack the structures tightly and to avoid damage to the archaeological remains, but gravel and stones were used to hold this filling in place and resist the wave action.

The soil and sand in direct contact with the structures was humidified and compressed in order to avoid shrinkage when soaked with water. Special care was taken to fill all cisterns, covered rooms and areas to prevent cave-ins and settling. These operations were carried out using wheelbarrows and shovels whilst the remaining backfill operation was carried out with mechanical diggers wherever possible. Occasionally, in some particularly difficult terrain situations, the materials were transported as close as possible to the trenches and then distributed manually. Once all the structures in the trench were covered with soil/sand, the backfill was sealed with gravel and stones. Generally, the soil fill was covered with at least a further 0.50m on top. In the case of gentle slopes, the

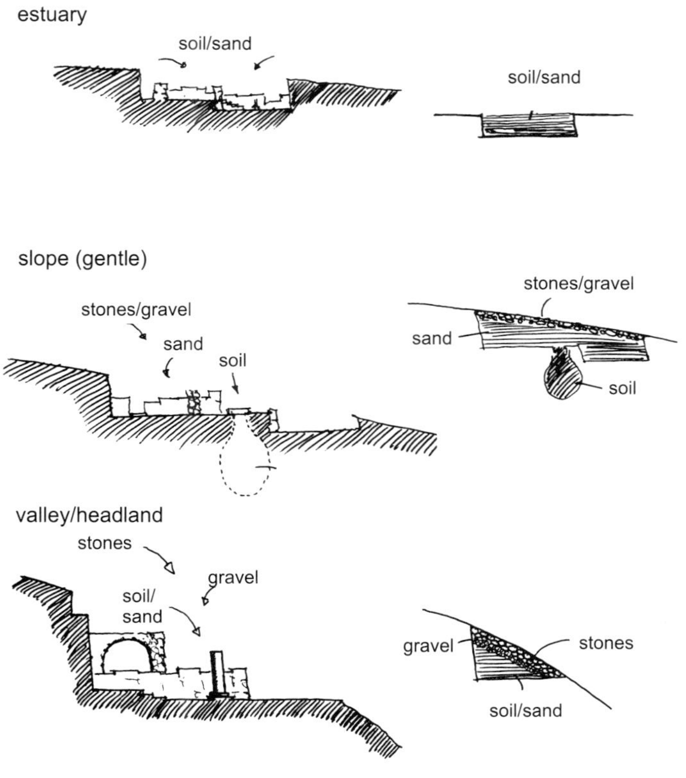

Fig 21.5 The different topographical features on site with the different layers of fill

gravel and the stones were mixed, while on steep hillsides the stones were distributed as a separate layer above the gravel. This had the effect of a natural filter where the bigger pieces held the smaller particles in place, keeping them from being washed away. The porous structure of the stone covering, similar to a sponge, helps to disperse the wave energy by forming numerous eddies.

A total of 8700m^2 spread over 20 areas was reburied. The backfilling volume was of 10,500 m^3, including 630 truckloads of sand (1400m^3), gravel (2200m^3) and stones (2700m^3) as well as the excavated soil. This enormous quantity of additional material was generously provided by the dam company - a silent sponsorship of the utmost importance for the success of the backfilling programme.

Post-excavation site protection and shoreline maintenance programme

The first post-excavation season at Zeugma has shown that due to the particular juxtaposition of the Birecik dam basin and the prevailing winds, the shoreline of the former Zeugma 2000 excavation is exposed to major wave activity especially on windy days. The winds are channelled by the Euphrates river canyon and then build up even further once they reach the enlargement of the basin in the 2–3 km before the dam. As the prevailing wind direction coincides with the inlet to the canyon, the waves can splash up to 3.0m. This results in major erosion activity along the exposed shoreline, especially the parts directly facing the oncoming waves.

A few months after the end of the excavation, there was clear evidence that in some areas the protection measures implemented during the excavation alone could not withstand the continued erosion in the fluctuation zone (283–5m). The amount of energy built up by the waves was great enough to remove and/or disrupt the heavy pebble covering (average stone size between 0.10–0.20m diameter) put in place. Furthermore, the wave action continued to disturb the unexcavated areas of the shore, continually bringing up archaeological remains. In some places, artefacts, such as wall plasters and mosaic floors protected during the excavation, were uncovered. While the protective lime-based mortar layer resisted such threats quite well, the structures supporting the artefacts had little or no resistance to the erosion activity on the waterfront. The mud stone walls (ashlar) started crumbling away and the floors were often laid on unreinforced ground undermined by the water. It was clear that further protection measures were necessary to minimise the damage to the archaeological layers in the water fluctuation zone.

To improve the efficiency of the protection system established in October 2000, rows of woven plastic (nylon) bags filled with gravel (10, 20 or 30mm in diameter) were arranged parallel to the shoreline. These were piled onto each other between 2–4 bags high depending on the degree of slope of the shore. The nylon bags have a very limited durability in outdoor conditions as they suffer from UV radiation, so the uppermost row of bags was filled with gravel mixed with 10% cement. Behind this barrier, a fill of heavy pebbles (100–200mm diameter) was put in up to the high water mark under storm conditions. Where the shoreline configuration allowed, a double row of bags was arranged about 2–5m apart. The space between the two bag walls was filled with pebbles (Fig 21.6)

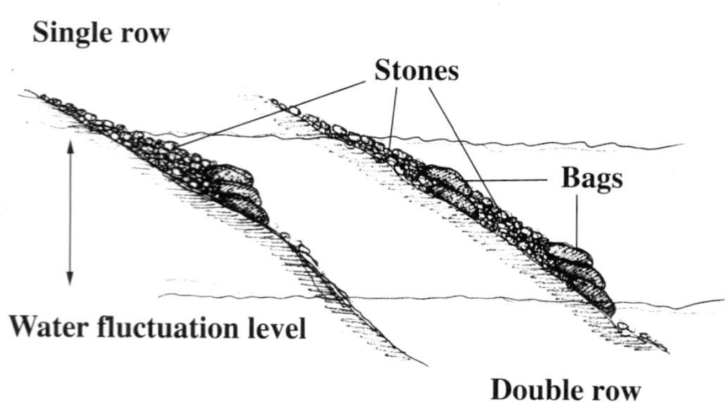

Fig 21.6 *A scheme showing the disposal of the gravel bags along the shoreline*

After one month, it was clear that the plastic bags were not suitable, as they had deteriorated very quickly, distributing their contents in the water and polluting the environment. It was therefore decided to use local traditional sacks made of completely biodegradable canvas. The sacks needed to be durable for long enough for the gravel/cement mixture to dry in its final location and to its final shape. This material was put in place using local manpower aided by tractors equipped with dump-load trailers under the supervision of conservation professionals. A total of 6000 bags and 100 truckloads – equivalent to 1000m^3 – of stones were used.

To maintain an area of 8000m^2 (the shoreline) in good condition during the first year after excavation took 600 working days plus materials and machinery, equivalent to US $20,000, that is $2.50 per m^2 per year. The maintenance will continue until evidence shows that a new equilibrium has been reached. The need for continued maintenance could be seen as evidence that the reburial has not been successful, but this kind of comment reflects a widespread attitude which views conservation as a one-off job. In practice the profession is becoming more conscious of the need for maintenance being an integral part of conservation. No one would doubt the necessity to maintain the reinforcements of a medieval castle on the coast, in a similar way the Zeugma site has become an excavation on the coast of the Birecik Dam Lake.

Conclusions

The Zeugma Archaeological Project 2000 was a huge excavation, organised and carried out in a very short time with multiple needs for conservation during and after the excavation; the ambient conditions were extremely harsh and unfavourable both for the operators and for the archaeological remains; moreover the water, rising according to an inexorable schedule, permitted no delays. The methodological approach chosen allows no access to the submerged area or to any artefact contained therein. Some particularly exposed, unexcavated areas on the steep slopes facing north-west have been badly damaged by waves on windy days. And finally the area within the 2.0m deep fluctuation level below the high water mark runs a high risk of complete destruction over time.

However, on the positive side we can say that all the excavated areas have been protected. Finds and threatened structures have been removed and conserved. The techniques used and the materials applied are inoffensive and completely compatible with the archaeological remains. The unitary contexts have been almost fully respected and the operations carried out are completely reversible. Archaeological and technical information has been collected and circulated, and at the same time the material evidence (currently underwater) will be preserved for future generations.

Despite the number of difficulties, an impressive area has been protected and has a good chance of surviving the coming centuries unharmed in a safe underwater 'Dornröschen Schlaf'. It is likely that a certain amount of minor damage may occur to some artefacts due to settling, dissolving mudbrick fills or similar. But it would be naïve to assume that an excavation of this size could not be undertaken without some damage, not to mention the extent of destruction that would have ensued had the wall paintings and plasters, mosaics, columns and paved areas been completely lifted. We would have been left with a similar choice in having to select the most spectacular pieces out of their context and exhibit them like rare gems. The mere fact that in an extreme situation our 'primitive' treasure hunter instinct did not overcome a profoundly meditated response is a success: the archaeological structures of Zeugma have returned, after protective measures have been carried out, to a stable underwater environment, which will protect them from the dangers of exposure. Accurate, archaeological documentation will enable them to be studied and the results disseminated.

Our solution was just one of several possible approaches; others would have been possible and have been proposed and discussed, with their advantages and disadvantages. In our opinion we made the best judgement possible under the given circumstances. Many people have disagreed with our choice and most probably many more will continue to do so. If nothing else this project has set a landmark, which will serve to develop any further discussions on the subject in the future.

Bibliography

Costanzi Cobau, A, 1986 Excavated wall plasters: conservation problems, in *Preventive measures during excavation and site protection*, International Centre for the Study of the Preservation and the Restoration of Cultural Property rep, 103–8, Rome

Costanzi Cobau, A, 1990 The Roman Forum. On-site conservation of floor surface during excavation, in *Mosaicos 5: conservación* in situ, International Centre for the Study of the Preservation and the Restoration of Cultural Property rep, 127–35, Palencia

Guichen, de G, 1984 Object interred, object disinterred, in *Conservation on archaeological excavations*, International Centre for the Study of the Preservation and the Restoration of Cultural Property rep, 21–9, Rome

Kennedy, D L, 1998 *The twin towers of Zeugma on the Euphrates: rescue work and historical studies*, J Roman Archaeol Sup Ser 27

Nardi, R, 1986 Conservation on the excavation: the Crypt of Balbus in Rome, in *Preventive measures during excavation and site protection*, International Centre for the Study of the Preservation and the Restoration of Cultural Property rep, 91–7, Rome

Shilston, D T, and Fletcher, S L, nd [1998] Geotechnical engineering for the *in situ* preservation of archaeological remains, in *Preserving archaeological remains* in situ: *proceedings of the conference of 1st–3rd April* 1996 (eds M Corfield, P Hinton, T Nixon and M Pollard), 60–5, London

22

Measuring the burial environment at archaeological sites to judge the *in situ* conservation potential: a Planarch project to facilitate planning decisions

ALEXANDER SMIT

Alexander Smit
Paleo Terra
Lamarckhof 10-I
NL – 1098 TK Amsterdam, The Netherlands

Abstract

Over the course of one year a set of parameters likely to contribute to the conservation potential of the burial environment was measured at 12 different sites at Voorne-Putten. The research is part of the Dutch National Service for Archaeological Heritage regional project at Voorne-Putten, made possible by the EU Interreg IIc program PLANARCH. The parameters measured are the groundwater level, pH, reduction–oxidation potential, oxygen content, total salt content, temperature and nitrate and sulphate content (all measured via monitoring dipwells), and also the reduction–oxidation potential via *in situ* platinum electrodes. In addition as a part of the Voorne-Putten project, the archaeological quality of several sites was measured by studying botanical remains, pieces of bone and the microstructure of the soil and soil layers. This latter part of the project is not discussed here.

The first results lead to the (preliminary) conclusion that the reduction–oxidation potential of the burial environment is a key factor. It seems that the reduction–oxidation potential is best measured using *in situ* electrodes, but using a dipwell is much cheaper and possible when strict procedures are followed.

It is suggested that in future monitoring work, at least the reduction–oxidation potential and pH of the burial environment are monitored. Also, the survival of different archaeological materials (eg at a scale of 1 to 5) should be recorded. When this is done at a sufficiently large number of sites it should be possible to point out favourable burial environments with greater reliability, even before degradation processes are fully understood.

Introduction

In recent years archaeological policy is changing from excavation of archaeological remains towards conserving these remains *in situ*. However, it is not always clear whether the remains will actually be preserved *in situ*. Therefore, an active policy of protecting and preserving archaeological remains in the soil almost automatically leads to the consequence of monitoring these remains (Kars 1997).

This monitoring of archaeological sites can consist of two different aspects, firstly monitoring the quality of the archaeological remains themselves and secondly monitoring of the burial environment.

The results of monitoring the quality of archaeological remains can be used in a direct way; so that when it is monitored through the years the results on an annual basis can have direct bearing on the decision whether to excavate or to leave the remains *in situ*. Some aspects which can be monitored are the quality of seeds and pollen, the histology of bone or the stratification of the soil via micromorphological analysis. Although this kind of archaeological monitoring does not require knowledge of the degradation processes of the archaeological remains, some disadvantages will result from the lack of understanding of such processes.

Without understanding the damaging processes going on in the soil, it is hard to predict the timescale of the deterioration of the remains. Also, it is rarely possible to take measures to prevent further deterioration. A third disadvantage of monitoring the quality of archaeological remains is the diversity of results which can be expected at short distances within a site, so that it is hard to tell whether a well-preserved piece of bone is representative of the whole site. In order to obtain a reliable picture of a complete site, it would be necessary to sample quite large amounts of archaeological materials; but when this is done annually, it hardly constitutes non-destructive research.

Research goal

The only way to overcome the disadvantages mentioned above is to understand the degradation processes of the archaeological remains. Once it is possible to measure the state of the burial environment and to understand the degradation processes going on in such an environment, the monitoring of archaeological sites need not be so dependent on archaeological material.

The first research goal is to find out which burial environments have good preservation potential. This can be found by comparing the characteristics of a sufficiently large number of burial environments. Not only should the soil chemical characteristics of the burial environments be monitored, but also the quality of archaeological remains. The final objective is to identify a set of environmental parameters which should be monitored in order to predict the rate of degradation of different archaeological remains. This, however, requires detailed knowledge of the degradation processes, something which still remains to be achieved.

Planarch opportunity

In 2000, the Dutch National Service for Archaeological Heritage started a regional project at Voorne-Putten as part of the EU Interreg IIc program Planarch. One of the goals of the Planarch project is to improve expertise to reduce the effects of town and country planning on archaeology. As part of this project, the archaeological quality of several sites is measured by studying botanical remains, pieces of bone and the microstructure of the soil and soil layers.

In order to use the results for the purposes of planning procedures, more information is needed. It is essential to measure (as well as possible) the *in situ* conservation potential of the archaeological sites and use this information in planning procedures. Perhaps it is even possible to improve the conservation potential at several sites, eg by raising the water table.

Therefore, it was decided that groundwater levels should be monitored for at least one year to obtain accurate information for the exact locations of the archaeological sites. Moreover, it was felt that more parameters which may be contributing to the conservation potential of the burial environment should be measured, such as the reduction–oxidation potential of the burial environment, microbial activity and basic chemistry of the soil water (Van de Noort et al 2001).

Research monitoring programme

A programme was set up to monitor several parameters of the burial environment at 12 different sites at Voorne-Putten. The sites were selected according to the magnitude of the difference in the degree of degradation of organic remains. It is assumed that the rate of degradation of organic remains depends mainly on the availability of oxygen and perhaps other oxidising agents. It is recognised that the presence or absence of microbes acting as catalysts might be of equal importance. However, their role is not investigated here due to practical constraints.

The parameters chosen are mainly characteristics of the groundwater: its level; temperature; salt content (electric conductivity); acidity (pH); reduction–oxidation potential (Eh); oxygen content; concentration of nitrates and sulphates. The measurements are taken by allowing groundwater into dipwells. The dipwells are purged a few days before taking measurements, to allow fresh water into the wells. Temperature, Eh and oxygen content are measured by lowering electrodes down the dipwells. This method is chosen to prevent changes in concentrations due to sampling and transportation which usually create greater inaccuracies than variations in the field. EC, pH, nitrate and sulphate are measured in the laboratory because significant changes due to sampling and transportation are not expected. In addition, the reduction–oxidation potential of the soil is measured directly as well. This is done by means of Pt electrodes which are installed *in situ*.

Groundwater level readings are taken every two months and all other parameters are measured every four months. These frequencies are a compromise between the desired number of readings and budgetary constraints.

Results

Three research questions underpin the analysis of the results. What relationships can be found between characteristics of the burial environment? How would any subsequent monitoring project be set up based on the experience so far? How does the actual quality of archaeological remains relate to the preservation potential of the burial environments? At present only the first two questions can be addressed. The third question cannot be answered yet, as results are not available. It is however one of the main goals of this research project to identify possible relationships between the quality of archaeological remains and the preservation potential of the burial environments.

Relative importance of oxygen to redox potential

Under the influence of oxygen and other oxidising agents, archaeological organic remains degrade. The process of degradation can be purely chemical or can be performed by microbes 'eating' into the remains. The overall ability of the burial environment to oxidise materials is expressed by the reduction–oxidation potential (Eh). A high Eh (+400mV) stands for a highly oxidising environment, a low Eh (-400mV) indicates preserving conditions.

Prior to this research, it was thought that the presence or absence of oxygen would be the main factor in determining the Eh of the burial environment. The first results indicate that a high oxygen content in the groundwater does indeed result in a high Eh (Fig 22.1). More than 20% saturation with oxygen results in an Eh greater than 0mV. However, low oxygen content does not always result in a low Eh. At 5% oxygen saturation, Eh values of –400mV up to 0mV are recorded. This means that oxidising agents other than oxygen (perhaps nitrates and sulphates) are equally important to the preservation potential of the burial environment.

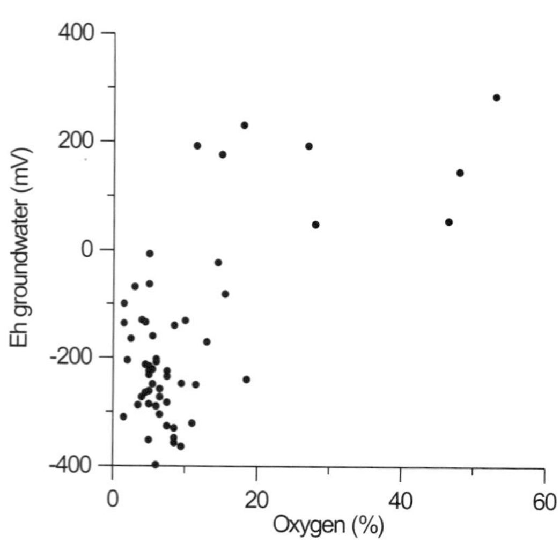

Fig 22.1 The Eh measured can only partly be explained by the concentration of oxygen

Redox potential depends on the length of time water stays in the soil

When measurements were taken in March, the outside temperature was around 5°C. The temperature of the deeper soil layers was 10–12°C. For March, the redox potential of the groundwater is plotted against its temperature (Fig 22.2). It shows that the higher the water temperature, the lower the Eh of the groundwater. This means that the longer the water stays in the soil, the lower its Eh, and thus its oxidising strength. This observation is supported by Chadwick and Chorover (2001) who give a hypothetical plot of the sequence of redox reactions, showing Eh falling steadily for a soil containing a finite supply of organic matter and oxidation capacity.

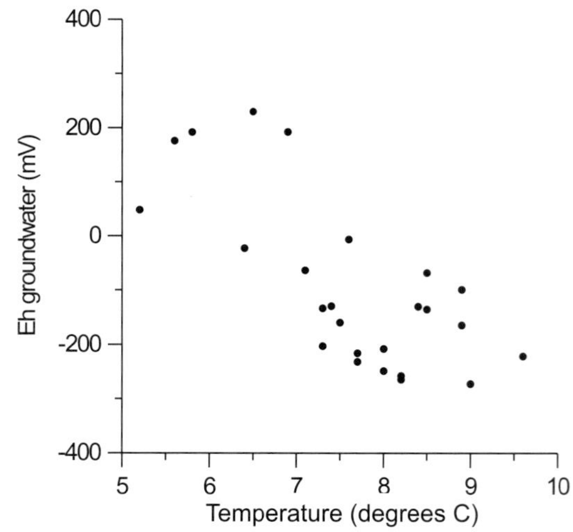

Fig 22.2 Eh being a function of water temperature indicates that the length of time water stays in the soil influences the Eh

Redox potential as measured in soil and groundwater

In this research, the Eh is measured in two different ways. First, the Eh of the groundwater in the dipwells is measured by lowering an electrode down the dipwell each time a reading is taken. Secondly, the Eh is measured in the burial environment directly by means of an *in situ* installed platinum electrode. Apart from several specific cases, the readings of both methods compare well (Fig 22.3), considering that it is known that different methods of measuring Eh often show poor agreement (Lindberg and Runnells 1984).

Exceptions here are cases where the *in situ* platinum electrode is installed at a depth above the groundwater table. In these cases, Eh readings from the *in situ* electrode are much higher than Eh readings from the water in the dipwell. This is not surprising, as more oxygen is expected above the groundwater table than below. Other exceptions are cases where the dipwell is not purged a few days prior to the readings; the Eh of the water inside the dipwell can be much lower than the Eh of the surrounding soil. This can be

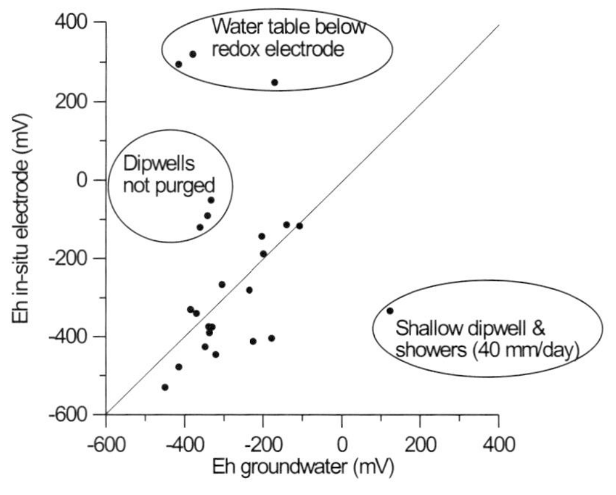

Fig 22.3 The Eh can be measured directly via an in situ *soil electrode and indirectly via the groundwater; most readings compare well*

explained by 'rotting' of the water inside the 'dead end' of the dipwell (a small reservoir of water inside the dipwell which does not easily exchange with the environment).

The last exception encountered during this project concerns a very shallow dipwell (150–400mm below the surface) where the water inside the dipwell showed much higher Eh readings than the surrounding soil. This can probably be explained by the ingress of oxygenated rainwater into the dipwell via a shortcut after heavy showers (*c* 40mm) the day before taking readings. This exception emphasises the importance of sealing the space between the upper part of a dipwell and the surrounding soil resulting from the auger hole made to install the dipwell. Sealing can be done with bentonite grains or rods.

It can be concluded that it is possible to measure the Eh via the groundwater inside a dipwell and directly by using an *in situ* platinum electrode. As the Eh readings via a dipwell are only correct when several precautions are taken, *in situ* electrodes are preferred. However, *in situ* electrodes cost over six times as much as a dipwell.

Future monitoring

In situ electrodes

The work undertaken to date has demonstrated several points. First of all, it is better to use *in situ* electrodes where possible instead of lowering electrodes down a dipwell. However, apart from platinum electrodes to measure Eh and perhaps solid-state electrodes to measure pH, suitable *in situ* electrodes are not readily available. Another disadvantage of *in situ* electrodes is cost, especially when monitoring equipment can easily get lost or damaged. During the Voorne-Putten project so far, about half of the monitoring dipwells in agricultural fields have got lost or become damaged, in spite of agreements with the local farmers.

Redox potential (and pH) as key factors

Prior to the Voorne-Putten research project, it was thought that oxygen was the major oxidising agent threatening archaeological remains. However it has been demonstrated that in many cases the Eh cannot be explained by the concentration of oxygen alone. Apparently, other oxidising agents can be equally important. This shows that in the first instance, monitoring Eh is the best way to gain insight into the preservation potential of the burial environment. Only when the Eh appears to be unfavourable, is it necessary to measure the concentrations of the different oxidising agents. In this way, it can be ascertained which oxidising agent is causing the unfavourable Eh and what planning or civil engineering measures should be taken to prevent the presence of that agent.

In addition to the Eh, the pH should also be monitored. From chemical theory, it follows that the strength of many chemical reactions (eg the degradation of archaeological materials) depends on both the Eh and pH (Caple and Dungworth 1998; Pollard nd). When the Eh and pH of the burial environment are known *as well as* the nature of the main degradation reactions, the preservation potential of a burial environment can be given for different archaeological materials.

Stagnant burial environment

It has also been found that the oxidising strength of groundwater decreases with the greater the length of the water stays in the soil. This conclusion points to what might be the key factor in assuring a burial environment with high preservation potential. By creating a burial environment with minimal transport of oxidising (and other harmful) agents, the preservation of archaeological remains might be assured. To create such a stagnant burial environment, the flow of water and air through the soil should be minimised, as the flow of water and air are the main ways of transportation through the soil. This conclusion implies that the flow of water and air should be monitored as well.

Microbiology

As has been mentioned above, microbes probably play an important role in catalysing reduction–oxidation reactions in the soil. By 'eating' into organic archaeological remains, microbes can speed up the degradation of these remains compared with purely chemical decay. Therefore, the presence and/or activity of microbes should be part of a complete monitoring programme.

Future monitoring

Given our present knowledge and experience, any future monitoring should include clearly defined monitoring points which may vary between one and ten depending on the size and nature of the archaeological site. Each basic monitoring point would consist of an *in situ* platinum redox electrode; an *in situ* solid-state pH electrode and one monitoring dipwell serving several goals. The latter would monitor the groundwater level and measure temperature and salt content (these sometimes valuable parameters are very easy to measure once a monitoring dipwell has been installed and should therefore be included). The *in situ* redox electrode needs a reference electrode preferably in contact with soil water; when felt necessary, the dipwell makes it possible to check the readings of the *in situ* Eh and pH electrodes.

Ideally, the *in situ* electrodes and the groundwater level meter can be read off site either by GSM technology or else they should be equipped with dataloggers. When such technology is installed, it is advisable to take readings every day. Without such technology, the desired frequency of taking readings is dependent on financial constraints. Depending on the expected stability of the burial environment, it is suggested readings are taken at least once every two months, up to at most once a week.

Based on data from the basic monitoring point, it should be possible to estimate the preservation potential of the burial environment. When needed, a more extensive set up of monitoring points could provide further insight into the factors causing an unfavourable burial environment. Additional dipwells can be installed to different depths to estimate vertical groundwater flow and (complying with strict specifications) can serve to lower an apparatus capable of measuring horizontal groundwater flow. In addition apparatus exists to measure oxygen flow through the soil and groundwater samples can be analysed for oxidising agents.

Directions for future monitoring work

In my opinion, the two different types of monitoring archaeological sites cannot be viewed separately over the next decade. Until such time that the degradation processes are understood, monitoring of the quality of archaeological remains should be accompanied by monitoring of the burial environment. Only when a clear relationship between burial environment and degradation processes is established, can they be viewed separately. When more conclusions are based on measurements of the burial environment, archaeological monitoring can become increasingly non-destructive.

Until the relationship between burial environment and degradation processes is sufficiently understood there are three reasons for continued monitoring. Firstly knowledge of the state of degradation of the archaeological remains is needed to facilitate planning decisions to prevent putting too much effort into worthless remains. Secondly knowledge of the preservation potential of the burial environment is needed to facilitate the design of civil works. Finally combined knowledge of the burial environment and the state of preservation of remains is needed to help elucidate degradation processes. Thus it will, in future, be possible to conclude what kinds of burial environments are favourable for archaeological remains.

This last reason is based on my view that it will still take a long time before degradation processes are fully understood. I think that by combining monitoring of the burial environment with monitoring the preservation of remains it will be possible to distinguish favourable burial environments many years before degradation processes take place. At the same time, the data gathered can help unravel these degradation processes.

However, a substantial data-set is needed in order to indicate favourable burial environments reliably. Only when it is established at several sites that a certain burial environment results in well-preserved archaeological remains, can that burial environment can be considered favourable. Future monitoring in different countries should be gathered into one database to prevent information being lost. Moreover, at least the key parameters of pH and reduction–oxidation potential should be measured at each monitoring project and recorded together with the survival of different archaeological materials (eg at a scale of 1 to 5). Following this conference an agreement on a minimal package of key parameters to measure (and also how to measure them) in order to ensure the formation of a valuable database could be established. Apart from this, a well-defined research project focusing on microbiology and groundwater flow needs to be initiated.

Bibliography

Caple, C, and Dungworth, D, 1998 Waterlogged anoxic archaeological burial environments, unpub Engl Heritage Ancient Monument Lab rep 22/98

Chadwick, O A, and Chorover, J, 2001 The chemistry of pedogenic thresholds, *Geoderma* 100, 321–53

Kars, H, 1997 Conservation science and the archaeological property: a review of current research, needs, and opportunities, in *Archaeological heritage management in the Netherlands: fifty years state service for archaeological investigations* (eds W J H Willems, H Kars and D P Hallewas), Assen

Lindberg, R E, and Runnells, D D, 1984 Groundwater redox reactions: an analysis of equilibrium state applied to Eh measurements and geochemical modelling, *Science* 225, 925–7

Pollard, A M, nd [1998] The chemical nature of the burial environment, in *Preserving archaeological remains* in situ: *proceedings of the conference of 1st–3rd April 1996* (eds M Corfield, P Hinton, T Nixon and M Pollard), 60–5, London

Van de Noort, R, Chapman, H, Cheetham, J, 2001 Science-base conservation and management in wetland archaeology; the example of Sutton Common UK, in *Enduring records: the environmental and cultural heritage of wetlands* (ed B A Purdy), Oxford

23

Cawthorn Camps (North Yorkshire): damage and survival

GRAHAM LEE and

PETE WILSON

Graham Lee
North York Moors National Park Authority
The Old Vicarage, Bondgate, Helmsley
North Yorkshire YO62 5BP, UK

Pete Wilson
Senior Archaeologist
English Heritage Centre for Archaeology
Fort Cumberland, Fort Cumberland Road
Eastney, Portsmouth PO4 9LD, UK

Background

The earthwork complex known as Cawthorn Camps (NGR SE 784 900) lies just within the southern boundary of the North York Moors National Park, some 6km to the north of Pickering (Fig 23.1). They comprise a group of closely adjacent Roman military earthworks, located in a strong defensive position on the northern scarp edge of the limestone Tabular Hills (Fig 23.2).

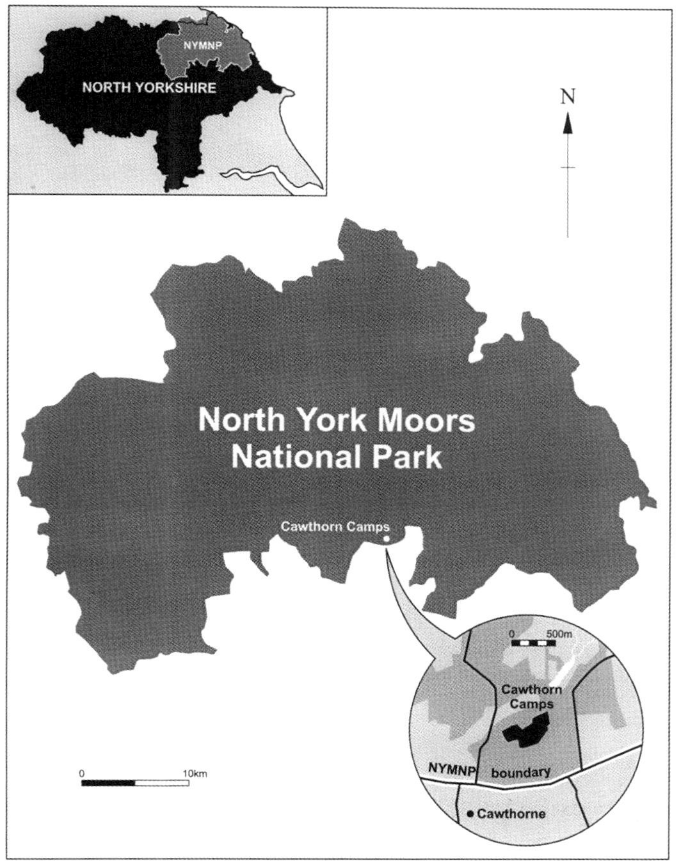

Fig 23.1 Location map of Cawthorn Camps site

The site first enters the archaeological literature with its publication by Drake (1736, 36) whose plan shows considerable detail with regard to the form of the defences. In the later 18th century General Roy published a plan which includes internal features in the south-east corner of the complex (Roy 1793, pl xi). Other accounts were published by Young (1817, 694), who labelled the camps from west to east D, C, A and B, and in the *Proceedings of the Oxford Architectural and Historical Society* (1861), both of which incorporated plans.

In the early 20th century the site was the subject of excavations undertaken by Sir N Boddington and S D Kitson. Their work included trenching of the west rampart of earthwork D, location of the *via principalis* and a 'paved area', but little else is known (Simpson 1923, 29–30). Far more extensive excavations were undertaken by F G Simpson, Sir Ian Richmond and a local archaeologist, Dr J L Kirk, between

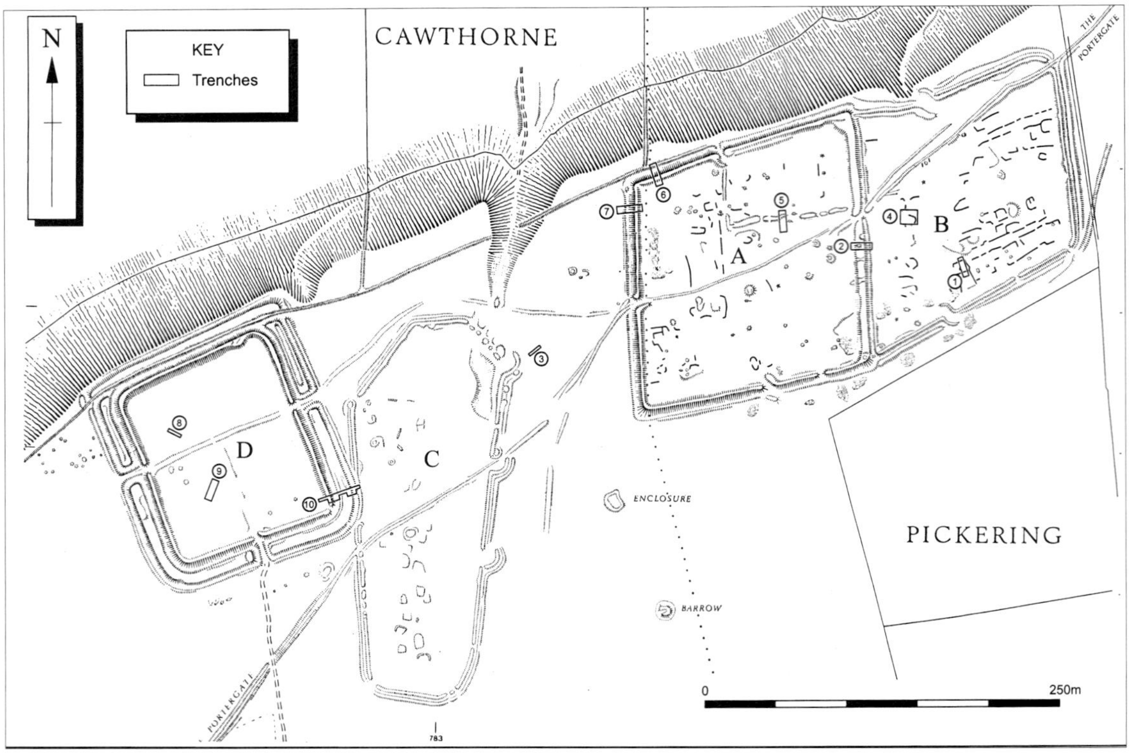

Fig 23.2 Site plan showing trench locations (after Welfare and Swan 1995, fig 115)

1923 and 1929 (Richmond 1932; Simpson 1923). Unfortunately the reports do not include trench plans and therefore our knowledge of the extent of the excavations is limited since the reports concentrate on what was considered interesting. Richmond saw the site as a series of practice works belonging within the period AD 70–120, that had '…no connexion with the permanent occupation of Yorkshire…' (Richmond 1932, 78). Richmond's report stood as the definitive account of the site until the publication of the Royal Commission on the Historical Monuments of England (RCHME) study of *Roman camps in England* (Welfare and Swan 1995, 137–42) and subsequent work by the present authors (Lee 1997; Lee 1998–1999; Wilson and Lee 2000; Wilson and Lee 2000–2001). The former RCHME work led to the reinterpretation of the earthworks as a camp (Camp C), two forts (Forts D and A), the latter associated with an annexe (Annexe B). Richmond had suggested that Fort A was of two phases, and the former RCHME work, along with Lee's subsequent reconsiderations of the site, led to the suggestion that Fort D might also be of two phases. The most recent work has also incorporated an evaluation of all available aerial photographs by the English Heritage Aerial Survey Team (Stone 1999).

Aerial photographic evidence

The general plan used by Simpson (1923, 31) and Richmond (1932, 18) derives from the 1912 edition of the Ordnance Survey Map, Yorkshire, North Riding, lxxv, 10. It shows the majority of the site, with the exception of Fort D, as planted up within a conifer plantation. The portrayed location of forestry tracks help to explain several significant breaks within the earthworks, particularly at the south-west corner of Camp C.

The locations of these tracks are confirmed on aerial photographs of 1925, which also usefully indicate no disturbance in the vicinity of a putative postern gate at the north-west corner of Camp C, linking it to the eastern gate of Fort D. Importantly these photographs show the complex as clear of trees, which are thought likely to have been cut down during the First World War, and illustrate a number of Richmond's trenches, together with the extent of cultivation within the northern portion of Fort D (photographs NMR SE 7890/5/2; NMR SE 7890/8/1; Stone 1999, figs 2a, 2b and 9).

The aerial photographs further illustrate the nature of the 1920s excavations, comprising a great many long narrow trenches, and also serve to confirm the fact that Richmond's report is selective in terms of the information presented. The aerial photographs reveal a number of trenches in significant locations which are not referred to in the report.

By 1945 the site was open heather moorland with a scattering of trees. A number of shallow craters are visible on aerial photographs, which are interpreted as due to a bomber dumping the remains of its bomb load (photograph CUCAP K22, 02/08/45; Stone 1999, fig 5). Otherwise scars relating to the use of the site for military training or other forms of erosion appear to be slight.

Aerial photographs from 1968 show the line of a firebreak which was bulldozed in 1959 (photograph CUCAP RC8-E/1, 31/05/68; Stone 1999, fig 6). Some remodelling/ reinstatement of the affected earthworks was apparently undertaken prior to 1968, as the damage to the ramparts had been made good. Also visible is the increase in tree cover due to natural regeneration (self-seeding), particularly in Fort D and Annexe B, as a result of removing sheep flocks from the moor and the decimation by myxomatosis of the local rabbit population in the 1950s.

By 1975 the extent of tree cover had increased significantly (photograph NMR SE 7890/14; Stone 1999, fig 7) and many of the trees were reaching maturity. The need for their management was regarded as an urgent priority for the newly formed National Park Authority, both due to the continuing penetration of roots into the earthworks and the risk of more serious damage should trees be uprooted during storms. Inappropriate access, by horse-riders and motorcyclists, and bracken growth were compounding the problems. An agreement was initially negotiated to allow the felling of dangerous trees but in the course of further negotiations 42 acres of Cawthorn Moor were offered for sale to the Park Authority. Purchase was completed in 1983.

Subsequently a management plan (NYMNP 1985) was drawn up for the estate and timber carefully cleared from the site. Much of this was done by hand by the British Trust for Conservation Volunteers to avoid any damage to the earthworks. Bracken was controlled by aerial spraying. This clearance allowed a topographic survey of the site to be carried out by the RCHME, which was further updated in the 1990s. Fort A and Annexe B were then resurveyed in greater detail by Ed Dennison Archaeological Services in work sponsored by the National Park Authority.

The excavated evidence

In all ten trenches were excavated during 1999 and 2000 by the English Heritage Centre for Archaeology (Fig 23.2). They were targeted to test key relationships within the monument complex and examine areas that might help address management issues, such as the erosion referred to above and previously discussed in print by Lee (1994). Despite the catalogue of damage set out above, in general the site survives remarkably well. Within Fort A, Annexe B and probably Camp C, there are the upstanding remains of early Roman military turf-walled buildings, including features noted by Roy (1793). Although the ditches of Fort A contain 0.70–1.0m of fill on the northern and western sides of the fort (Trenches 6 and 7) they remain open to a depth of 1.0m fronting ramparts that survive to 1.5m. The outer ditch of Fort D (Trench 10) contains no more than 0.3m of fill, possibly because it is not directly associated with a rampart that could be subject to erosion. This suggests that the earthworks achieved a degree of stability fairly quickly after their abandonment.

Under the ramparts of Fort A (Trenches 2, 6 and 7) we, and Richmond before us, recorded the existence of a layer of bleached turf (Fig 23.3), it was also noted in patches elsewhere on the site. This layer clearly demonstrated that when Fort A was constructed the area underlying its defences was not cleared of turf, but that the dump construction rampart was constructed directly on top of it. Indeed the first action of the rampart builders, best seen in the eastern rampart but visible in the northern and western ones, was to clear the ditch of turf and dump that material, which survived as recognisable turves, as the core of the new rampart. Turf was incorporated in the defences of Fort D, but appeared very different in the excavated section which showed as a near-classic Roman turf rampart, albeit without any evidence of timbering or a palisade (Fig 23.4).

Fig 23.3 Re-excavation of the southern half of Richmond's 'Officer's dug-out' number 2; a Grubenhaus *cut into the eastern rampart of Fort A (view to west) Note the layer of bleached turf underlying the rampart which is visible in the southern and western edges of the feature, also the dark modern fill of a posthole excavated by Richmond, located in the foreground on the left hand edge of the photograph. This posthole lay some 0.25m outside the pit of the* Grubenhaus *when excavated in the 1920s; it now lies within the eroded edge of the feature*

The second phase rampart in Fort A, which was examined in Trenches 6 and 7, and was absent on the eastern side (thought to be due to the addition of Annexe B, which removed the need to strengthen the defences), produced no

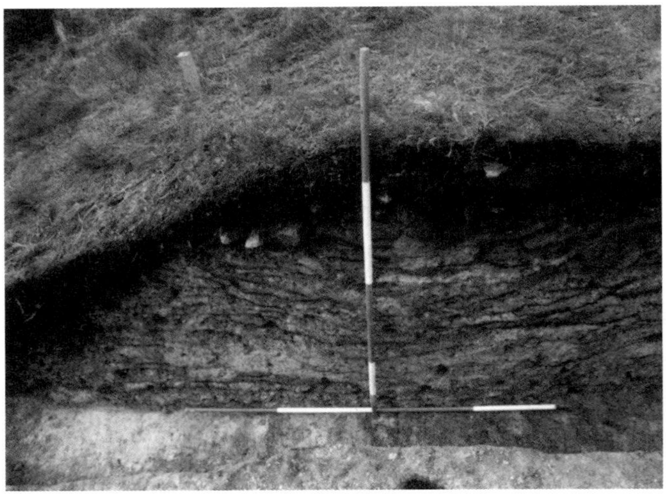

Fig 23.4 Rear portion of the turf-built eastern rampart of Fort D showing clear evidence of individual turves; the earlier phase of the rampart is represented by the lower c 0.20m of the structure where fewer turves are recognisable (view to north)

Fig 23.5 Stone foundation of south wall and remnant of stone surface within Building 5 in Trench 5; the turf mound of the south wall of Building 4, which Building 5 replaced, is visible against the far baulk (view to east)

evidence of a palisade. This and the lack of a palisade in both ramparts associated with Fort D could be taken to suggest that the ramparts had seen considerable erosion. However, this was not necessarily supported by the extent and character of the filling of the ditches of Fort A. The filling of the ditches was fairly limited and appears largely to consist of redeposited natural. This suggests that the material may not have been derived from the upper part of the ramparts (which appear to have incorporated considerable quantities of turf and soil).

The most obvious 20th-century impact on the site is that of forestry and woodland regeneration, although overall the level of impact appears to have been less than might have been expected. A number of tree stumps were encountered in the trenches excavated and in general they appear to have been fairly shallow-rooted with only occasional major roots penetrating into archaeological deposits. However, in Annexe B (Trench 4), a tree stump was located on the turf mound that represented the upstanding wall of the early Roman military structure investigated. Due to the extent of its root system, combined with the presence of a 1920s excavation trench, a quadrant of the building had to be excluded from the area investigated. There was also evidence of tree-throw holes elsewhere on the site, including in Trench 9.

No evidence of Roman period structures survived in Trench 9, although there was a spread of stone in the northern part of the trench that, given the evidence elsewhere on the site, could have been associated with a structure, or represent the remnants of a surface. The fragility of the structural remains is emphasised by buildings found in Fort A (Trench 5). Building 4 had a degraded turf north wall, which stood to *c* 0.35m, and in which the individual turves could no longer be recognised. This wall was either lined with or replaced by a wattle and daub structure. On the south side of the building the opposite wall only survived to about 0.10–0.15m in places. Building 4 was replaced by Building 5 which incorporated a stone surface and stone foundations which survived to less than 0.10m (Fig 23.5) and were set over an irregular trench.

There was no direct evidence of the agents of destruction that might have impacted on the features in the central part of Fort D, but Trench 8 in the northern part of the fort was much easier to interpret. As noted above, the 1925 aerial photographs clearly show ploughing, probably by a steam plough, in the area and excavation demonstrated that any features to a depth of 0.35m would have been destroyed. In the case of Trench 8 this meant everything that may have existed in the area, although there remains the possibility of deeper cut features surviving elsewhere in the northern part of the fort.

Another significant 20th-century impact has been the archaeologist. As previously stated, we lack a trench plan for the work in 1908 and in the 1920s and wherever you go on the site you have a chance of finding Boddington and Kitson, Richmond, Simpson or Kirk. In Annexe B (Trench 1) a well-preserved early Roman turf building was examined, the walls surviving to *c* 0.35m and incorporating well-defined turves (Fig 23.6). Fig 23.6 also shows accumulated material, presumably turf eroded from the wall, to either side of the core of the wall and serving to stabilise it. However in the southern part of the building we located one of Richmond's unrecorded trenches (Fig 23.7), the soft fill of which served as 'planting hole' for a self-seeded tree cleared in the 1980s. The nearest feature recorded on Richmond's plan lies some 6.0m to the south-east (Richmond 1932, pl xx, pit B1). A further legacy arises from the fact that some, at least, of the 1920s trenches were not backfilled. Spoil heaps are recognisable on the aerial photographs (Stone 1999, figs 2a/b, 3a/b). This was confirmed by the re-excavation of Richmond's 'Officer's dug-out' number 2 (Richmond 1932, 66). This feature was cut into the eastern rampart of Fort A and was recorded by Richmond as a pit incorporating a posthole on its western edge, with three more set *c* 0.25m to the east of it. On re-excavation it was shown that the eastern and southern edges, at least, had eroded back to the extent that the eastern postholes now lie within the edge of the feature (Fig 23.3). The interpretation as an 'Officer's dug-out' appears as an example of the excavator's First World War experiences conditioning his interpretation. It is clear that the feature was a *Grubenhaus*, a feature type not known at the time, as were probably the other four that Richmond illustrates (Richmond 1932, 66), and possibly some of the 'many pits [which] were identified, but [of which] few were of interest' (Richmond 1932, 68).

Fig 23.6 Turf wall of Building 1 in Trench 1; the individual turves that form the 0.90m wide wall are visible, as are dark areas to either side of the wall that represent material eroded from its top which now serve to protect and stabilise the in situ turf structure (view to east)

Fig 23.7 Unrecorded 1920s trench with Trench 1; the softer material accumulated within the 1920s excavation has served as a 'planting hole' for a self-seeded tree which was cut down in the 1980s (view to south)

Little direct evidence of Second World War military activity was encountered in the areas investigated, other than spent 0.303 rounds in the turf and topsoil over the inner ditch of Fort D, which appears to have been used as a rifle pit. However, one substantial feature that was cut into the area between the inner and outer ditches of Fort D (Trench 10) can be assigned to this period. It took the form of a trench aligned on a slightly oblique angle to the ditches of the fort. At 1.5m deep it incorporated steps on its eastern side and was too deep for a rifle pit, or slit trench, and may represent a fire control position for mortar or explosives training.

Conservation and present-day management

Vegetation management is a continual requirement – particularly in the case of trees self-seeding back onto the site. In this way birch, in particular, is the woodland equivalent of garden weeds. Initially, encroaching vegetation was controlled by sheep grazing, the totality of the site having been fenced (Lee 1994, 90). The drawback to this, if the sheep are left on the site for long periods, especially from early on or until late in the season when ground conditions tend to be soft, there can be increased erosion of specific areas in which the sheep gather or along the trails which they tend to establish.

Sheep are best used for only a limited, *c* two month period when ground conditions are firm at the peak of the growing season; the alternative is labour-intensive hand-spraying/weed-wiping. There is also the need to control the development of bracken, which obscures the earthworks and causes ground disturbance through root (rhizome) growth, which tends to be very extensive. Control is by hand-spraying, also used to deal with rhododendron and other invasive plants, or by beating down.

Erosion control and management is an important aspect of sustainable access (Lee 1994, 90–1). In order to deal with human erosion a hard surfaced trail has been created over the years in a circuit around the complex, with the exception of 'soft surfacing', turf over soil and terram, through the centre of Fort D.

This has still not particularly helped to resolve 'desire line' problems; the site is locally important not for its internationally significant preservation of early Roman military earthworks but as an open space and dog-walking circuit. The latter brings associated hygiene issues, which are significant in relation to the educational use of the site.

Erosion has also increased along the northern ramparts of Forts D and A, after the felling of trees on the scarp slope below the earthworks opened up the moorland vista, with the most significant impact being on the north rampart of Fort A. This problem is being monitored but the National Park Authority is likely to need to consider hard-surfacing (limestone crusher-run surfaced with compacted dust) in order to create a sacrificial layer to limit further erosion of the archaeologically significant layers in specific locations such as the north-west corner of the fort and its northern gateway.

Summary

While the obvious problems associated with forestry on fragile earthworks within the interior have been addressed and large-scale uncontrolled archaeological exploration is no longer likely, the site remains subject to pressures that threaten its continued preservation *in situ*. Efforts to balance the competing demands that arise from visitor access, as an ancient monument that is promoted in a low-key way, as locally important public open space, and as a site with a high nature conservation interest, require sensitive management. The earthworks appear to have achieved stability, with little evidence for erosion that cannot be explained as a result of visitor use, or in recent times the stocking regime, but as an unmanned site in a relatively remote moorland setting, it remains potential vulnerable. The site is protected as a Scheduled Ancient Monument and is owned by a conservation minded authority, but continued monitoring is required.

Acknowledgement

The drawings were prepared by Vince Griffin of the English Heritage Centre for Archaeology Graphics Studio.

Bibliography

Drake, F, 1736 *Eboracum: or the history and antiquities of the City of York*, London

Lee, G, 1994 Erosion control on Cawthorn Roman Camps, in *Erosion on archaeological earthworks: its prevention, control and repair* (eds A Berry and I W Brown), 87–92, Mold

Lee, G, 1997 Cawthorn Roman military complex: an update, *Archaeol J* 154, 260–7

Lee, G, 1998–1999 Cawthorn Roman military complex: an update, *Ryedale Hist* 19, 6–13

NYMNP, 1985 North York Moors National Park, 1985 *Cawthorn Roman Camps: a management plan*, Helmsley

Proceedings of the Oxford Architectural and Historical Society 1861, third meeting Michaelmas term, *Proc Oxford Architect Hist Soc* 1, 76–80

Richmond, I A, 1932 The four Roman camps at Cawthorn, in the North Riding of Yorkshire, *Archaeol J* 89, 17–78

Roy, W, 1793 *Military antiquities of the Romans in north Britain*, London

Simpson, F G, 1923 The Roman camps at Cawthorn, near Pickering, *Yorks Archaeol J* 28, 25–33

Stone, J L, 1999 Cawthorn Camps, North Yorkshire: air photograph evaluation, unpub Engl Heritage rep

Welfare, H, and Swan, V, 1995 *Roman camps in England*, London

Wilson, P R, and Lee, G, 2000 Cawthorn Camps 2000: summary interim report, *CBA Forum* 2000, 23–7

Wilson, P R, and Lee, G, 2000–2001 Cawthorn Camps; trial excavations 1999, *Ryedale Hist* 20, 5–8

Young, G, 1817 *History of Whitby and Streoneshalh Abbey*, Whitby

24

An integrated monitoring strategy at Chedworth Roman villa (Gloucestershire)

JOHN STEWART, SOPHIE JULIEN and SARAH STANIFORTH

John Stewart
(formerly) Adviser on the Conservation of Archaeological Sites and Monuments

Sophie Julien
Environmental Monitoring and Control Assistant

Sarah Staniforth
Adviser on Paintings Conservation and Environmental Control

The National Trust
36 Queen Anne's Gate
London SW1H 9AS, UK

Abstract

Chedworth Roman villa (Gloucestershire) consists of extensive residential remains excavated in the 19th century and partly protected under a range of different shelters, all creating variable internal environments. An integrated monitoring strategy based on radio telemetry is successfully collecting data from the disparate monitoring sites for precipitation, internal and external temperature and relative humidity, internal surface temperature, external surface temperature of reburied mosaics, soil moisture content and, less effectively, masonry moisture content. This methodology attempts to interrelate ground and ambient conditions in order to achieve an understanding of the factors contributing to the stability or deterioration of the mosaics and wall plasters at their interface. The ultimate objective is to mitigate deterioration by means of enhanced preventive conservation, through improvements to the environments within existing buildings, or their replacement with a new structure built to a dedicated environmental design.

Introduction

Exposed archaeological structures of stone, brick and mortar are generally subject to more aggressive forms of deterioration by environmental mechanisms than similar features buried in the ground. The surface of exposed monuments, at the interface of two differing dynamic environments, constitutes the zone in which most deterioration is concentrated. Associated loss is all the more serious when the surfaces are decorated or have other markings of intrinsic historic interest.

Buried building materials are also vulnerable, either through progressive chemical leaching of mineral constituents, ground frost or mechanical damage (by geological subsidence, roots, rodents, invertebrates). Although these specific materials may be more stable in a buried environment than in an exposed one, the rate of deterioration is less easily characterised and monitored.

Strategies to mitigate the causes and the rate of deterioration need to be based on the identification and assessment of all possible risks. However, only some of these can be easily measured. At Chedworth Roman villa in Gloucestershire, in the south-west of England, an integrated monitoring programme is attempting to identify the most important environmental factors contributing to the deterioration of the ancient fabric, both exposed and buried. This is part of a more comprehensive evaluation aimed at formulating a conservation plan for the site.

The villa dates largely from the 4th century AD, and has been progressively excavated, following its discovery, in 1861. It contains one of the largest collections of *in situ* ancient mosaics in Britain, some wall plaster, and excellently preserved hypocausts, or under-floor heating systems. The vulnerability of the mosaics and wall plaster was certainly recognised at the time of their discovery. Some of these were selected for presentation and protection under three purpose-built shelters. Mosaics that were not to be presented to the public were reburied under turf, or asphalt footpaths. A dedicated site museum was also erected.

In the mid 20th century, a fourth enclosure was built, along with a temporary marquee or tent, in the 1980s, anticipating the construction of a fifth. This did not occur, and the marquee is still in place. The four proper buildings have timber frames on stone walls, approximately 1.0m high. Three of these are enclosed on four sides, whilst one is open to the elements on its southern elevation.

Conservation surveys

The geographical components of the monitoring programme are identified in Fig 24.1. However, monitoring is only one part of a larger strategy, encompassing an inventory of the natural and historical resources of the site, and a condition survey of the archaeological fabric, leading ultimately to the diagnosis of causes of deterioration (Stewart et al in prep). This latter, of course, is intimately influenced by dynamic environmental factors.

Three surveys undertaken as part of this strategy are particularly relevant to the monitoring programme. A geophysical survey of buried archaeological features sought to identify possible obstacles to the site's natural hydrology, or unknown and unrecorded drainage systems. Second, a hydrogeological survey was commissioned to characterise site geology, hydrology, and the extent and condition of buried drains: ancient, 19th century, and modern. The site sits on geology of variable limestone with layers of clay or siltstone, which deflect groundwater to the surface in the form of a series of springs. Third, a survey was undertaken of mosaics excavated and reburied in the 19th century and intermittently exposed and recovered. These had never been properly recorded. Representative sample areas were excavated, confirming the scope of surviving pavements, their condition and burial stratigraphy. This also provided the opportunity to upgrade reburial cover, where feasible, pending a decision on future presentation.

The condition surveys identified that the most serious natural risks to the ancient building fabric are from soluble salts and microbiological growth affecting stone, ceramic and plaster substrates variably. There was also strong evidence of frost damage to external masonry, and possibly also to insufficiently buried mosaics.

Environmental factors

Moisture in liquid or vapour form is a primary agent of these processes, along with ambient temperature, relative humidity, and light in the case of photosynthetic micro-organisms.

Moisture originates from superficial condensation, from local springs or groundwater, and from precipitation. Rainwater reaches the foundations of the building either directly through the soil, or indirectly, via roof gutters and downpipes which feed into stone-filled soakaway pits relatively near each structure (Fig 24.2).

Capillary moisture from ground sources does not have equal access to the ancient building fabric. In the case of mosaics, one pavement has discontinuous contact with the ground only through the piers of its hypocaust, which were cleared of invasive soil (Fig 24.2). Parts of two other mosaics are atop more substantial hypocaust supports, whose soil-filled channels remain unexcavated. Another mosaic has been lifted and re-laid on a bedding layer impervious to groundwater penetration.

The internal environment within each protective structure is further defined by their different designs, and the degree of ventilation. Air flow is due to public access in the months during which the site is open to the public (March to November). Characteristics of each shelter are as follows.

Triclinium

Fully enclosed, with insulation, winter heating, seasonal ventilation, and clear glazing in windows.

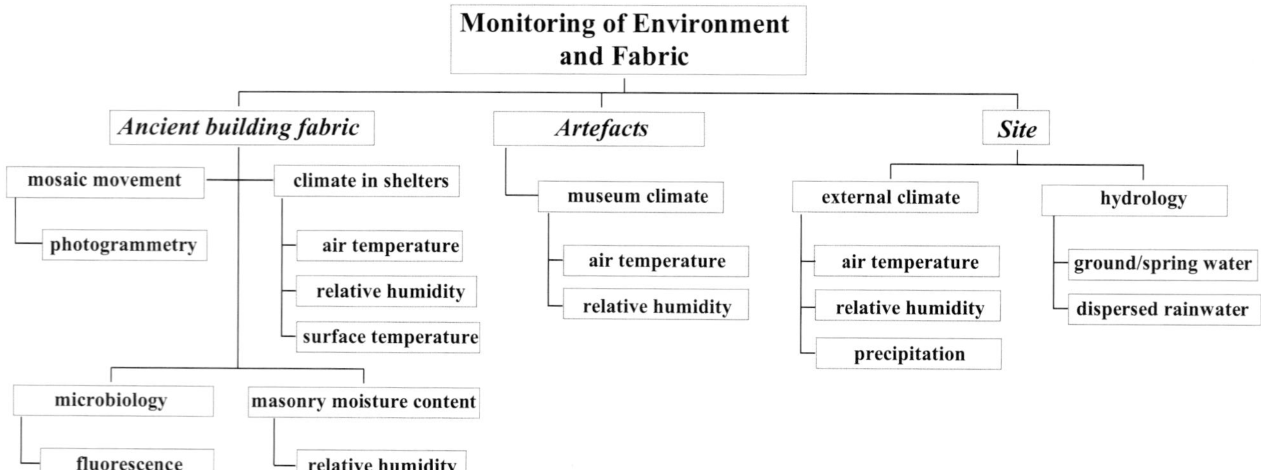

Fig 24.1 Monitoring sectors and parameters, highlighting those described in this article

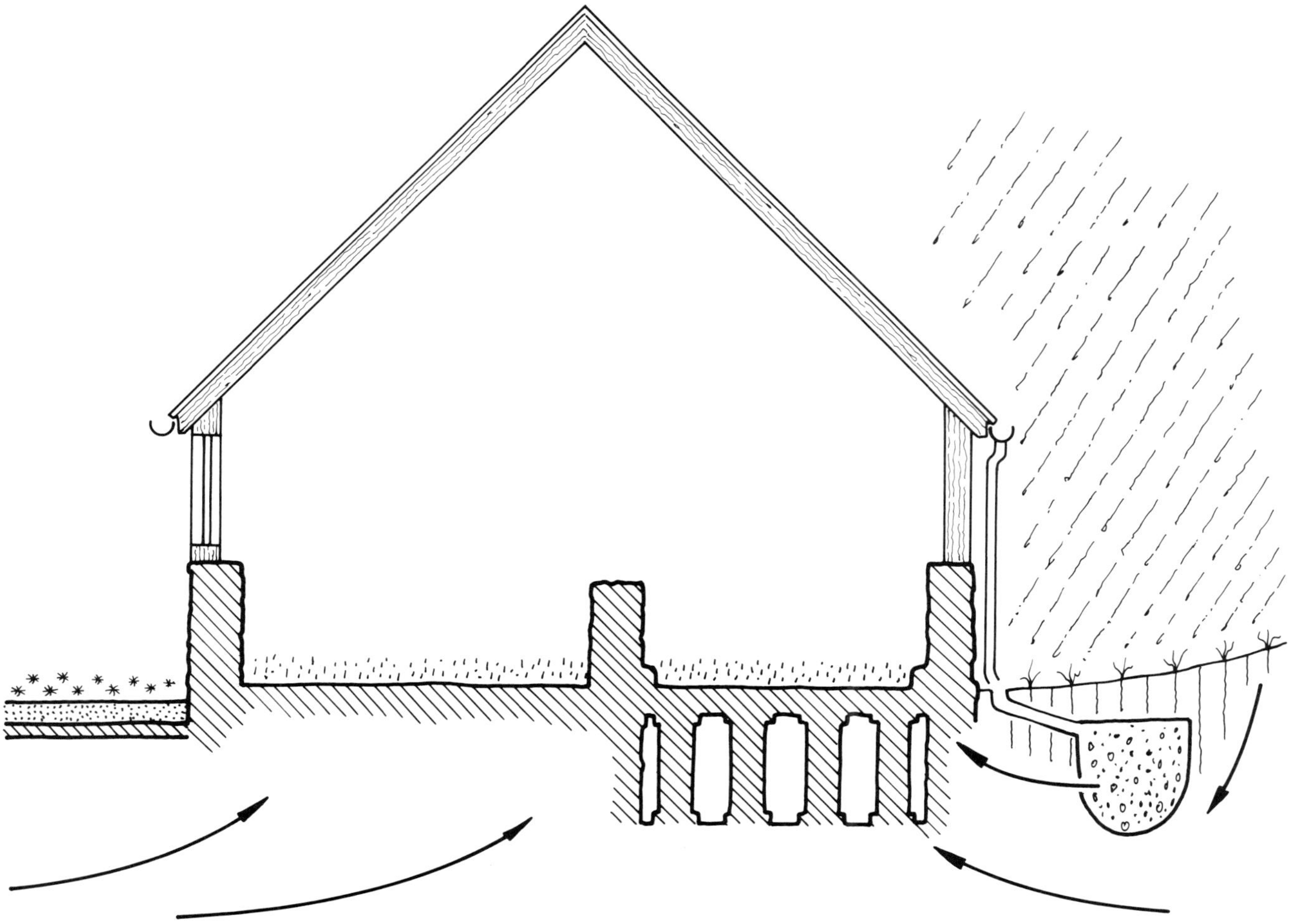

Fig 24.2 *Schematic section through one shelter building with floor mosaics, and an external reburied mosaic, showing typical sources of moisture (Graham Reed)*

West bath house
Fully enclosed, without insulation but with winter heating, seasonal ventilation, and frosted glazing in windows.

North bath house
Fully enclosed, with no insulation or winter heating, very limited ventilation throughout the year, and large-scale clear glazing.

Cold plunge bath
Semi-enclosed, with no insulation or winter heating, but full ventilation throughout the year.

Living room (marquee)
Semi-enclosed, with no insulation or winter heating, but full ventilation throughout the year.

In order to quantify the relative influence of different moisture sources, as well as other factors influencing deterioration processes, an environmental monitoring programme has been progressively implemented around the site. There are few other sites with shelters on which internal environment has been characterised (Agnew et al 1996).

Two of the protective structures are now over 140 years old. Direct public access has been gradually restricted in response to increasing conservation awareness and concern. The early buildings are certainly of historical interest, but all of the shelters fall short of satisfying the higher standards of preventive conservation that are now both feasible and considered fundamental to good curatorship.

The environment containing the ancient fabric can certainly be improved, either by the modification of existing structures or by replacing them with a new comprehensive enclosure with an enhanced environment. In either case, building design must be based on an understanding of the existing environment and synergistic deterioration processes.

Environmental monitoring framework

Telemetric system and dataloggers

The nature of the site and the scope of monitoring that was envisaged demanded an integrated system that would encompass the different areas, but use the minimum amount of cabling and dataloggers. A system based on radio telemetry had obvious advantages, as data collected at separate sites can be transmitted back to a central receiver, which then transfers the data to a single logger. An additional benefit is that all data can be downloaded simultaneously, and exported to a monthly spreadsheet for processing. The various items of equipment of the system are illustrated in Fig 24.3. Monitoring parameters and equipment are described below.

Air temperature, relative humidity and surface temperature

The internal environment of the shelters is monitored using three parameters, ambient temperature (AT), relative humidity (RH), and surface temperature (ST) of mosaics and wall plaster.

Relative humidity is the ratio of the amount of water vapour contained in a given volume of air at a certain temperature over the total amount of water vapour that volume could contain at saturation (100% RH). The sensors used for monitoring are Vaisala Humitter 50-U capacitive sensors, with a specified accuracy of ±5% over a range of 0–90% RH, combined with thermistors for air and surface temperature measurement. From the AT and RH values, one can calculate the absolute humidity (AH) value, in g/m^3. This value is by itself of limited interest, but is significant when compared to the AH of outdoor or neighbouring environments. The indoor AH usually follows the outdoor AH values very closely, perhaps with some buffering from the building structure. However, when internal values are found to be consistently higher, it can be assumed that there is an additional source of moisture within the building, possibly combined with poor ventilation. Another value which is calculated from the AT and RH values is the dew point temperature (DPT) of the air, the temperature to which the air would need to be cooled to reach 100% RH and form condensation (liquid moisture).

Surface temperature measurements (ST) have two purposes. The first is to give an indication of possible instances of condensation. When ST falls below the DPT of the air, condensation can be expected on this surface. In fact, the accepted margin is a difference of 1°C for the formation of condensation. The calculation of ST-DPT is called the dew point margin and can be plotted on a graph. The second purpose of ST measurements is to monitor variations, possible freezing events, localised heating due to solar gain, or cooling due to moisture evaporation at the surface.

At Chedworth, AT/RH probes are placed in each shelter, with surface temperature sensors on the mosaic floors or wall plaster, and in the case of the living room under the marquee, attached to the side of a stone *pila*. Sensors were fixed by

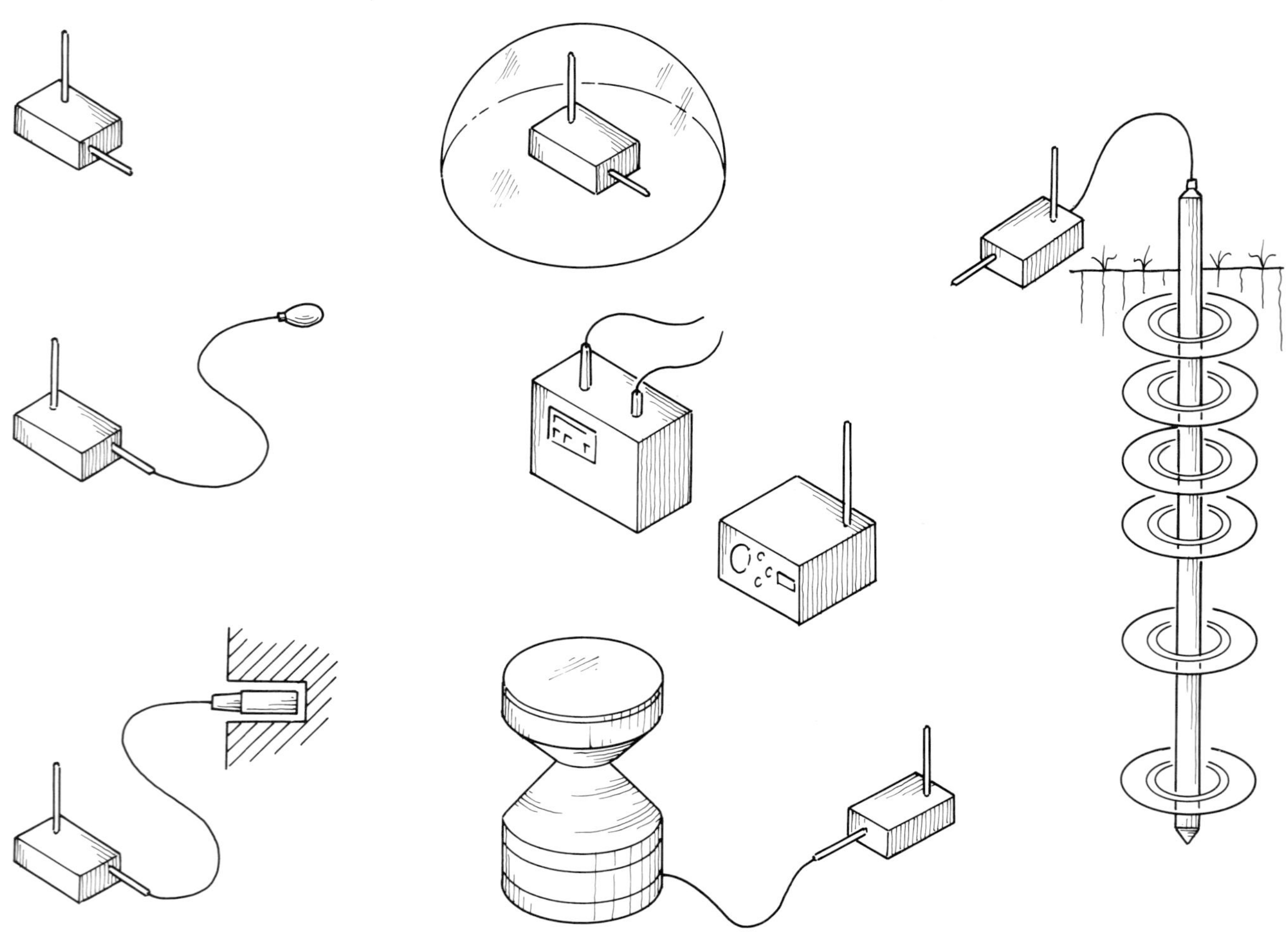

Fig 24.3 The range of monitoring equipment integrated into the radio telemetric system (Graham Reed)

means of small pads of Plastazote, polyethylene foam adhered with Paraloid B72 acrylic co-polymer adhesive and a separation layer of cotton gauze (this method of adhesion was devised by Dr Robyn Pender, formerly of the Wall Paintings Conservation Course, Courtauld Institute, University of London). One dedicated AT and RH probe has been placed on a mosaic, and enclosed by a plastic hemisphere. This is intended to confirm that moisture is being released from the floor surface.

A further AT/RH probe, placed in a special housing to protect against rain and vegetation, monitors external conditions. Additionally, two ST probes are used outdoors to monitor the temperature of mosaics reburied under different materials and at different depths (100mm and 400mm respectively).

Precipitation

A tipping rain gauge is used to precisely record precipitation at Chedworth, so this can be related to each shelter's interior climate and soil moisture content. The instrument consists of a small bucket contained within a funnel-shaped box, designed to minimise turbulence effects. The bucket collects precipitation and tips every time it becomes full (0.2mm per tip). Each action is logged and the number of 'tips' per hour relayed to the datalogger. This gives an accurate pattern of rainfall which can be directly related to the hourly readings of other environmental conditions.

Masonry moisture content

The measurement of moisture variations within masonry and their relation to ingress of moisture from ground sources is an important parameter in the monitoring programme. However, this remains problematic. The traditional method of determining moisture content is by the extraction of samples and measurement of water content at a specific point in time. For monitoring of moisture content in real time there are a number of techniques available commercially, but they all have their particular drawbacks of accuracy, coverage, applicability or expense (Dill 2000). Furthermore, not all of the available methods are easily compatible with datalogger technology. Monitoring at another site in a similar context utilised the measurement of electrical resistance through an electric current passed between two metal electrodes inserted into the masonry (Knight 1994). This was not considered appropriate here, as it is known to be affected by salts which are present in abundance at Chedworth.

As there are no accurate and economical models for monitoring moisture it was decided to apply technology developed to measure the drying of concrete for the construction industry. It was hoped that this would be able to cope with the particularly high moisture values found at Chedworth. The probe utilised is a Vaisala HMP44 capacitive sensor, similar to an AT/RH probe, with a specified accuracy of ±2% in the 0–90% range and ±3% in the 90–100% range. The sensor fits into a hole in the masonry to a depth of 80mm, and is held in place with a plug that also insulates the environment of the hole from ambient conditions. The measurement of RH in the air space within the hole (in equilibrium with the masonry) represents a relative value of moisture content as a percentage rather than an absolute g/m^3 value. In theory this can be directly compared to readings taken at other times or at other monitoring sites. Probes were deployed at three different locations, with the intention of relating masonry moisture to different moisture sources from the ground.

Soil moisture content

The latest addition to the system, installed on a trial basis, is a probe designed to measure the moisture content of soil at varying depths, and does so to an accuracy of ±3%. This specific probe is relatively new technology used principally in agricultural applications. The PR1 Profile probe, manufactured by Delta-T, consists of six rings at depths of 100, 200, 300, 400, 600 and 1000mm. Each ring generates an electromagnetic signal and records how much of this signal is reflected by the soil. Due to the high dielectric constant of water, the higher the moisture content of the soil the stronger the return signal will be. Two probes have been inserted at different locations thought to be affected by water run off from an adjacent hill slope, and one possibly by natural spring water related to the Roman nymphaeum, or water feature. Precipitation is detected by the upper sensor rings, and ground or spring water primarily by those at the bottom of the probe.

The insertion of the probes is done through access tubes. Auguring needs to be very precise in order to obtain an even gauge along the entire profile. The data from the probes is in the form of an electrical signal, and needs to be converted through an equation relating the amount of water (m^3) contained with an equal volume of adjacent soil, or percentage volume reading. A highly accurate soil calibration was not considered necessary in this case since the main interest is to identify trends and relationships between the readings obtained.

Integrated monitoring

The monitoring of multiple parameters simultaneously on a unitary system enables an integrated approach to the diagnosis of mechanisms of deterioration by identifying specific relationships between environmental factors and their interdependency (Fig 24.4). The ingress of moisture, seen as a rise in internal absolute humidity, can be related either to dispersed rainfall and/or to spring or groundwater entering foundations from below. Instances of condensation can be related to outside conditions, such as precipitation or damp air. The occurrence and frequency of frosts and solar gain can be set against the other ambient conditions.

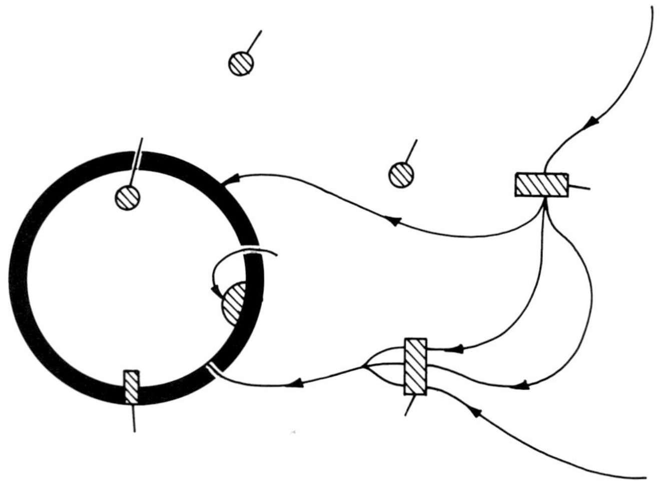

Fig 24.4 Schematic illustration of the relationship of monitoring equipment and environmental parameters (Graham Reed)

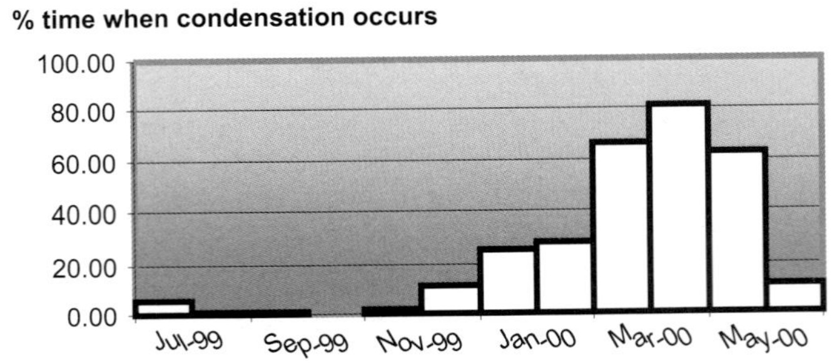

Fig 24.5 The frequency of condensation occurrences within the marquee tent over a period of one year

Discussion of results

Up to three years' of data from parts of the system have now been collected and analysed. A number of useful conclusions can be drawn for the performance of each building in variable climatic conditions and levels of moisture ingress.

The three fully-enclosed shelters each create their own particular environment, which is dependent on the ventilation, moisture ingress and insulation characteristics of the building. RH levels in these buildings tend to be high and stable, but are subject to rapid change upon opening of the windows/door for public access. This can result in condensation. Both the triclinium and north bath house show evidence of an internal moisture source from the ground. The north bath house, which is permanently closed, has the highest and most stable RH levels. The radiant heating in the triclinium and west bath house is effective at preventing frosts and also reducing RH levels, but not sufficiently to prevent condensation. A trial using humidistatic control of the heating is now under way.

The two other shelters that are open (cold plunge bath and marquee), as expected show a much closer relation to outside daily and seasonal variations. The stone and timber cold plunge bath has a higher thermal mass than the marquee, and is a better buffer against extreme conditions. Condensation in these shelters is closely related to rainfall and damp air. There are periods of constant condensation in winter (Fig 24.5).

The surface of the wall plaster in the semi-enclosed stone cold plunge bath is partly protected from nightly frost by the thermal mass of the soil behind it, but can still be subject to prolonged freezing conditions. The stone *pilae* under the marquee are exposed to daily freeze-thaw cycles in the winter that often coincide with condensation.

Although there is evidence of moisture movement through the mosaic floors and plaster-covered walls, this does not increase visibly following rainfall. Rather, it is constant and appears to be greater during periods of temperature rise.

Morning sunlight entering east-facing windows causes thermal gain on the surfaces of all mosaics, though frosted glass windows reduce this effect. Midday sunlight also affects the surface temperature of south-facing wall plaster in the semi-enclosed cold plunge bath (Fig 24.6).

In the upper layers of soil, moisture can remain high for several days following rainfall. In the deeper layers however, moisture content is higher and extremely constant, indicating the presence of stable ground or spring water levels during the limited period of monitoring. The soil moisture profile is different for the two locations currently being monitored. Further information on specific geological and archaeological characteristics of these areas is required to better understand this data (Fig 24.7).

Areas of masonry monitored for moisture content have for most of the time been either saturated or have had a value beyond the range of the probes. However, recent data suggests that the walls begin to dry out on very hot, dry days when the internal conditions are also hot and dry. There does not seem to be any direct correlation with rainfall, although prolonged periods of rain do seem to result in an overall rise in absolute humidity within the walls.

In the two reburied mosaics where surface temperature is being recorded, the one covered with the greater bulk of soil overburden clearly has a more stable temperature, even during nightly and prolonged freezing conditions. The temperature varies only by 1°C a day. However, the mosaic buried under the asphalt footpath and a shallow layer of soil shows a temperature gain of up to 5°C a day in summer. Winter conditions are of course of greater concern when freezing conditions may occur, but monitoring here has not yet been carried out over a full winter.

When enclosed within a hemisphere, the lack of air movement at the surface of the mosaic results in RH levels between 90 and 100%, with a corresponding rise in AH. This implies that there is moisture movement through the mosaic floor, despite the presence of an external French drain near the monitored area. The hemisphere is essentially replicating on a smaller scale those conditions when the shelter is kept closed and unventilated.

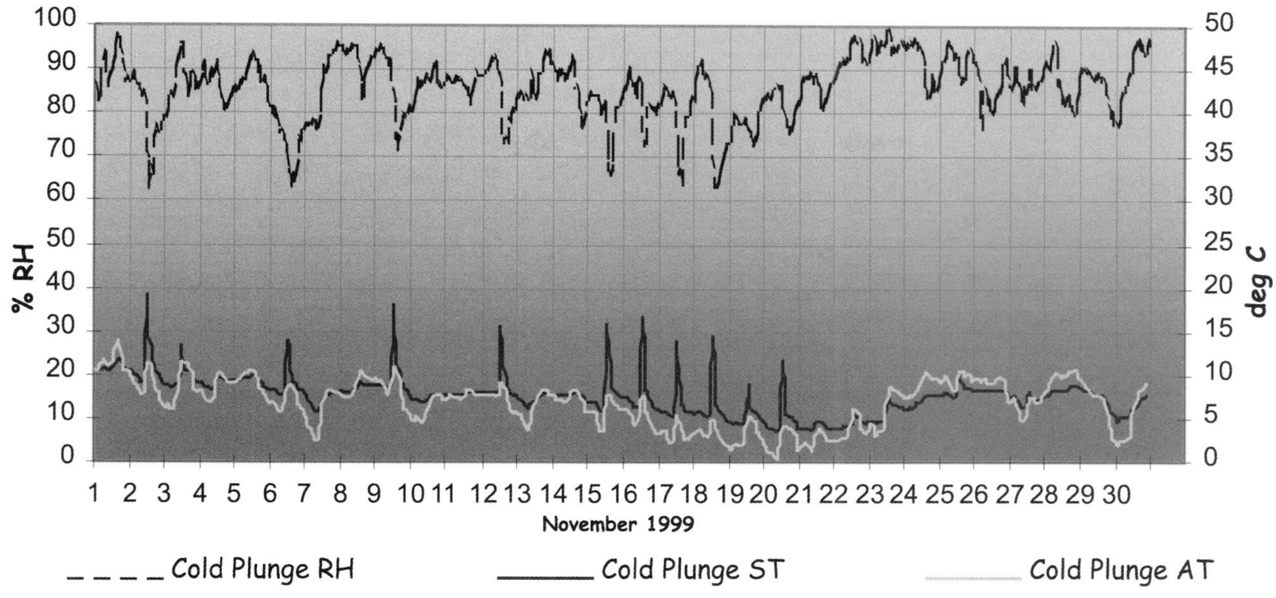

Fig 24.6 *Graph showing the effects of direct sunlight on surface temperature of the cold plunge bath*

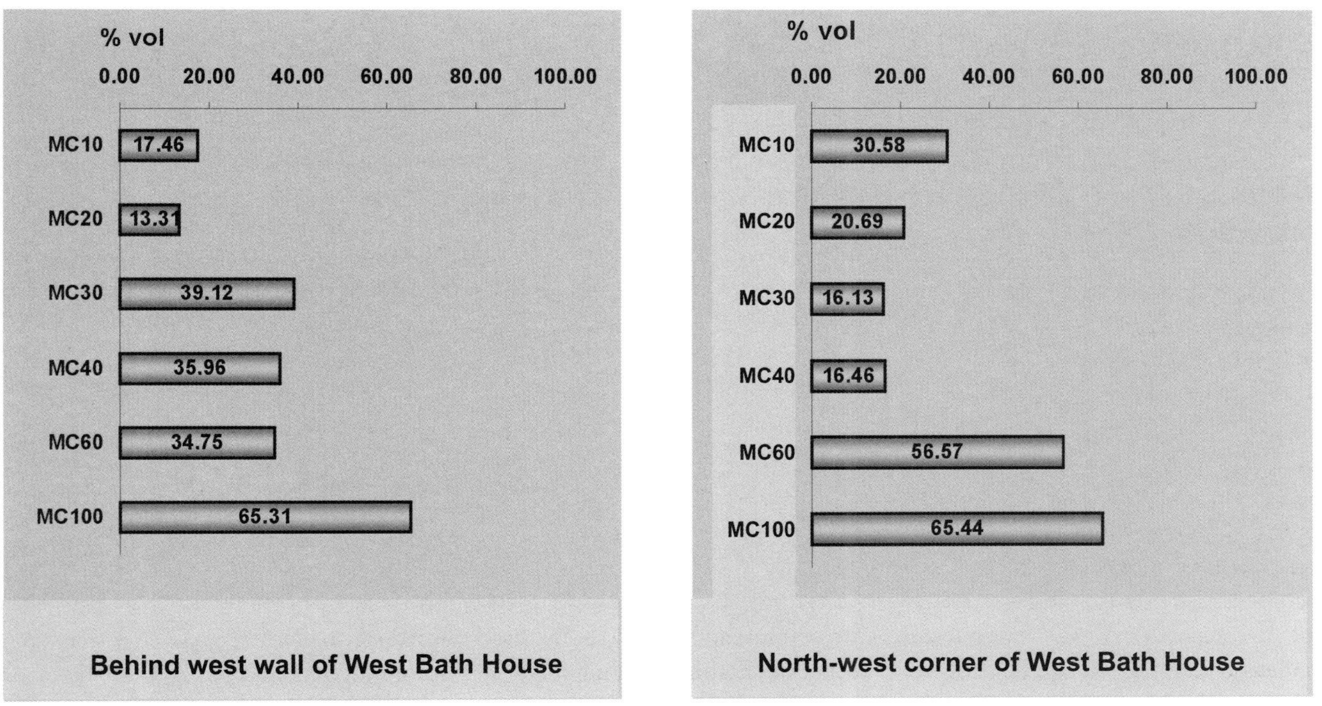

Fig 24.7 *Soil moisture (MC) profiles, in descending depth, of two probes in different locations*

Although a number of the principal trends could easily have been surmised, monitoring has provided tangible data to corroborate them. This permits a highly detailed evaluation of individual shelter performance which will inform the design brief and environmental specification for new or altered shelters. There are still outstanding questions relating to precipitation, ventilation and moisture ingress.

Modification of the internal environment of one shelter is now also under way. The heating system is being fitted with a relative humidity control to determine if condensation can be prevented by maintaining the RH below a critical level. More reliable data also needs to be gathered from the masonry and soil moisture probes to make their contribution even more significant to the interpretation of ambient data.

Challenges

An important consideration has been the duration of monitoring. It was initially envisaged that one year of uninterrupted data would provide an accurate picture of environmental conditions. Substantially more time was needed, not only because technical faults resulted in loss of sometimes crucial data, but also because conditions did vary significantly from year to year, and it was important to identify accurate ranges and average values. Three years is now considered the minimum period and of course, monitoring must be continued to assess any modifications undertaken to the buildings.

Although the overall system has worked well, there were a number of inevitable and unpredictable problems that arose in the course of the project from which lessons were duly learnt. Many of the difficulties encountered were due to the lack of local staff on site, with time available to monitor potential problems in the system. Regular site visits were usually carried out every three months, with supplementary visits when required.

Loss of data

Intermittent loss of data was a result of occasional detachment of surface temperature probes, the gnawing of cables by rodents (which was resolved by providing armoured cables), and clogging of the rain gauge by leaves and debris. Frequent maintenance inspection of the rain gauge was impossible due to the long intervals between site visits.

High levels of humidity and condensation, and calibration of AT/RH probes

The AT/RH probes suffered from prolonged high humidities in the winter of 2000–2001, which resulted in condensation on the probes and inaccurate readings of nearly 100%. Calibration of some of the probes during this event also proved problematic. Since this time, the probes have dried out and recovered their initial accuracy following a second calibration. The limitations of the Humitter 50-U Vaisala probes in the conditions of Chedworth thus became apparent, as the probes can drift in periods of very high humidity. The product data does not give a specified accuracy above 90% RH, though they do work above this level. Although these conditions were infrequent at Chedworth, and insufficient to warrant exchanging the probes to better performing ones, this is a consideration for future systems.

Masonry moisture probes

The masonry moisture probes were chosen specifically because of their application in the monitoring of drying concrete. However the walls and floors monitored at Chedworth were so near saturation that the probes gave constant readings of 100%. This may be because the reading is taken at depth rather than at the surface where wetting and drying cycles may be more perceptible. Repeated calibration of these probes did not solve this problem. Therefore, this type of probe is not ideally suited to the monitoring of moisture in conditions for which it was not designed, such as this consistently very damp historic masonry of Chedworth.

Other available and economically viable monitoring technologies will be reassessed.

Soil moisture probes

Although the two soil moisture probes have provided interesting readings during an initial trial, from their very limited area of moisture catchment, data cannot be considered characteristic of the site as a whole. A more thorough invasive survey of the site's geology should enable more calculated positioning of the probes, but in this initial trial period it was desirable to minimise disruption of archaeological stratigraphy.

Data processing and interpretation

Processing the great amount of data generated by monitoring proved to be painstaking and its accurate interpretation complex. However, this level of analysis has only recently been possible with the availability of fast computers with large memories. A more powerful and flexible data manipulation and statistics programme, able to quickly handle such amounts of information, would be a useful supplement in the future.

Relationship of data to site events

Because of an infrequent presence on site, it was difficult to relate features in the data to actual events such as, for example, malfunction of the heating.

Future perspectives

Remote downloading of data would be a useful facility in the future. However, this could never be a substitute for regular site visits. Rather, it would enable closer scrutiny of conditions at the site, detect any system failures in good time, and help to schedule site visits around these events.

The development of such a monitoring programme and addition of new equipment inevitably was dependent on the limitations of available technology. The rain gauge and soil probes were later additions, and there was an inevitable delay in their configuration into the network.

Despite these limitations, the core system has been highly appropriate for this specific application. Its flexibility and its software enabled a number of components to be added after the initial design, thus permitting the incorporation of new technological developments.

The monitoring campaign is ongoing. Results gathered to date have characterised the internal environment within each type of protective structure; confirmed the interrelationship between dynamic environmental conditions and the deterioration of the ancient fabric; and provided a sound basis for modifying internal environmental conditions and appraising the efficacy of such changes.

Acknowledgements

The authors are indebted to technical consultants for their contributions to the development of the monitoring system: Lucien Hatfield and Mike Bass of Eltek Ltd; David Cridlsand of Vaisala and staff at Delta-T. Illustrations are by Graham Reed. Current or former National Trust colleagues also assisted in the project: Janet Berry, Katy Lithgow, Philip Bethell and Mark George.

Bibliography

Agnew, N, Maekawa, S, Coffman, R, and Meyer, J, 1996 Evaluation of the performance of a lightweight modular site shelter, *Conservation Management Archaeol Sites* 1, 139–50

Dill, M J, 2000 *A review of testing for moisture in building elements*, London

Knight, B, 1994 Environmental monitoring at Brading Roman villa, Isle of Wight, unpub Engl Heritage Ancient Monuments Lab rep 57/94

Stewart, J, Staniforth, S, Berry, J, in prep Chedworth Roman villa: a methodology for the monitoring of *in situ* mosaics, in *Mosaics: conserve to display? The VIIth conference of the International Committee for the Conservation of Mosaics, 22–28 November 1999, Arles – Saint-Romain-en-Gal, France* (eds E Chantriaux and P Blanc)

25

The management of archaeological sites in arable landscapes

GEORGE LAMBRICK

Abstract

The Management of Archaeological Sites in Arable Landscapes Project is being undertaken by the Oxford Archaeological Unit in conjunction with the Council for British Archaeology (CBA), Oxford University and leading Agricultural Consultants for the Department for Environment Food and Rural Affairs (DEFRA).

The fact that archaeological sites and landscapes can be progressively damaged or destroyed altogether by modern farming practices has been acknowledged for centuries, but the scale of destruction was brought home by English Heritage's Monuments at Risk Survey. The objective of this study is to establish the basis for developing a management strategy for preserving archaeological sites on arable land that focuses on where damage is most serious and which provides sustainable remediation of the problem. To be sustainable solutions must be both effective in substantially reducing the medium to long term threat of damage (within a clear perspective of where damage is most likely to occur), while also maintaining agricultural viability and minimising extra cost to the exchequer or loss of revenue to the farmer.

The project is divided into seven main stages all involving the production of interim working papers and ranging from the identification of the issues, evaluating and testing models to preparation of guidance for farmers.

The project is now well into testing the models and beginning to look at options for remediation schemes. The site-specific model attempts to predict, on a site by site basis, where plough damage is likely to be a problem. The development of a national model has the potential to predict, in broad terms, where plough damage to archaeological sites is likely to be most prevalent. This has significant potential for use in deciding at a regional level where resources devoted to pilot schemes for addressing the issue might best be concentrated, and possibly for longer term strategies, depending on more general trends in agri-environmental policy.

Remediation measures will focus on both mechanisms for delivery (good practice codes, agri-environmental schemes, ancient monuments management agreements) as well solutions (reversion to grassland, direct drilling, minimum cultivations etc). Consideration is being given to how issues such as farm business type, capital investment strategies and crop rotation systems affect the viability of different schemes as well their technical value.

George Lambrick
Director
Council for British Archaeology
Bowes Morell House
111 Walmgate, York
North Yorkshire YO1 9WA, UK

Introduction and background

The Management of Archaeological Sites in Arable Landscapes Project is being undertaken for the Department for Environment Food and Rural Affairs (DEFRA) by Oxford Archaeology (OA) in collaboration with the Council for

British Archaeology (CBA), the Institute of Environmental Change at Oxford University and Reading Agricultural Consultants.

Archaeologically we know that cultivation damage to *in situ* archaeological remains of earlier human activity is as old as ploughing itself. The fact that archaeological remains can progressively be damaged, or destroyed altogether by ploughing has been acknowledged by antiquaries and archaeologists since at least the early 17th century. The famous early 18th-century antiquary, William Stukeley (quoted in Lambrick 1975) described the Roman town of Alchester in the following terms:

> Everyone has a little portion of it to plough up; whence we may well imagine the land is racked to the last extremity and no great care taken in the management of it; yet it bears very good crops of wheat. As I traversed the spot at every step I saw pieces of pots and vessels, of all sorts of coloured earth … I perceived them strown very thick over the whole field together with bits of brick of all sorts: the husbandmen told me they frequently break their ploughs against foundations of hewn stone and brick; and we saw upon the spot many paving stones with a smooth face, and laid in a very good bed of gravel, till they drew them all up by degrees, when the plough chances to go a little deeper than the ordinary.

One of the most passionate accounts of plough damage was by Colonel Lane-Fox (later General Pitt-Rivers) who was so moved by the destruction of Dorchester Dykes in Oxfordshire, that in 1870 he wrote in the Saturday Review (2 July, quoted in Cook and Rowley 1985):

> … the fortress at Dorchester and the fortress on Sinodun (Castle Hill) are among the most speaking monuments of the earliest history of our island, and till lately they were among its most perfect monuments. But it is a grievous truth that while we are writing the dykes at Dorchester are being levelled. Hitherto the neighbouring ground has been grazed, and the harmless sheep is no foe to history; but it has lately occurred to the owner of the ground that a few shillings more of yearly profit might be gained by turning pasture land into arable; and to such a sordid motive as this, these precious antiquities are at this very moment being sacrificed. At least a third of the dyke has already been lowered, and will gradually be utterly destroyed beneath the yearly passage of ruin's merciless plough share.

The concern raised by cases like Dyke Hills contributed to the passing of the first Ancient Monuments Act in 1882. A hundred years later the issue again came to the fore, culminating in a conference and publication in 1980 of *The past under the plough* (Hinchliffe and Schadla-Hall 1980), in which Professor Colin Renfrew spoke passionately of the seriousness of plough damage as a threat. Andrew Saunders, the then Chief Inspector of Ancient Monuments, spoke for many in summing up the general feeling: 'There can be few, if any archaeologists who are not aware that arable farming and forestry are the agencies of the greatest destruction of the material evidence for the understanding of our past …'

The CBA has continued to highlight the problem over the last 25 years and more (eg CBA 1988; CBA 1993), but until the last few years there was very little progress in addressing the problem. Archaeological conservation became a statutory concern for the then Ministry of Agriculture in 1986, and was incorporated into the development of agri-environment schemes (for example the West Penwith Environmentally Sensitive Area was designated partly for its archaeological value). However, these schemes remained focused on wildlife and attractive landscapes within a political context where lobbying on these issues was much more powerful and effective than could be achieved by the archaeological community. This included better national data-sets on loss of wildlife than there were for archaeology – despite several regional surveys from the 1970s onwards that have quantified the problem of archaeological plough damage (Drinkwater 1972; Gingell 1976; Schadla-Hall 1977; Drewett 1980; Saville 1980; Allden 1982; Lawson et al 1986; Barnatt and Collis 1996; Dunkin nd). For much of this time land under intensive arable was seen as having very little general environmental value, while in the archaeological world the last 15 years have been much more concerned with addressing the pressures of building and infrastructure development, which have always been seen as both a more drastic and more immediate threat. The growth and new challenges of professional contracting archaeology have thus been almost exclusively concerned with development under the planning system, to the detriment of dealing with the more insidious and uncontrolled damage caused by agriculture.

But with an increasingly conservation conscious climate developing within the former Ministry of Agriculture Fisheries and Food (MAFF) and continued persistence from the CBA and English Heritage during the late 1980s and 1990s, ongoing concern about the issue was given vital underpinning by the results of English Heritage's Monuments at Risk Survey (MARS, Darvill and Fulton nd). This major study brought home the scale of destruction. Agriculture has been responsible for 10% of all cases of destruction, and 30% of all piecemeal, cumulative damage to ancient monuments in the last 50 years. Of monuments in arable areas 65% are at medium or high risk of damage. According to figures from Monuments at Risk Survey data provided by English Heritage for rural monuments only, 32% of all rural archaeological sites and 21% of rural sites protected as Scheduled Ancient Monuments were still under arable cultivation in 1995 (S Trow, pers comm).

After not just two decades, but arguably two centuries of increasing concern, MARS was thus a crucial catalyst in finally putting the issue of archaeological damage in arable landscapes seriously onto the agri-environmental agenda. In these days of evidence-based policy development, it effectively enabled English Heritage and the CBA to press for the project that forms the subject of this paper.

Getting a better understanding of the scale of the problem through MARS may have done most to trigger a realisation that something needs to be done about it, but it is worth noting some other crucial characteristics that make this distinctive as an environmental issue for arable landscapes.

1) Archaeology is the only source of information for 80% of human history since the introduction of farming, and a vital source of evidence for the rest.

2) Rural areas represent the most extensive and varied source of this evidence, and much of Britain's countryside is in arable production.

3) Archaeological sites – in contrast to most wildlife habitats – are entirely non-regenerating: a Bronze Age barrow cannot be recreated.

4) Much of the resource is in the middle of intensively cultivated fields; you cannot choose where to apply remediation measures as you can in using field margins for habitat recreation, they must be where the resource is.

5) Most of the resource is buried – albeit often at shallow depths – which makes reliable assessment of damage intrinsically difficult and potentially expensive. Thus an approach based on risk assessment is likely to be more realistic than one based on solid evidence of damage.

Project objectives and approach

Taking account of these broad issues, the objective of this three-year study, which is due to be completed in April 2002 is to establish the basis for developing a management strategy for preserving archaeological sites on arable land that focuses on where damage is most serious and which provides sustainable remediation of the problem. To be sustainable, solutions must be both effective in substantially reducing the medium to long term threat of damage (within a clear perspective of where damage is most likely to occur), while also maintaining agricultural viability and minimising extra cost to the exchequer or loss of revenue to the farmer.

The project is divided into several stages all involving the production of interim working papers:

Stage 1: bibliographic review and questionnaire-based gathering of existing evidence.

Stage 2: review of the issues that affect how plough damage occurs.

Stage 3: creation of site specific and national models of risk of plough damage.

Stage 4: evaluating and testing the models.

Stage 5: developing remediation scheme options for conserving archaeological sites in arable.

Stage 6: report preparation.

Stage 7: preparation for guidance for farmers.

The project is now well into testing the models and beginning to look at options for remediation schemes.

Issues and the development of risk models

Issues

The range and character of damage to archaeological sites in arable landscapes is well-attested and varied. They include the cultivation of previously unploughed sites or parts of sites – including lateral encroachment; the gradual planing flat of upstanding earthworks surviving within arable; the deeper cultivation or regular subsoiling of flattened sites in existing arable; the effective deepening of cultivation caused by water and wind erosion, and peat shrinkage, encouraged by ploughing to a constant relative depth; the desiccation of organically preserved sites caused by drainage; breakage and chemical deterioration of portable antiquities; physical damage caused by ancillary works such as farm tracks, barns, irrigation lagoons etc; damage to uncultivated sites within arable landscapes caused by external factors such as burrowing animals and visitor wear.

Recent work commissioned by English Heritage has started to quantify the rate of loss of ridge and furrow earthworks in the East Midlands, and is indicating that desiccation due to drainage poses the greatest threat to wetlands (Van de Noort et al 2002).

The example at Knowlton, Dorset illustrates several of these issues, including the ecologically rich grassland of the main surviving henge; partly upstanding barrows closely ploughed round; entirely ploughed over barrows whose truncated ditches are visible as cropmarks; and a modern farm building complex occupying part of another large prehistoric enclosure.

Where earthworks do survive in arable, they are not immune from damage. For example, the presence of an extensive rabbit warren and badger setts in the ramparts of Cherbury Iron Age fort, a site where typically the surviving earthworks stand within a sea of arable cultivation.

A recent condition survey of the Avebury World Heritage site, Berkshire (Fig 25.1a) has shown that in this predominantly arable farmland, cultivation accounts overall for about 46% of threats to the archaeological monuments, but this figure varies from one landscape management zone to another (Fig 25.1b).

However, while it is often easy to demonstrate that damage has occurred, it is often a great deal more difficult to show whether it is active. While it may be clear when a previously uncultivated site is ploughed up for the first time, it is much harder to judge whether sites already in cultivation are being damaged – or how fast. Often there is a threshold effect – as illustrated by the Raunds Roman villa in Northamptonshire. Nearly ploughed out barrows are potentially far more vulnerable than those surviving to a greater height, as indicated by a small barrow excavated at Rollright in the 1980s (Lambrick 1988). This barrow stood only about 100 mm high – virtually nothing remained of the

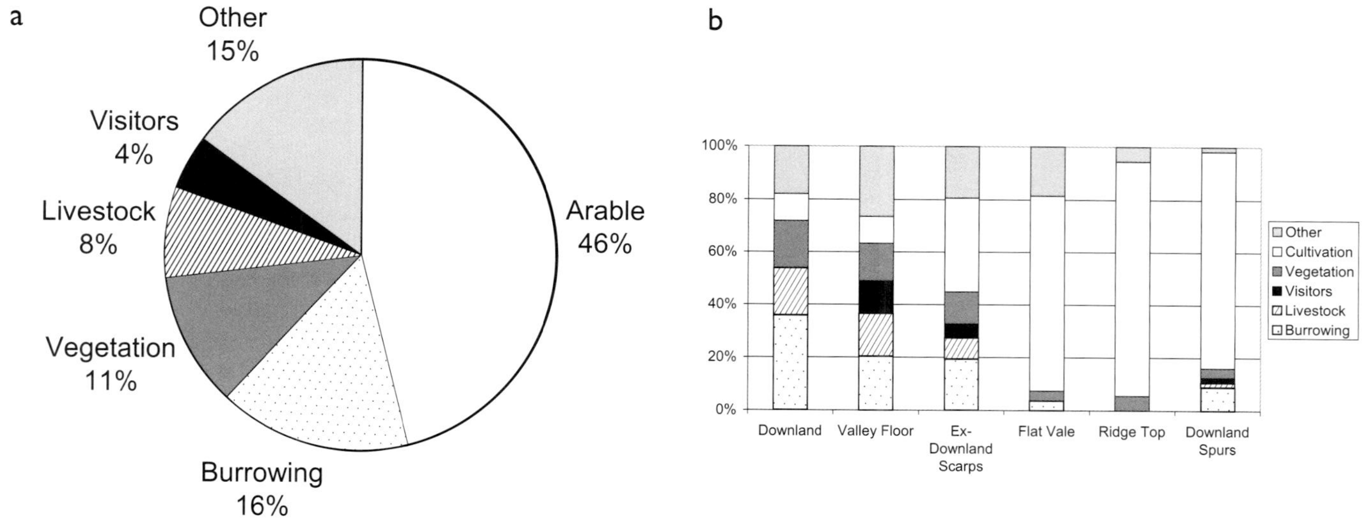

Fig 25.1 Condition survey of Avebury World Heritage site, a – types of damage to archaeology; b – types of damage to archaeology grouped by landscape zones

barrow mound but a thin layer of flat stones – but in the old ground surface there were not only cremations, but also a most unusual scatter of Mesolithic bladelets and valuable palaeoenvironmental evidence. In other instances the threshold of damage may be buffered by a sealing layer of older ploughsoil, colluvium or alluvium – as repeatedly found on floodplain and wetland sites. Here it is not continued cultivation so much as subsoiling, drainage and peat shrinkage that poses the threat (Lambrick 1992; Van de Noort et al 2002).

The MARS assessment of risk to archaeological sites was based on a fairly simple assessment of the scale and likelihood of damage that was applied to all kinds of rural and urban land use, not just arable agriculture. If a remediation scheme specifically to deal with arable damage is to be developed in which individual decisions can be based within a general strategy for addressing the problem, then a more tailor-made approach is needed, geared to the key factors that affect whether or not damage is likely to occur.

A review of issues likely to be important in determining whether plough damage occurs came up with two broad categories – those factors that are intrinsic to the site, and those which represent aspects of how the land is managed. Intrinsic site factors include the nature of the archaeological remains and quality of their survival; the depth of soil including depth and extent of any previous cultivation soils, colluvium and alluvium relative to the level of surviving archaeology; soil characteristics (erodability, drainage requirements, susceptibility to compaction etc); issues of slope influencing the likelihood of erosion. Management factors include cropping patterns and rotation; cultivation methods and depths and timeliness; background issues relating to farm structure and economics.

Site specific risk model

The idea of developing a site specific model to assess whether sites are at risk of damage is intended to provide a reasonably straightforward framework for predicting, on a site by site basis, whether and where plough damage is likely to be a problem. Ideally, with access to relevant information and some training, it will become usable by farmers and non-archaeological environmental advisors as well as by archaeologists.

Two approaches have been developed. The first is a scoring method that effectively is designed to provide a means of scoring each of the main factors outlined above, guided by basic criteria to assign a very high, high, medium, low or negligible risk score to each factor, which can be amalgamated and weighted to provide an overall risk score. The other approach being examined is a decision-tree model based on working step-by-step through various intrinsic site factors to lead out to some management scenarios that could be suitable for remediating the risk identified. A decision on what management prescriptions might be needed to address the problem for the site would be based on seeing how well the option indicated by the assessment matched with actual practice – and what could be made realistically practicable within the context of that particular farm business.

At present, neither model is particularly simple, but both seem to work reasonably well when tested as a desk exercise on published cases. Currently they are being field-tested by archaeologists in DEFRA and local authorities and through application to archaeological evaluation projects being undertaken for other purposes. It remains to be seen which is more effective and/or easy to use, and how they may need to be modified and improved for practical use.

National model

The development of a national model has the potential to provide an alternative approach to assessing the overall extent and scale of the risk, and thereby predict in broad terms where plough damage to archaeological sites is likely to be most prevalent. This has the prospect of being able to be used at a regional level to aid decisions as to where resources devoted to addressing the issue might best be concentrated, and possibly for longer-term strategies, depending on more general trends in agri-environmental policy. This includes the potential to identify suitable areas to pilot any remediation schemes.

In considering how to develop a model that would map these factors at a national level, it was realised that several of the critical issues would be reflected in characteristics of different 'Soil Associations' as summarised in the legend of the Soil Survey's 1:250,000 map of England and Wales, and by Evans (1990). Furthermore, very broad general land use would be reflected in cropping records provided by farmers' regular normal returns to DEFRA in claiming production support. The approach has therefore been to develop the national model on the basis of using documented characteristics of different Soil Association in terms of depth, erodability and drainage, combined with cropping patterns derived from DEFRA land use data that was especially prepared for this project. These data-sets are available in the form of the predominant characteristics per 1km or 2km square, and by scoring and weighting these four factors it has been possible to derive a single combined score that maps plough damage risk on a 1km square basis. This can then be tested by overlaying the incidence of sites recorded as having being affected by plough damage in the MARS study. As the resulting preliminary plot shows, this does indicate a reasonably good general correlation between recorded damage and the higher risk areas (Fig 25.2). A perfect correlation is not to be expected: whereas the risk model is based on the predominant conditions for each kilometre square, the archaeological sites represent point data (usually to within 100m). It is thus reasonable to expect both that undamaged sites may occur in pasture fields or woods in areas that are predominantly used for intensive arable (some such sites may be protected monuments); likewise it is to be expected that cultivation damage occurs in arable parts of areas that are predominantly pasture.

In addition it can be tested against older, but more detailed regional data-sets (Schadla-Hall 1977; Drewett 1980; Saville 1980; Barnatt and Collis 1996; Dunkin nd). A set of preliminary plots again at first sight suggests a reasonable general level of correlation, but they also seem to indicate that the greatest incidence of damage appears to correlate with kilometre squares scored for a high, but not the highest risk categories. This may suggest that the scoring of risk has given too much weight to one factor compared with others; but it could also be a genuine trend, perhaps suggesting that the areas mapped as highest risk are where sites have already been destroyed.

Having established what appears to be a reasonably convincing model, some further refinement is in progress. For example the problem of peat shrinkage needs to be incorporated into the national model, and more statistical work is needed to refine the scoring and weighting of different factors and quantify apparent correlations.

Remediation measures

In practical terms there are broadly two means of stopping, or at least slowing down the problem of cultivation damage. The first is to revert land to grassland or other benign non-intrusive land use. The second is to adopt archaeologically benign methods of cultivation that are not damaging or at least substantially postpone the point when damage will start to occur again. Either can be combined with other measures to address the problems that arise from other threats to archaeology in arable landscapes, such as drainage and farm infrastructure requirements or burrowing animals and visitor wear.

Reversion

Taking reversion first, as Lane Fox said with reference to Dyke Hills (Cook and Rowley 1985), 'the harmless sheep is no foe to history'. There can be little doubt that reversion to permanent, preferably sheep grazed, grassland is the ideal solution to the insidious degradation of sites being damaged by cultivation. A recent first estimate of the cost of taking all Scheduled Monuments in England out of cultivation is about £2.5m per year (S Trow, pers comm). Both the Avebury and Stonehenge World Heritage Site management plans envisage extensive reversion to ecologically rich grassland, and a special pilot Countryside Stewardship Scheme is being developed to look at this, including the level of compensation payments required.

Purchase of key sites for grassland reversion and leasing out to graziers under a Countryside Stewardship Scheme is an approach that has been developed – and shown to be financially sustainable – by the Norfolk Archaeological Trust (Wade-Martins 1996). As yet this type of approach is fairly small scale and limited to only a few counties, and the logistics and cost would seem likely to be prohibitive in terms of addressing the generality of the problem.

Targeted set-aside is another real possibility (long since advocated by the CBA when set-aside was first introduced 15 years ago). Although it does not necessarily offer a permanent solution, it is increasingly widely recognised that long term set-aside offers valuable means of achieving other environmental objectives such as supporting breeding birds such as the curlew and skylark; mammals like the brown hare and can also be beneficial for game.

Reversion is already possible under agri-environmental schemes but is not always practicable or attractive from a farming point of view. In predominantly arable areas where high yields are possible the available compensation based on flat rate national rates for income foregone is seldom realistic. Moreover many farms in predominantly arable areas simply do not have stock.

Archaeologically benign cultivation

Work carried out for the CBA back in the mid 1970s on the archaeological effects of different cultivation systems

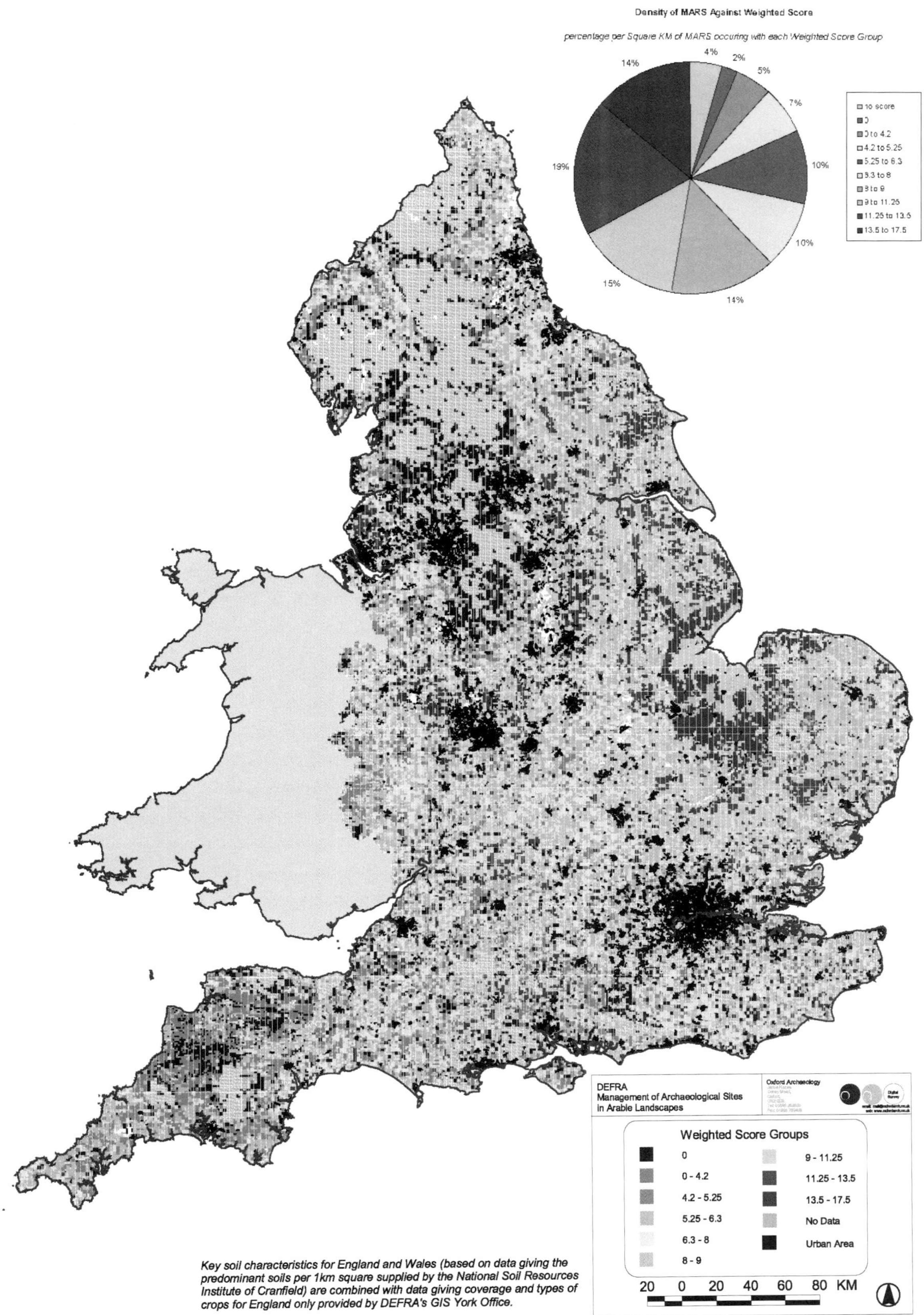

Fig 25.2 Map of the risk of plough damage to archaeological sites (from DEFRA 2002, fig 4)

(Lambrick 1977) showed clearly that forms of minimal cultivation, and especially direct drilling – or so-called 'no-till' cultivation – even then offered real potential for conserving archaeological sites within ongoing arable management. This is simply because they offer means of reseeding grass and growing crops that can be combine harvested (mainly cereals, oil seeds, peas and beans) with either much reduced soil disturbance or none at all (Christian and Ball 1994; Rasmussen 1999). There still remain some serious concerns that an entirely no-till cultivation system is not a long-term viable option, and that occasional subsoiling or pan-busting is needed. But some long-term trials have proved successful (MAFF 1998). In any case tackling this shortcoming in any scheme promoting such techniques for conservation purposes would potentially be a matter of assessing the loss of yield – and therefore income likely to be forgone – if conservation interests demand that such associated disturbance should not be allowed.

From an archaeological point of view, a key feature of direct drilling is the geographical pattern of its applicability. The areas where direct drilling – and by inference other minimum cultivation techniques – are most readily applicable generally coincide with the areas assessed as being at highest risk of archaeological damage.

As with reversion, the environmental attractiveness of no-till and other minimum cultivation techniques (which come in various guises as integrated crop management conservation tillage, ECOtillage) is not just archaeological. Indeed they offer potential benefits in terms of economic and agronomic performance as well (MAFF 1998; Monsanto 1999; Tebrugge and During 1999; Soil Management Initiative 2002). Economic advantages include reduced fuel costs (eg no-tillage only 17% of conventional ploughing); reduced labour costs (eg ECOtillage only 48–60% of conventional ploughing); reduced use of expensive agri-chemicals (eg 15% reduction in fertiliser costs; 35–40% reduction in herbicide fungicide and insecticide costs); reduced maintenance and wear costs of machinery. These advantages generally - and sometimes significantly – outweigh the disadvantages of maintained or somewhat increased seed costs (eg +12%) and maintained or somewhat reduced yields (eg -15% to +5% change).

Overall, with various 'no till' and 'minimum cultivation' systems it is possible in the right conditions to reduce inputs, deliver environmental benefits and still maintain profitability. While output may reduce, so do the costs of production. Furthermore, as grain prices fall, margins are not hit as hard with integrated crop management as with conventional systems.

Agronomic advantages that have been claimed for conservation tillage systems include better soil structure and organic content (eg 43% increase in infiltration rate and 18% increase in moisture retention); improved soil fauna and flora (eg 44% increase in soil organisms, 63% increase in earthworm biomass, and increased natural predators on slugs and aphids); reduced machinery passes leading to less soil compaction; better timeliness of operations reducing risks of damage to soils.

Environmental advantages that have been claimed include significantly improved archaeological conservation in the right conditions; maintained potential in the right conditions for habitat creation for arable weeds, conservation headlands, arable farmland bird habitats; reduced erosion (eg 48% reduced run off and 68% less sediment loss); improved soil chemistry (eg 81% reduction in loss of phosphate and 69% reduction in soil nitrogen depletion); reduced carbon dioxide and other emissions to air; less use of fossil fuels (eg 35% reduction in energy consumption).

Even without the benefit of improved long term conservation of non-renewable archaeological assets, there thus appears to be a strong case, where soil and other physical conditions allow and in relation to combine harvested crops, for arable farmers to be moving their production systems towards the minimising of cultivation operations. However, their ability to do so will be governed by a number of other factors including farm size, labour availability and skills, existing cropping and rotation patterns, future expansion plans, machinery running costs and the residual value of existing machinery (Park et al 1997; Monsanto 1999; University of Cambridge 1998). The relative versatility of traditional cultivation techniques (eg for roots as well as combine harvested crops), coupled with natural conservatism with regard to common, tried and tested techniques, represent further barriers to acceptance of the new reduced tillage systems.

As the need to reduce basic costs of arable farming becomes more critical, there may be more inclination for farmers to move to reduced tillage systems, though this is likely to be tempered by implications of having to put extra investment into capital equipment. Any support mechanism designed to promote the adoption of these techniques for environmental reasons must thus address not just their technical merits and performance in relation to running costs and yields, but also how the practical viability of different schemes for the farmer will relate to types of farm business, capital investment strategies and crop rotation systems. However, moves towards more cultivation through contractors, and more farm cooperatives and partnerships sharing the costs of investment in capital, suggest that some of the current barriers to their adoption may be weakening. This has recently been exemplified by a consortium of arable farmers in Suffolk who have jointly invested in a reduced tillage system purely on a business basis.

Limitations

In general all this is very encouraging, but we do also need to be aware of some of the limitations of this apparently rosey outlook. Reduced cultivation systems may be only a stop-gap measure, especially if hampered by restrictions on dealing with compaction issues; and given that benign cultivation systems are likely to have to be applied to whole fields rather than just the most archaeologically sensitive areas, apparent cost savings compared with reversion to grass more tightly focused on key areas, may prove illusory in many cases.

The practical viability and economics of different remediation measures will vary from one region to another. Of all areas under threat of damage, the East Anglian Fens and other areas of lowland peat in arable production probably present the greatest challenge. Well-preserved upstanding earthworks and gradually decaying waterlogged structures and deposits continue to emerge from beneath peat shrinking in the Fens at a rate of 2.0m to 3.0m per century. However, the economics and logistics of exclusively arable farm enterprises engaged in high-value root crops, unsuitable for minimum cultivation methods, mean that neither reversion nor benign cultivation systems offer obviously practicable or cheap solutions in agricultural terms. On the other hand, excavation, which is often thought of as the cheaper alternative to preservation *in situ* in a development context, would be even more costly on the scale involved over large tracts of wetland areas, and the polluter pays principle of *Planning policy guidance note 16* (PPG16, DoE 1990) would have to apply at a strategic, rather than individual level. From this perspective, relatively costly reversion targeted at key areas, perhaps combined with wetland habitat regeneration, could seem a good deal.

An emerging vision for archaeological conservation in arable landscapes

A new, more integrated strand of agri-environment support for archaeological conservation in arable areas could offer both reversion and direct drilling or minimum cultivation techniques as viable options for greatly enhanced preservation of archaeological remains *in situ* in arable areas, which could also include measures to deal with other, non-cultivation threats to archaeological sites in arable landscapes. Relocation of medium to long-term set-aside could potentially provide another means of stemming the threat.

Such provisions would focus primarily on conserving non-renewable fixed assets such as archaeological sites, but would offer additional incentives to use these areas to deliver locationally less constrained benefits of extended habitat recreation. This could have the great advantage of linking archaeological conservation to achieving key biodiversity targets such as reversing the decline in farmland bird species like the stone curlew. Research (eg Asteraki 2000) is demonstrating the increased biodiversity benefits of habitat creation in wider rather than relatively narrow field margins, but the potential role of archaeological conservation to maximise such environmental benefits on an even larger scale through set-aside or agri-environment schemes has tended to be ignored. This is partly because while Countryside Stewardship Schemes and Environmentally Sensitive Areas are fundamentally integrated approaches to conservation (and Stewardship in particular includes valuable archaeological provisions that are increasingly being used to help address this issue), tackling the primary conservation issue of arable damage to non-renewable archaeological assets has not been a principal objective that triggers environmental grant-aid – even in the Stewardship pilot arable scheme. In addition, archaeological conservation has not been integrated into previous agri-environmental research, and so its needs and the potential opportunities that it offers are relatively poorly understood in formulation of policy.

More needs to be done to integrate these different environmental goals and use them to promote other good practice requirements to ensure soil conservation and control of agro-chemical pollution. This would ensure that the taxpayer gets better value for money from agri-environment support.

Policy context

The foot and mouth outbreak in 2001 has dramatically shown that it is tourism rather than food production that is the real backbone of the rural economy. But farming remains the only means of sustaining an attractive countryside that produces healthy food, attracts tourism and protects the environment, and the emerging policy shift of agricultural policy from commodity support towards environmental support has been given a major push forwards. This offers an important opportunity to ensure that expanded agri-environment schemes or other incentives should be used to ensure much better preservation of archaeological remains in arable areas. What emerges from the forthcoming review of agri-environment schemes in 2003, or from the potentially even more significant review of the whole of the Community Agricultural Policy (CAP) in a few years' time remains to be seen. What this project should do is at least to provide a steer on how the centuries-old problem of agricultural damage to archaeological sites in arable landscapes might start to be tackled in earnest.

There is a need for further research to back up this work and to start to test and deliver the real conservation results that could provide a substantial basis for addressing the most extensive challenge for preserving archaeological remains *in situ* that we face. The issue is bound up in much wider major policy issues for the future of the rural economy and reform of CAP – not to mention World Trade Organisation pressures on European farming. It is an issue that needs to start moving onto the European stage to grasp an opportunity to make common the cause for conserving our shared culture and archaeological heritage.

Bibliography

Allden, A, 1982 *An analysis of central government (DAHMB) support in 1982/3 for the recording of archaeological sites and landscapes in advance of their destruction*, London

Asteraki, E J, 2000 Cereal field margins: enhancement of plant and invertebrate biodiversity through management and targeting of ELMS in MAFF, *Review of MAFF's R and D on biodiversity and conservation management*, review document for conference 14 July 2000

Barnatt, J, and Collis, J, 1996 *Barrows in the Peak District: recent research*, Sheffield

CBA, 1988 Council for British Archaeology, A policy for the

countryside, *Brit Archaeol News* 3, 57–60

CBA, 1993 Council for British Archaeology, *The past in tomorrow's landscape*, York

Christian, D G, and Ball, B C, 1994 Reduced cultivation and direct drilling for cereals, in Great Britain, in *Conservation tillage in temperate agroecosystems* (ed M R Carter), 117–140, London

Cook, J, and Rowley, T, 1985 *Dorchester through the ages*, Oxford

Darvill, T, and Fulton, A K, nd [1998] *MARS: the monuments at risk survey of England, main report*, Poole

DEFRA, 2002 Department for Environment, Food and Rural Affairs, *The management of archaeological sites in arable landscapes (BD1701)*, London

Drewett, P L, 1980 Sussex plough damage survey, in Hinchliffe and Schadla-Hall (eds), 69–74

Drinkwater, J, 1972 Barrows in Gloucestershire, patterns of destruction, in *Archaeology and the landscape* (ed P Fowler), 129–56, London

Dunkin, D, nd [1996] The East Sussex plough damage survey 1995–6, unpub Institute of Archaeology, London rep

Evans, R, 1990 Soils at risk of accelerated erosion in England and Wales, *Soil Use Management* 6(3), 125–31

Gingell, C, 1976 *Archaeology in the Wiltshire countryside*, rep for Wilts Archaeol Natur Hist Soc and DoE, Devizes

Hinchcliffe, J, and Schadla-Hall, R T (eds), 1980 *The past under the plough*, DoE Directorate of AMHB, Occas Pap 3

Lambrick, G, 1975 Ploughing at Alchester, in The defences of Roman Alchester (J Young), *Oxoniensia* 40, 155–60

Lambrick, G, 1977 *Archaeology and agriculture*, CBA and Oxford Archaeol Unit, London and Oxford

Lambrick, G, 1988 *The Rollright Stones: megaliths, monuments and settlement in the prehistoric landscape*, Engl Heritage Archaeol Rep 6, London

Lambrick, G, 1992 Alluvial archaeology of the Holocene in the Upper Thames Valley 1971–91: a review, in *Alluvial archaeology in Britain* (eds S Needham and M Macklin), 209–26, Oxbow Monogr 27, Oxford

Lawson, A J, Brown, J E, and Healey, F, 1986 *Barrow excavations in Norfolk 1950–82*, East Anglian Archaeol Rep 29, Dereham

MAFF, 1998 Ministry of Agriculture, Fisheries and Food, *Integrated farming-agricultural research into practice*, report from the Integrated Arable Crop Production Alliance for Farmers, Agronomists and Advisers, London

Monsanto, 1999 *Progressive crop management: ECOtillage reduces inputs and maximises crop yields*

Park, J R, Farmer, P, Bailey, A, Keatinge, J D H, Rehman, T and Tranter, R B 1997 Integrated arable farming systems and their potential uptake in the UK, *Farm Management J* 9, 483–94

Rasmusssen, K J, 1999 Impact of ploughless soil tillage on yield and soil quality: a Scandinavian review, *Soil Tillage Research* 53, 3–14

Saville, A, 1980 *Archaeological sites in the Avon and Gloucestershire Cotswolds: an extensive survey of a rural archaeological resource with special reference to plough damage*, Comm Rescue Archaeol Avon Gloucestershire Somerset Occas Pap 5, Bristol

Schadla-Hall, R T, 1977 *Winchester District: the archaeological potential*, Winchester

Soil Management Initiative, 2002 *Improved soil management for agronomic and environmental gain*, Chester

Tebrugge, F, and During, R A, 1999 Reducing tillage intensity: a review of results from a long-term study in Germany, *Soil Tillage Research* 53, 15–28

University of Cambridge, 1998 *Report on farming in the eastern counties of England 1996/97*, Cambridge

Van de Noort, R, Fletcher, W, Thomas, G, Cartsiars, I, and Patrick, D, 2002 *Monuments at risk in England's wetlands*, Exeter

Wade-Martins, P, 1996 Monument conservation through land purchase, *Conserv Bull* 29, 8–11

III

ISSUES IN PRESERVATION and CONSERVATION STRATEGY

26

From arena to art gallery: the preservation of London's Roman amphitheatre *in situ*

HELEN GANIARIS

and NICK BATEMAN

Helen Ganiaris
Senior Conservator
Museum of London
London Wall
London EC2Y 5HN, UK

Nick Bateman
Senior Project Manager
Museum of London Archaeology Service
Mortimer Wheeler House, 46 Eagle Wharf Road
London N1 7ED, UK

Abstract

London's Roman amphitheatre was first discovered during excavations in 1988 on the site of what was to become the new Guildhall Art Gallery. Masonry walls around the arena, entrance passages and side-chambers were found, along with associated timber structures such as drains and thresholds. The site was scheduled as an ancient monument, and redevelopment plans were revised so that the remains could be preserved *in situ* at a basement level, with two further basements beneath. This short paper outlines the processes involved for piling, underpinning and protecting the masonry during building works, drying the walls, and conserving the timbers. Plans for the display of the arena, including 'below ground' recesses for the timber drains, are described.

Introduction

In spring 1988, the Museum of London was carrying out excavations just to the south of the City of London's Guildhall when it made the unexpected discovery of part of London's Roman amphitheatre. The excavations were being carried out in advance of construction of the new Guildhall Art Gallery for the City, which houses the collections of the Corporation of London. Following designation of the site as a Scheduled Ancient Monument, proposals for the new building were reassessed and the Museum of London Archaeology Service (MoLAS) returned to the site between 1992 and 1999. Further excavations revealed more of the substantial timber and masonry remains of the amphitheatre, exposing most of the east end of the elliptical structure, about half of the arena, and entrance passages and various rooms (Figs 26.1–26.2) (Bateman 1997; Bateman 2000).

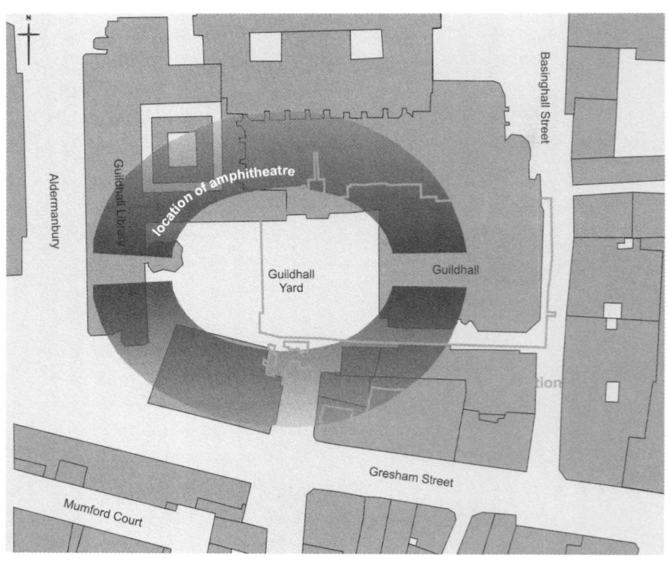

Fig 26.1 Plan showing the amphitheatre and the Guildhall Yard area with the extent of archaeological excavations (© Crown copyright. All rights reserved. Corporation of London 087254-2004)

Fig 26.2 *Archaeologists recording a length of arena wall; the walls stand up to 1.5m high in places*

The design of the new Art Gallery building – in particular piling locations and floor levels – was modified to allow retention of the amphitheatre remains *in situ* (Fig 26.3). With two further basements planned beneath the Roman levels, provision had to be made for supporting and maintaining the Roman masonry and for removal of the timber elements for conservation off site. The Art Gallery opened to the public in 1999, and plans for public access to the amphitheatre were proposed in 2000 when the Corporation decided to develop part of the basement as their new marketing suite. At the same time, the Roman remains would be prepared for public display.

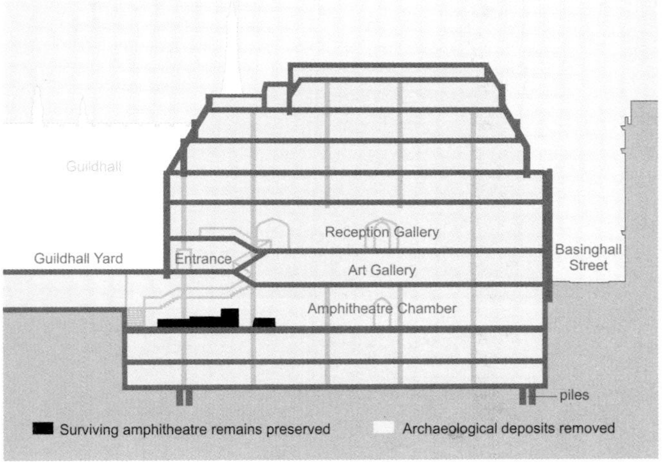

Fig 26.3 *Cross section through the new Guildhall Art Gallery, showing the level at which the amphitheatre was found and the two levels beneath*

Piling and underpinning

Before long-term conservation *in situ* could be considered, it was necessary to protect the amphitheatre walls during the initial construction work in 1994. As a first step, plywood boxing was constructed around them, with an outer 'tell-tale' of fine plastic netting to act as a hazard warning to drivers (Fig 26.4), while the space between the walls and the boxing was filled with spray foam (Nixon nd). The main piling operation was carried out from above a massive steel transfer deck (at new ground level), which was itself suspended above the amphitheatre remains on clusters of temporary mini-piles. With the main piles and the transfer deck in place, work started on raising the building's superstructure.

The amphitheatre walls, together with a block of underlying natural ballast, were then underpinned, extending the new basement floor beneath them section by section. After trial runs and many unforeseen problems, a successful system was devised. The first stage involved strengthening the gravel below the wall with an epoxy grout injected via tubes. The gravel foundations were then supported externally while the wall and underlying grouted gravel was undermined in sections to allow props, lintels and floor connectors to be placed within the void beneath the grouted gravel. Finally, the void was filled with reinforced concrete creating a 900mm thick support connected to the main floor slab (Fig 26.5). The underpinning process is described in greater detail elsewhere (Crooks 1995).

Fig 26.4 *The amphitheatre walls with plywood boxing for protection during construction work; the timber struts support the outer 'tell-tale' of wire netting*

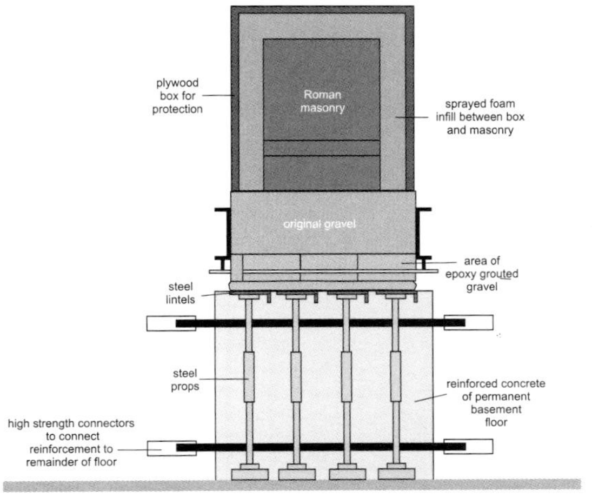

Fig 26.5 *One stage in the underpinning of the masonry; epoxy grouting was injected within the lower area of the Roman gravel, steel props supported the walls while the underlying soil was removed and concrete was poured to create the permanent floor*

To give more permanent protection to the walls before the next stage of construction work, the plywood boxing around the masonry walls was replaced with breeze-block enclosures. With the amphitheatre suspended and protected, it was possible to start excavating the ground beneath for the construction of the two further basement levels.

The masonry

Prior to the underpinning work described above, the entire area had to be dewatered to reduce the water table by 2m. It was initially planned to do this by pumping directly from sumps slowly enough to prevent settling of the Roman foundations. In the end a 'well point' system was used (Crooks 1995, 47–8). After dewatering, the masonry was still damp and it was anticipated that it might crack if allowed to dry too quickly. Heavy gauge polythene sheeting was placed loosely over the walls to slow down the drying rate (Fig 26.6). Ridout Associates, environmental consultants for the project, carried out monitoring of the environment within the entire display area, the plywood enclosures and the space beneath the polythene sheeting.

Fig 26.6 The masonry walls protected within breeze-block and plywood enclosures; heavy gauge polythene sheeting was draped loosely over the walls to slow down drying

Initially, this was carried out with sensors, placed in 28 locations, cabled to a datalogger. When preliminary building work began in 2001, these were replaced with ten stand alone dataloggers, not requiring cabling. Core samples and/or internal probes to monitor the moisture content of the masonry were not used because of possible damage to the masonry.

In advance of the development of the marketing suite scheduled to begin in late summer 2001, the Corporation's Department of Technical Services, together with English Heritage, prepared a method statement detailing plans for drying the masonry. As ambient conditions had stabilised to 50%RH (±5%), it was deemed safe to expose the masonry in stages over a period of 15 weeks. First the breeze block walls were lowered and new polythene lids were made to replace the plywood ceiling. One section of sheeting was removed from the masonry as an initial trial; as no cracking was observed, all the sheeting and eventually the lids were removed.

Throughout the process, the condition of the walls was monitored by conservators from Museum of London Specialist Services and Cliveden Conservation Workshop Ltd. During the final stages of drying, eight monitoring points were set up and measurements were taken at regular intervals. No damage from the final drying has been noted. When the remaining breeze-block walls are removed, some remedial work will be required to prepare the walls for display.

The timbers

The original amphitheatre (dated *c* AD 70) was built entirely of wood, though it was later enlarged and rebuilt incorporating both masonry and timber structures in the early 2nd century. Over 500 timbers were found relating to the amphitheatre including beams, posts, drains, silt traps, thresholds and gateposts. On the whole the timbers were found in very good condition, with many clearly showing toolmarks and an unprecedented range of carpentry techniques. The degree of survival had not been predicted as the site is neither low-lying nor directly adjacent to the course of an extinct river. The bowl-shaped depression created by the amphitheatre, and subsequent silting up of the drains, seem to have created localised waterlogged conditions - which also preserved later Saxo-Norman timber structures. In total, from all phases of the site, over 1000 timbers were found (Goodburn 1995).

Many of the timbers were retained for study and dating and, of these, a large proportion was stored wet in tanks for future display (Fig 26.7). The final selection for display *in situ* includes a 6.0m section of the main axial drain under the entrance ramp, a 3.0m section of the same drain beneath the arena, an associated silt trap, and two thresholds for gateways leading to side chambers. Other timbers will be retained for Museum of London collections and displays.

Fig 26.7 *One of the large timber beams from the arena entranceway being winched off site*

The selected timbers are being treated at the Museum of London and the York Archaeological Trust. Immersion in polyethylene glycol (two stage process using PEG 200 and PEG 4000) will be followed by freeze-drying. Installation of the timbers on site will take place in 2004.

The display

It was decided that, rather than create a conventional museum, the display would emphasise the architecture itself and recreate the excitement and drama of entering a Roman amphitheatre. The design, by Branson Coates Architecture Ltd, evokes this with a theatrical use of light, projection and sound highlighting the masonry and timber features.

Visitors will be able to walk through the arena entrance, which will also be visible from above through a viewing window in the Art Gallery. The timber drains and silt trap will be mounted, as found, below floor level, within two long recessed chambers covered by glass. Visitors will then be able to walk over, and look down on, the conserved timbers. Since the amphitheatre will be open to the public (summer 2002) several years before conservation of the wood is complete, timber structures will be represented at first by archaeological line drawings placed just below the glass.

A series of graphic panels along the route will give visitors key information. Guided tours will be given by Museum of London staff and the Museum will display selected finds from the site.

Postscript

Subsequent to this paper being presented the amphitheatre is now open to the public. The amphitheatre is entered through the Guildhall Art Gallery, for details of admission visit the Corporation of London website (http://www.guildhall-art-gallery.org.uk) and for organised tours contact the Museum of London (http://www.museum-london.org.uk).

Acknowledgements

The Corporation of London has generously sponsored the excavations and subsequent work on the site and English Heritage (London Division) has played a vital role in the archaeological process throughout.

We are grateful to the following who contributed information and images: Branson Coates Architecture Ltd, Cliveden Conservation Workshop Ltd, Ridout Associates, and J Crooks, construction manager for the site. Structural work on the building was carried out by Oscar Faber Consulting Engineers and DY Davies.

Bibliography

Bateman, N, 1997 The London amphitheatre: excavations 1987–95, *Britannia* 28, 51–85

Bateman, N, 2000 *Gladiators at the Guildhall: the story of London's Roman amphitheatre*, MoLAS, London

Crooks, J 1995 Monumental headache, *Building Magazine* 8, 45–50

Goodburn, D 1995 Summary of the finds of ancient woodwork made at the Guildhall excavations, unpub MoLAS rep

Nixon, T, nd [1998] Practically preserved; observations on the impact of construction on urban archaeological deposits, in *Preserving archaeological remains in situ: proceedings of the conference of 1st–3rd April 1996* (eds M Corfield, P Hinton, T Nixon and M Pollard), 39–46, London

27

Between a rock and a hard decision: the role of archaeology in the conservation planning process

KATE CLARK

Kate Clark
(formerly) Head of Historic Environment
English Heritage
23 Savile Row
London W1X 1AB, UK

Abstract

Significance lies at the heart of all conservation decisions, and archaeology has an important role in both understanding fabric and assessing significance. Yet many decisions about the conservation of buildings, landscapes or places are taken without any understanding of the potential role of archaeology. The result is, quite simply, that places are damaged. At the same time, many archaeologists are locked in the 'mitigation ghetto' where they see their role as one of recording damage rather than informing positive decisions. They work within a relatively narrow 'bandwidth' of values, failing to recognise the wider significance of the historic environment.

The twin processes of conservation planning and heritage impact assessment are now becoming central to managing the historic environment. They are in fact a codification of what has always been implicit, but not explicit, in how we make decisions about the past. These processes are generating a new and enhanced role for archaeology in the process, but only if archaeologists are prepared to respond to the challenge by rethinking how they collect data, how they assess significance and how they work with other professions.

Introduction

A lot of the debate on the preservation of archaeological remains *in situ* has focused on whether or not buried remains will still be there for archaeologists to excavate in years to come. But the preservation of archaeology for archaeologists is only a tiny part of a much wider, and much more important, debate about the conservation of the historic environment as a whole. That process – conservation – involves handing on what we value to future generations, whoever they may be, not just to excavate but also to enjoy. Archaeology has not always played a very large role in this wider debate, for two reasons; partly because other environmental professionals do not always see a potential role for archaeology, and partly because archaeologists themselves remain in a sort of 'mitigation ghetto' tightly focused on preserving buried remains.

In this paper I want to challenge archaeologists to move away from this narrow perspective, and to begin to take up a role as stewards and interpreters of the historic environment, whose work could underpin the whole process of sustaining that historic environment.

Conservation: first principles

In order to understand the potential role of archaeology in conservation, it is first useful to explore current definitions of conservation. Although the word conservation itself often acts as a barrier to communication between professionals, it is one in common use in the UK (as opposed to heritage management or cultural resource management). Its meaning today is almost akin to 'sustainability'.

What does conservation involve? It has almost as many definitions as there are practitioners in different fields. Within heritage, there is a very narrow definition which sees it very much as relating to the physical preservation of buildings or objects – especially their repair and restoration. The broader definition, set out in the Australian Burra Charter, defines conservation as 'All the processes of looking after a place so as to retain its cultural significance' (Australia ICOMOS 2000).

'All the processes' can be interpreted to include almost anything concerned with the management of historic places – from repair and maintenance through to access, visitor management, the management of the organisation or commercial activities, all of which will have an impact on significance. This definition has emerged as the idea of conservation has moved from isolated monuments or objects in museums, to embrace the wider historic environment, including buildings, landscapes, buried remains and historic city centres. By necessity, conservation in these areas has come to encompass a wider range of activities than simply intervention into fabric.

The second important thing about this definition is that it places retaining cultural significance at the centre of conservation activities. Critics have suggested that this could be taken to mean that values should be retained rather than fabric; this is not the case. Judgements about significance simply provide a framework for making individual decisions about places, rather than confining the processes to a series of traditional ethical recipes such as 'restoration', 'repair', 'reinstatement' and 'reconstruction', many of which would be inappropriate in the context of, for example, managing London.

The National Trust stresses that conservation does not mean preserving or fossilising things, which is expressed in the following definition: 'Negotiating the transition from past to future so as to ensure the transfer of maximum significance' (Holland and Rawles 1993, quoted National Trust 1995, 21). This can be paraphrased as, managing change in order to hand on what we value to future generations.

It is important to follow these changing definitions, as they encapsulate some of the ways in which conservation practice has changed. Conservation professionals today are not simply technical conservators – they may be architects, engineers, archaeologists, conservators, gardens or landscape specialists, museum curators, planners, or any one of many professions. They will need to understand not only the technical specifications of the things they deal with, but the wider economic, social, political and policy context of conservation.

Our concept of what we conserve is also moving on. Although we tend to retain our professional areas of expertise, we are increasingly moving towards an idea of historic environment which moves away from the focus on individual sites, buildings, objects towards the idea of a historic environment, which encompasses all of the physical remains of the past. The move is akin to that in ecology from a focus on species, to a concern for first habitat, and then the idea of environment as a whole.

This change in attitude is perhaps best seen in the contrast between *Planning policy guidance note 16* (PPG16) published in 1990, and *Planning policy guidance note 15* (PPG15), just four years later (DoE 1990; DoE 1994). Although PPG16 pays lip service to historic landscapes, the fundamental purpose of archaeology is to determine 'whether archaeological remains exist on a site where development is planned and the implications for the development proposal'.

PPG15, in contrast, deals with the historic environment as a whole, and asks us to consider the implications of any proposal for the special interest of the site. These are two very different questions, and imply two different approaches – the one emphasising 'finding remains' the other, considering the impact on value.

More recently, the idea of the historic environment as a whole was set out in *Power of place; the future of the historic environment* (English Heritage 2000), a sector-wide statement on the future of heritage conservation, and was picked up again strongly in the governments response, *The historic environment; force for our future* (DCMS/DTRL 2001).

This general move towards a wider definition for conservation, as well as a greatly expanded idea of what is to be conserved, has meant a change in the way that those professions and individuals who practice conservation operate.

Significance

Although we are increasingly interested in the conservation of the historic environment, and that historic environment can encompass almost anything which has been made or influenced by people, this does not mean that everything should be preserved. Decisions are made all the time about what will or will not be kept. As noted above, the basis of every such decision is, somewhere along the line, an assessment of significance or value. Value lies at the heart of all conservation decisions.

Not everything in the historic environment is of equal value, and it is notions of value which justify protection, designation, regulation and the expenditure of public money. Conservation is at its most basic, the declaration of a public interest in private property, and the role of those of us who practice conservation is to defend and if need be, help to articulate, that public interest.

The historic environment can be valued in a number of different ways. The Burra Charter notes that cultural significance can mean 'aesthetic, historic, scientific, social or spiritual value for past, present or future generations' (Australia ICOMOS 2000).

For archaeologists, the value of the historic environment lies largely in its 'ability to demonstrate' past events. Physical remains of buildings, landscapes or places are a source of evidence which enable us to create historical narratives and give meaning to modern places. We use the techniques of field archaeology and a range of analytical tools (such as the idea of stratigraphy) to interpret physical evidence, and borrow a range of interpretive tools from anthropology and other disciplines in order to give meaning to that evidence. What we often seek to preserve is that evidential value – whether the sequence of changes in a building, the buried deposits beneath a London basement or the earthwork remains of a motte and bailey castle.

Of course judgements about value are enshrined in legislation which enables sites to be designated or set aside for special treatment. For buildings, in England at least, the 1990 *Planning (Listed Buildings and Conservation Areas) Act* enshrines 'architectural and historic interest' as a key value against which buildings are listed. This in turn is broken down with age, with more rigorous tests applied to more modern structures. For historic areas, the legislation refers to the idea of 'character or appearance which it is desirable to preserve or enhance'. 'National importance' is the basis on which Scheduled Monuments are designated in the 1979 *Ancient Monuments and Archaeological Areas Act* (Darvill et al 1987).

There are of course, a much wider range of values for the historic environment than are set out in the legislation. Understanding those wider values is an essential part of the management process (English Heritage 1997; GCI 2000).

Economic value

There is increasing evidence for the economic value of the historic environment. The conservation of the historic and natural environment takes a decisive share in the dispersed economic flows from tourism, although this can be hard to capture. Investment in conservation enhances the value of a property and makes it more attractive for the visitor, either as the object of a visit or as the backdrop to other leisure activities. The conserved landscapes of the south-west, for example, attract a spend of £2354 million from holiday trips in the region, and activities and spending by the National Trust contribute £248 million to the local economy.

English Heritage has conducted research which shows that listed commercial buildings perform as well or better than their counterparts. Public funding acts as an economic multiplier: every £10,000 of English Heritage grant in historic areas generates an additional £48,000 of match funding from private sector and public sources. Together this delivers on average 177m^2 of improved commercial floorspace plus one new job, one safeguarded job and one improved home (English Heritage 1999). There are over 800 different companies represented in the annual Building Conservation Directory (Cathedral Communications 2001).

It is also often forgotten that over half of the £54 billion spent on construction each year is devoted to existing buildings. This means that the maintenance and management of existing buildings is a major element in the economy of the country and as important as, if not more important than, the construction of new buildings.

Obviously these are isolated statistics, but they are beginning to demonstrate the contribution that the historic environment makes to the national economy. Such figures matter because they help the sector to demonstrate that heritage is not a brake on development – as it is so often portrayed – but plays a positive role in the economy of Britain. Business leaders in some of the key growth sectors of the economy (eg knowledge based industries) repeatedly identify environmental quality as a major influence on locational decisions.

Aesthetic and amenity values

Archaeologists are often very bad at recognising the aesthetic, spatial and design qualities of the historic environment; but it is often these qualities which are most appreciated and for which a place may be conserved. The enjoyment of places, whether for their familiarity or for their ability to inspire, is critical to conservation and is part of what we seek to hand on to future generations. Acknowledging, understanding and working with aesthetic concepts is essential.

Social values

Most people would accept that the historic environment has social value. It is the place where people live, work and enjoy themselves, or a place of commemoration or memory. People are tied to their historic environment by memories and events, or simply by the comfort created by an old and familiar place. Yet many of the systems for designating sites acknowledge social values, but there has been relatively little thought amongst heritage practitioners about what social value involves or indeed how to capture it (Byrne et al 2001).

Cultural diversity

The Stephen Lawrence report (Macpherson 1999), published after the death of a young architect, highlighted the role of institutions in shaping perceptions about people and their culture. It asked that, Local Government and relevant agencies should specifically consider implementing community and local initiatives aimed at promoting cultural diversity and addressing racism and the need for focused, consistent support for such initiatives.

As Professor Stuart Hall (1999) has pointed out, 'national heritage is a powerful source of meaning: those who can't see themselves reflected in that mirror are therefore excluded'. Cultural heritage practitioners have a particular responsibility to ensure that the heritage they seek to identify and defend does reflect the full diversity of Britain. At the same time, the right to a heritage can be seen to bring with it a responsibility to respect the heritage of others. The recent Convention on Identity published by the Council of Europe notes that 'individuals and communities have a fundamental right to self-defined identities to know their history and to shape their future through their heritage', and that 'individuals and communities have a right to enjoy their heritage. They equally have an obligation to respect the heritage of others and the communal interest in all heritage'.

The issues set out here only touch on the growing literature on the subject of value and how it is defined, both in archaeology and in heritage conservation as a whole (Darvill 1987; Carmen 1996; Briuer and Mathers 1996). This literature shows that there are many other, more sophisticated ways of defining value, and indeed that the exploration of value and how it is defined is a critical part of heritage practice. For the moment, it is enough to recognise that values – however they are defined – are central to conservation.

Sustainable development

A parallel initiative which has shaped thinking has been the concept of sustainable development. Emerging out of the Rio Conference, and Agenda 21, the fundamental ideal that development and conservation can be reconciled (IUCN 1991) In other words, that they should not be seen as opposing forces, but that they can work together in order to bring both economic benefits and environmental benefits. With Agenda 21 came a raft of ideas, including the importance of bottom up as well as top down approaches to conservation and the need to integrate economics and conservation.

One of the greatest successes of the environmental movement has been the recognition that ecology cannot be sustained by the actions of ecologists alone. Environmental responsibility is something that we all share, and the choice we make will have some influence at least, on the environment. The appointment of government 'green ministers' shows also that environment conservation is not just seen as the responsibility of the government department in charge of the environment; it demonstrates that all areas of public life potentially have an impact on the environment, whether transport, employment, business or education.

All of these approaches apply as well to the historic environment as the natural one. If the word 'historic' is inserted in front of any of the Rio environmental recommendations the result is what seems to be an eminently sensible set of approaches which heritage conservation would do well to consider.

Indeed, the English Heritage leaflet on *Sustaining the historic environment* (1997) showed how the arguments about sustainability could equally apply to cultural heritage.

Vulnerability and loss

Green conservation has for decades collected and made strategic use of information not just about what they have got, but also about what is happening to it. Data on air quality, species counts and biodiversity enable them to make and support arguments for better conservation, through identifying key indicators which are then fed into government policy.

The historic environment is only just beginning to catch up. Australia and Canada are now into their second round of reporting on the State of the Historic Environment (Parks Canada 1998). Scotland has just published its first audit. In England data is provided in the Heritage Monitor, the Buildings at Risk work (English Heritage 2001) and the Monuments at Risk Survey (Darvill and Fulton nd) – but are only at the planning stage of comprehensive state of the environment reporting.

For the countryside, the Monuments at Risk Survey (Darvill and Fulton nd) has shown a significant decline in the condition and survival of rural monuments (which make up over half of all Scheduled Ancient Monuments), including: an estimated 22,500 monuments have been destroyed since 1945; 20% decline in survival of earthworks over the period; 2% of all monuments at risk of serious damage or destruction in the near future; 33% field monuments, 39% earthworks under cultivation; 68% of rural earthwork monuments destroyed or very poor.

Our understanding of patterns of loss in urban areas is therefore, at best, anecdotal. Backlands buildings, medieval plot boundaries, working class housing, small industrial buildings, upstanding and below ground archaeological remains (such as town walls) are all likely to be at risk. Mass produced building materials erode the quality of areas whose character and distinctiveness is based on local materials – the rich handmade bricks of the Midlands towns, the particular stone tiled roofs of the Cotswolds.

Quality of life

One of the fundamental planks of sustainability is that the environment is important because it contributes to the quality of our lives. In other words, it goes beyond the protection of individual species to make a contribution to how we live. This is important to counter arguments that it is development and by implication economic prosperity that generates quality of life, and only more development will create better lives.

We have been slow in the historic environment to counter these arguments with our own. We know instinctively that the historic environment contributes to pleasant places to live but as noted above, are only just beginning to gather the economic data that will demonstrate this.

Informed conservation: what is the role for archaeology in sustaining the historic environment?

In order to sustain the historic environment, we need to understand what is valued because it is arguments about value which inform the management of change. Two different tools can be used to ensure that value lies at the heart of decision-making. The first is the proactive approach – the conservation planning process which ensures that an understanding of what matters and why drive decisions about how it is managed. The second is the reactive approach – impact assessment – ensuring that an understanding of value informs decisions about change. The two cannot be separated.

Both proactive planning, and reactive decision-making depend fundamentally on a good understanding of what is valued, and that, in turn, depends upon an understanding of the site – in other words – its archaeology. Archaeological techniques can underpin both processes.

Conservation plans

Conservation (or value-led) planning is a technique for proactive management planning. The difference between conservation plans and the more traditional management planning process which has been in use for almost a century in national parks and other areas, is that conservation planning places significance at the heart of decision-making. This is not to say that other approaches have not done this. However conservation plans place a greater emphasis on understanding a place and recognising that there might be a variety of different values, than do other approaches.

Conservation planning emerged in Australia in the early 1980s in response to the publication of the *Burra Charter*. In 1982 James Semple Kerr published *The Conservation Plan*, a guide to preparing conservation plans for historic places. The aim of a conservation plan was to set out the significance of a site, and to show how that significance would be retained in any future use, alteration, repair or management. Kerr's approach involved four basic steps – understand the site, assess significance, define issues and set policies. Kerr's approach has evolved over the years. It is now recognised that one of the most important aspects of conservation planning is the process, and with the revision of the *Burra Charter* in 2000 it is now know as the Burra Charter Process. Greater emphasis is now placed on involving stakeholders from the outset, in implementation and on periodic review. The whole process looks broadly like this: identify stakeholders; understand the site; assess significance; identify vulnerability; set policies; implement; and review.

In 1998 the Heritage Lottery Fund (HLF) in the UK introduced conservation plans as a means of providing an assessment of places prior to considering grant aid (HLF 1998). The HLF had a wide remit, bringing together buildings, landscapes, collections, archives, archaeology and ecology, and needed a single approach to assessing value. The Kerr approach was introduced, but modified to recognise the wide variety of sites and values which needed to be considered. The HLF guidance was launched at a conference in 1998 (Clark 2001) at which there was much debate over the value of the process, including concerns that conservation plans would be used by developers to justify schemes.

Since then, conservation planning has been used widely in the UK, not only in conjunction with HLF funding, but also in association with the management of sites. English Heritage, the National Trust and a number of different organisations use conservation planning in order to manage their portfolio of properties. Value-led planning can also be seen to be part of a much wider general trend in managing cultural sites (Clark 2002).

Impact assessment

Many people have argued that conservation planning is not new but is simply a codification of existing best practice. This is also true of the process of impact assessment – the way by which conservation practitioners react to or make decisions about whether or not change is appropriate. But like conservation planning, there has been relatively little discussion of the process and what it involves. PPG15 asks that applicants provide local planning authorities with sufficient information 'to determine the impact of the proposal on the special interest of the building' (DoE 1994, para 3.4). So it is reasonable to see impact assessment as central to decision-making, however, the process is rarely explained (except in the context of formal Environmental Impact Assessments, required for a relatively small number of cases under separate legislation). Broadly, impact assessment involves assessing the need and justification for the work; identifying the potential impact of the work on the significance of the site; establishing whether any negative impacts can be mitigated through better design, choice of alterative locations or looking again at need and making a decision (which may involve modifying the original scheme). It is not a linear process but an iterative one, where one is often looking at need and justification in the light of impact, and seeking to use design as a way of reducing any negative impacts.

This whole approach will be very familiar to archaeologists who are used to assessing proposals and establishing mitigation strategies. But there is a very big difference here. In this case, the process involves looking at the whole project (not just the bits that affect below ground remains). It also involves assessing the impact of the work not just on buried remains, but on the whole significance of the site, and identifying mitigation strategies which are not just about choosing what trench to dig, but involve playing an active and positive role in the design process.

For example, assessing the impact of a new set of housing proposals; the archaeologist might devise an evaluation strategy which involves digging a series of trenches to establish whether buried remains survive. If they do, the mitigation strategy might involve some form of excavation prior to construction or some selective preservation *in situ*. Whereas a full impact assessment of that proposal might involve assessing the impact of new houses not just on buried remains, but on the historic landscape, on views, on community values, on existing services and on ecology. In exploring mitigation, one would explore design, materials, location, layout and of course need and justification. This approach means working with other professionals, and not in isolation from them. It means looking at a wider range of views and being prepared to explore a greater range of issues.

Yet although it is a much broader approach, the role of archaeology is enhanced. Archaeology is needed to understand the current layout and evolution of the area affected, whether hedgerows and boundaries, field patterns, past management regimes, buried and visible remains, trackways and the nearest settlement. An understanding of the whole character of the area and the impact of the proposals on that character is vital, an understanding which can begin with the archaeology of development in time and space.

The role of archaeology

Archaeologists should be good at two things – firstly understanding physical remains in time and space, their previous pattern of use alteration and repair (including the pattern of loss), which can be applied to any type of site, whether it is a historic landscape, a town, a stately home or a set of earthworks, The forensic analysis of fabric and how it has changed through time, as well as the implications context, forms the basis of understanding any historic place. The second thing that archaeologists – or at least those who were at university from about 1980 onwards – should be able to understand is the volatile and complex relationship between material culture and value, including the recognition of multiple values for places.

These two skills – understanding places and articulating their value – should underpin every decision about conservation, whether about a landscape, a building or a site. Unfortunately, years of rote evaluations and poor quality contract work, mean that they appear to have been bred out of much of the profession.

In arguing for the importance of understanding as the basis of decision-making, I am not seeking to suggest that archaeologists should become the universal all-purpose heritage experts. There is much more to valuing places than is taught in the average archaeology degree or experienced by the average working archaeologist. Few archaeologists know enough architectural history to understand the wider significance of buildings; our knowledge of formal landscape design and aesthetics is lacking; hardly any of us know why buildings stand up or how they were built. We know little of colour and often less of early technology, of ecology and ecological values or of the techniques of conservation. We do not know how to draw, or how to present information in a format that other conservation professionals can understand. Although excavators now deal confidently with rafts and piling, how many of us can read a repair specification intelligently? And despite lip service, archaeologists are poor at stakeholder participation and recognising social value.

Against that, there remains in conservation a fundamental problem with the quality of information on which decisions are based. In 1998, English Heritage, the former Royal Commission on the Historical Monuments of England and the Institute of Field Archaeologists commissioned a study of the role of information in PPG15 and found that most conservation officers found that their work was delayed by lack of appropriate information (Grover 2000). A rigorous assessment of a building – how it has changed through time and what survives – is rarely set out at the beginning of a conservation plan, and as a result many of the later assessments of significance and vulnerability are flawed. It is difficult to look after a place whose significance has not been adequately understood.

Conclusions

Significance lies at the heart of all conservation decisions, and archaeology has a role to play in both understanding fabric and assessing significance. Unfortunately, the majority of decisions which are made about the conservation of buildings, landscapes or places are taken without any understanding of the potential role of archaeology. The result, as the data on loss shows, is that places are damaged. At the same time, many archaeologists are locked into a mitigation ghetto where they see their role as one of recording damage rather than informing positive decisions. They work within the relatively narrow bandwidth of values, failing to recognise the wider significance of the historic environment.

The twin processes of conservation planning and heritage impact assessment are now becoming central to managing the historic environment (Clark 2001). Of course they are a codification of what has always been implicit, but rarely explicit, in how we make decisions about the past. These processes are generating a new and enhanced role for archaeology but only if archaeologists are prepared to respond to the challenge by rethinking how they collect data, how they assess significance and how they work with other professions.

There is a role for archaeology in many of the major initiatives in conservation. Urban spatial master-planning depends upon a good analysis of urban built form through time. The repair of a complex historic building depends upon value judgements about what matters and why and an understanding of how it has developed through time. The decision to reinstate a Repton pathway in Bristol will depend upon our archaeological understanding of what survives, and our ability to anticipate what impact the restoration will have.

Yet archaeologists are their own worst enemies. The perception that archaeology is about mitigating damage, rather than informing conservation persists. The PPG16 precept that we are there to find archaeological remains, rather than to assess impact on value remains in place. Because we have not made the case for proactive involvement early enough and well enough there is a danger of what I call the 'fabric backlash' in international conservation. A perception that values are intangible, that it is values we conserve, and that fabric is somehow not important. Yet even the values of an aboriginal community which may have no relation to man-made structures, will be tied in some way to places, and an understanding of those places, their extent and how they have used and evolved will be vital.

Peter Read, in his book *Lost places* (1996), writes about what it means to an individual to lose a place, whether as the result of environmental catastrophe or poor planning decisions. The statistics on loss show that we are losing aspects of the historic environment every day. It is as well for heritage practitioners to remind ourselves of what happens if we do our job badly.

Bibliography

Australia ICOMOS, 2000 International Council on Monuments and Sites, *The Burra charter*, [Canberra]

Briuer, F, and Mathers, W, 1996 *Trends and patterns in cultural resource significance: an historical perspective and annotated bibliography*, Institute for Water Resources, Alexandria VA

Byrne, D, Brayshaw, H, and Ireland, T, 2001 *Social significance: a discussion paper*, NSW National Parks and Wildlife Services

Carman, J, 1996, *Valuing ancient things: archaeology and law*, Leicester

Cathedral Communications 2001 *The building conservation directory*, Cathedral Conservation, Tisbury

Clark, K, 2001 The use of archaeology in building conservation at Down House, Kent, in *Archaeology and buildings: papers from a session held at the European Association of Archaeologists fifth annual meeting in Bournemouth 1999* (ed G Gunilla Malm), BAR Int Ser 930, Oxford

Clark, K (ed), 1999 *Conservation plans in action*, London

Clark, K, 2001 *Informed conservation: understanding historic buildings and their landscapes for conservation*, London

Clark, K, 2002 Preserving what matters: value led planning for cultural heritage sites, *Conservation – the Getty Conservation Institute Newsletter* 16(3), 5–12

Cooper, M A, Firth, A, Carman, J, and Wheatley, D (eds), 1995 *Managing archaeology*, London

Darvill, T, 1987 Valuing archaeological resources, in Cooper et al 1995, 40–50

Darvill, T, and Fulton, A K, nd [1998] *MARS: the monuments at risk survey in England, 1995 main report*, Poole

Darvill, T, Saunders, A, and Startin, B, 1987 A question of national importance, *Antiquity* 61, 393–408

DCMS/DTRL, 2001 Department for Culture, Media and Sport/ Department for Transport, Local Government and the Regions *The historic environment: a force for our future*, London

DoE, 1990 Department of the Environment, *Planning policy guidance note 16: archaeology and planning*, London

DoE, 1994 Department of the Environment, *Planning policy guidance note 15: planning and the historic environment*, London

English Heritage, 1997 *Sustaining the historic environment: new perspectives on the future*, London

English Heritage, 1999 *The heritage dividend: measuring the results of English Heritage regeneration 1994–99*, London

English Heritage, 2000 *Power of place: the future of the historic environment*, London

English Heritage, 2001 *Buildings at risk: the register 2001*, London

GCI, 2000 Getty Conservation Institute, *Values and heritage conservation research report*, Los Angeles

Grover, P, 2000 Local authority practice and PPG15: information and effectiveness, unpub Oxford Brookes Univ rep

HLF, 1998 Heritage Lottery Fund, *Conservation plans for historic places*, London

IUCN/UNEP/WWF 1991 *Caring for the Earth: a strategy for sustainable living*, Earthscan, Gland, Switzerland

Kerr, J S 1982 *The conservation plan; a guide to the preparation of conservation plans for places of European cultural significance*, 5 edn, National Trust of NSW, Sydney

Macpherson, W, Sir 1999 *The Stephen Lawrence enquiry*, London

National Trust, 1995 *Linking people and place*, Cirencester

Parks Canada, 1998 *State of the parks 1997 report*, Ottowa

Read, P, 1996 *Returning to nothing: the meaning of lost places*, Cambridge

28

Coastal change and the historic environment

PETER MURPHY

and STEVEN TROW

Peter Murphy
Regional Advisor for Archaeological Science
English Heritage, East of England
Brooklands, 24 Brooklands Avenue
Cambridge CB2 2BU, UK

Stephen Trow
Head of Countryside Policy
Department of Conservation
English Heritage
23 Savile Row
London W1X 1AB, UK

Abstract

Coastal erosion, relative sea level rise, and management schemes related to flood defence and nature conservation objectives all damage, destroy or submerge archaeological sites. In this paper the UK and EU policy frameworks for the management of coasts are outlined, UK heritage legislation and planning regulations are presented, and ways of integrating these disparate measures to the best advantage of the coastal historic environment are considered. The coastal archaeological resource in England is described. There is a consideration of the archaeological impacts of coastal management schemes. Methods for evaluating and recording sites and for developing mitigation strategies are outlined. The prospects for preserving coastal sites *in situ* are discussed.

Introduction

Coastal change is a pressing concern for all maritime nations, and increasingly for their archaeologists and heritage managers. Coastlines have always changed, and there is certainly nothing new about the loss of sites and historic buildings by erosion or submergence (Fig 28.1). However, current climate change models suggest that rates of change will increase over this century: see, for example, the website of the Intergovernmental Panel on Climate Change (http://www.ipcc.ch/). The 2002 United Kingdom Climate Impacts Programme report (UKCIP 2002) indicates that, by the 2080s, relative sea level in the London area might be between 0.26m and 0.86m higher than during the period 1961–1990. Extreme sea levels – as experienced during storm surges – are the cause of most damage at coasts, and these might be experienced between 10 and 20 times more frequently than at present, particularly in south-east England.

Around the low-lying coasts of the North Sea, rising sea levels and fixed coastal defences have already led to the intertidal zone being 'squeezed'. This has caused the loss of coastal habitats, such as salt marsh and mud flat, which are important not only for their ecological value, but also because they provide natural sea defences. Archaeological sites, stratified within intertidal sediments, or located immediately landwards of salt marsh, are being lost. Elsewhere, rates of beach and spit mobility, and cliff erosion are expected to increase. Damage or loss of sites as a result of these natural (albeit humanly-enhanced) processes are further exacerbated by construction projects on coasts – notably for ports. It is already apparent that archaeological projects associated with current commercial port developments in England will provide opportunities for much technical innovation. However, discussion of such projects would be premature, both for reasons of commercial confidentiality and because they are at an early stage.

This paper is based on *Coastal defence and the historic environment: English Heritage guidance* (now published 2003), and so focuses

Fig 28.1 *Walton Castle, Suffolk; this late Roman fort on the Suffolk coast near Felixstowe was part of a defensive system along the south and east coasts of England (the Saxon Shore), it was entirely destroyed by cliff erosion in the 18th century*

on the coasts of England, but it addresses issues relevant everywhere. Recent UK policy, strategy and consultative documents include *Safeguarding our seas* (DEFRA 2002a), *Modern ports* (DETR 2000), and *Seas of change* (DEFRA 2002b). In a wider context, the recommendation of the European Parliament and Council of 30 May 2002 on Integrated Coastal Zone Management (2002/413/EC) requires member states to take a strategic ecosystem-based approach to coastal management, in which the cultural heritage is an important consideration. This should lead to a reappraisal of the coastal heritage throughout Europe, and to new surveys along the lines of the current English Heritage Rapid Coastal Zone Assessment Surveys (RCZAS). Meanwhile, we hope that our experience in England will be useful to colleagues worldwide. The principal challenge facing heritage managers is to translate the references to cultural heritage in EU and UK documents into effective programmes designed to preserve sites and buildings.

Finally, in the context of this conference, we need to consider how and where coastal sites may be preserved *in situ*. This is considered in a final section.

Coastal management in England

In England, coastal defence was until recently the responsibility of the Ministry of Agriculture, Fisheries and Food (MAFF) – now the Department for Environment, Farming and Rural Affairs (DEFRA). Coastal defence schemes are undertaken by Operating Authorities (often the relevant local authority), using grants provided by MAFF subject to each scheme meeting technical, environmental and economic criteria. The Environment Agency has particular responsibilities in respect of flood protection for low-lying land and may, in some instances, be the relevant Operating Authority.

There is now a widespread recognition that defence can no longer be wholly dependent on traditional 'hard' defence works, accompanied by a change in policy towards sustainability and long term economic viability. It is also now recognised that the coastline is a series of interlinked physical systems, comprising onshore and offshore elements, known as sediment transport cells. Each cell contains a source of sediment (eg rivers or eroding cliffs): areas where sediment is moved by coastal processes and sediment stores or sinks (eg beaches or estuaries). Within each cell, a series of 'sub-cells' has been identified. Consequently, coastal defence works on one stretch of coastline may have a profound effect on the rate of erosion or likelihood of flooding elsewhere. Future shoreline management policy requires a high-level, strategic, and long-term approach to planning, involving groups of Operating Authorities acting together, rather than the uncoordinated and short-term approaches which have previously prevailed.

Shoreline Management Plans

In 1995 MAFF first published guidance on the preparation of Shoreline Management Plans (SMPs) by Operating Authorities. SMPs are now a key element in DEFRA's flood and coastal defence strategy. The aim of an SMP is to provide the basis for sustainable shoreline management policies over the next 50 years within a particular length of coastline. They provide a large-scale assessment of the risks to people and to environmental assets associated with coastal processes (flooding, coastal erosion, cliff instability etc) and present a policy framework to manage and reduce those risks in a sustainable manner.

The plans are intended to be working documents, which should be reviewed and, where appropriate, revised at regular intervals to incorporate new information and reflect changes in policy guidance. This will lead to the production of successive 'generations' of each plan. The 'first generation' of plans, covering the entire coastline of England, has been completed and DEFRA subsequently commissioned a review of their strengths and weaknesses. In some areas, preliminary work on 'second generation' plans has commenced. Second generation plans may incorporate data from supporting studies (DEFRA 2001, 27), including results from the Future Coastal Evolution Study, which will consider scenarios for longer-term changes (50–100+ years); additional assessment of sediment budgets; assessment of condition and performance of existing defences; Coastal Habitat Management Plans (CHaMPs), which will provide a framework for managing internationally important nature conservation sites – which are being produced by Operating Authorities and English Nature; flood risk maps, produced by the Environment Agency; flood and erosion risk studies; scoping studies for inclusion of estuaries within SMPs.

The coastal sub-cell has so far been the principal frame of reference for SMPs, with each plan considering a single sub-cell or groups of adjacent sub-cells. The term 'Coastal Process Unit' is also pertinent. These are lengths of shoreline (which may include an estuary) within which a set of physical processes operates relatively independently from adjacent units. Within these a series of Management Units are identified and, for each Management Unit, the SMP identifies a preferred strategic defence option. MAFF guidelines for Operating Authorities (MAFF 2000) and the new DEFRA document *Shoreline management plan: a guide for coastal defence authorities* (DEFRA 2001) recognise five generic policies: hold the line; advance the line; managed realignment; limited intervention and no active intervention. The new guidance also recognises that, over the 50-year period of the plan the preferred option may change.

Strategy Plans and Scheme Design

Coastal defence now comprises a hierarchy of plans, summarised in Table 28.1. Large scale plans such as SMPs will typically cover a whole coastal cell or estuary system. Strategy Plans will normally be prepared for coherent subsets of the large-scale plan. Scheme Design and appraisal will be undertaken for particular time-limited capital or maintenance projects required to deliver the Strategy Plan.

Like SMPs, Strategy Plans should be prepared by Operating Authorities, be long term and subject to review. They will provide a basis from which scheme-specific designs can be formulated. All of the generic coastal defence options can have implications for the historic environment (discussed further below).

Recently, MAFF has issued detailed guidance on project appraisal procedures in the form of *Flood and coastal defence project appraisal guidance notes* (FCDPAG, MAFF 1999a-d; MAFF 2001a; MAFF 2001b). These FCDPAGs include important references to the historic environment and will be cited in the text below as appropriate.

The coastal archaeological resource

Archaeological sites are common on all sections of the coastline of England (Fulford et al 1997). However, they are often poorly covered by local Sites and Monuments Records, and are under-represented in the lists of statutorily protected sites – mainly because systematic survey has not been undertaken in many areas. English Heritage is currently commissioning a series of Rapid Coastal Zone Assessment Surveys. These projects include a desk-based assessment of easily accessible sources including aerial photographic and

Table 28.1 Stages in the appraisal process (after MAFF 1999c Approaches to risk)

Stage	SMP	Strategy Plan	Scheme
Aim	To identify policies to reduce risks	To identify appropriate scheme types to implement policies	To identify the nature of works to implement preferred scheme
Delivers	Broad brush assessment of risks, constraints and opportunities, areas of uncertainty	Preferred approach (ie scheme type) including economic and environmental decisions	Comparison of different implementation options for preferred scheme type
Output	Generic policies (eg Hold the line, Managed realignment etc)	Type of scheme (eg beach recharge, seawall realignment etc)	Type of works (eg revetment, wall, recycling etc)

cartographic sources, followed by rapid 'walk-over' survey. They are intended to enhance the record of coastal archaeology in both local Sites and Monuments Records and the National Monuments Record.

The 'soft' coastlines of our major estuaries and of the south and east of England have a particularly complex and vulnerable archaeology. Sites on this type of coastline fall into several categories.

1) Prehistoric sites (usually Mesolithic–Neolithic) on the basal surface. These were usually dry land sites when occupied, but due to later sediment cover are surface-intact and very well preserved. The palaeotopography and prehistoric archaeology of the former landscape now beneath the North Sea has been reviewed by Flemming (2002).
2) Peats formed on the former dry land surface, often with associated 'submerged forests' (Fig 28.2).
3) Wooden structures, such as fish weirs, trackways, houses etc, of Bronze Age to post-medieval date within intertidal clays, originally formed on mud flats and/or salt marsh (Fig 28.3). In some instances prehistoric human and animal footprints survive intact within these clays. Occasional stone structures, such as fish weirs, also occur at some locations.
4) Peats within the mineral sediment sequence. These may include wooden structures such as trackways.
5) Saltern sites (Bronze Age to medieval) and other site categories associated with coastal economies (eg oyster pits, decoy ponds).
6) Sea walls and structures relating to drainage and reclamation. These are of various dates, but include some medieval or earlier defences (Crump 1981). Sea walls may just be earthen banks, but at some locations include timber structures, which can date back at least to the medieval period.
7) Defensive structures, and remains of aircraft, such as V2 rocket engines in the Essex estuaries (Fig 28.4). Military remains range in date from Roman forts to Cold War installations, though most date from the 1940s.

Sites are also common along other types of coastline. Palaeolithic coastal sites occupied at times of high interglacial sea levels are now well inland on raised beaches (Roberts 1998), but Palaeolithic landscapes and sites have also become submerged by rising relative sea level. An example is the valley of the former 'Solent River' (Wymer 1999, 105). Sites are exposed in cliff sections and foreshore platforms at many locations.

Fig 28.2 Peat rafts on the beach, at Walberswick, Suffolk, indicate erosion of near-shore sediment outcrops

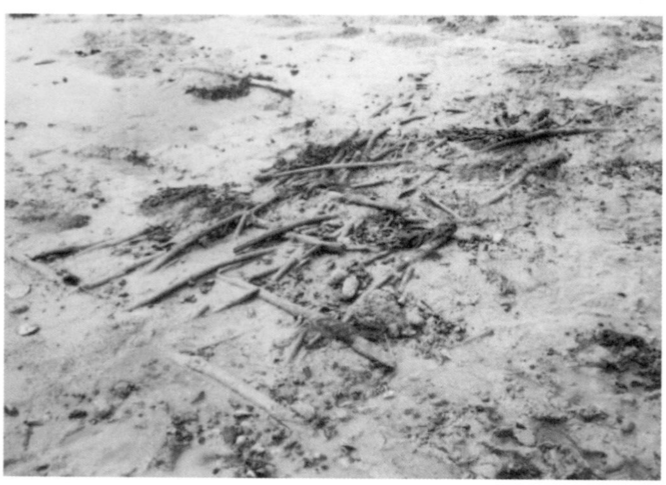

Fig 28.3 Bronze Age hurdle, part of a trackway, exposed on the foreshore by natural erosion at Rolls Farm, Tollesbury, Essex

Fig 28.4 Part of a post-medieval timber building, exposed on the beach adjacent to a 16th-century coastal fort at Cudmore Grove, Mersea Island, Essex

Shipwrecks and hulks occur in inter-tidal to sub-tidal situations all around the English coast (RCHME 1996). Sand dune coastlines and shingle banks are inherently mobile. Both dunes and shingle banks are vulnerable to breaching during storms, and dunes vulnerable to erosion resulting from human activity. There may also be long-term dune recession. Archaeological sites in semi-terrestrial situations landwards of coastal barriers are vulnerable. A well-known example was the Bronze Age timber structure dubbed 'Seahenge' at Holme-next-the-Sea, Norfolk (see paper 20). It was constructed on salt marsh behind a dune barrier, but due to dune recession was later exposed on a tidal beach. It was then directly subjected to wave erosion, necessitating archaeological intervention.

Cliff coastlines fall into two categories: 'hard-rock' or consolidated cliffs, developed from resistant sedimentary, igneous or metamorphic bedrocks; and 'soft-rock' or unconsolidated cliffs composed of easily eroded materials such as shales or glacial/fluvioglacial deposits (typically till or sand). The latter are characterised by a short cycle (4–5 years) of cliff failure, removal of collapsed sediment by erosion, cliff steepening and renewed failure (Pethick 1992). Erosion leads to loss of archaeological sites located in cliff top positions, particularly where the cliffs are of unconsolidated material. The almost complete loss of the medieval town of Dunwich, Suffolk, is a well-known example (Linford 2000; West 1970).

The policy framework for managing the historic environment in England

The historic environment comprises landscape features, archaeological sites and historic buildings. These assets are unique and, once damaged or destroyed, cannot be re-created. Perception of the breadth of the historic environment continues to develop, and in recent years more recent industrial and military sites have come to be included. There is also now a more holistic view, including surviving field and settlement patterns. However, only a very small proportion of recorded historic sites – less than 5% – have any form of statutory protection, through the Ancient Monuments and Archaeological Areas Act 1979, the Protection of Wrecks Act 1973 and the Town and Country Planning Act 1990.

Government Planning Guidance on the historic environment is set out in detail in *Planning policy guidance note 15: planning and the historic environment* (PPG 15) and *Planning policy guidance note 16: archaeology and planning* (PPG 16). This guidance (under revision at the time of writing) emphasises the desirability of protecting historic remains, whether designated or not, and recognises that this should be a material consideration in the planning process. It also stresses that planning decisions should always be informed by an understanding of their impact on historic assets. The guidance recognises that as these assets are non-renewable, they should wherever possible be preserved.

This approach is echoed in MAFF's *Flood and coastal defence project appraisal guidance* (MAFF 1999a–d; MAFF 2001a). The need to consult with local authority conservation and archaeological officers, (mostly based in planning departments), is emphasised, to ascertain implications of strategic plans for the historic environment.

Government advice in PPGs 15 and 16 is that, wherever practicable, important historic sites should be preserved. This may not be achievable or sustainable in the long term in dynamic coastal environments, but the objective of preserving nationally important sites *in situ* should always be fully considered. FCDPAGs 3 and 5 (MAFF 1999b; MAFF 1999d) offer advice on the valuation of historic assets as part of this process, in terms of costs of protection from erosion, excavation and recording or moving listed buildings to another location. Nevertheless, FCDPAG 3 (MAFF 1999b) also acknowledges that it may not always be possible to place a monetary value on all assets.

Integrating the historic environment in the shoreline management process

Statutory background

Section 7 of the Environment Act 1995 and sections 61A and 61B of the Land Drainage Act 1994 require the Environment Agency, Internal Drainage Boards (IDBs) and Local Authorities respectively to contribute to the conservation of nature and heritage when carrying out their flood defence functions under the relevant Acts. In respect of the historic environment the main duties are to have regard to the desirability of protecting and conserving buildings, sites and objects of archaeological, architectural or historic interest; to have regard to preserving public access to places of natural beauty, and to buildings, sites and objects of archaeological or historic interest; to take into account any effect which proposals would have on the preservation of such public access or on the beauty or amenity of any rural or urban area, or on any flora, fauna, features, buildings, sites or objects.

Procedures

The key to ensuring proper consideration of the historic environment within the shoreline management planning process is to ensure that adequate and properly interpreted information is integrated into all stages of the shoreline management planning process, as well as into operational procedures. Recent guidance on the compilation of SMPs (MAFF 2001b; DEFRA 2001) and guidance on project appraisal (MAFF 1999a–d; MAFF 2001a) provides a good basis for achieving this integration at all levels of activity.

The key to this is a staged, proportionate and iterative approach to the gathering of historic data which fits closely into the hierarchy of plans and appraisals advocated by MAFF. This will ensure that the significance of historic assets is fully recognised at the high level and strategic planning stage, avoid unnecessary damage to sites, and reduce the risk of unexpected – and consequently costly – discoveries during scheme implementation.

This is particularly an issue in terms of archaeological sites, which are less likely to be recognised, or the significance of which is less likely to be proactively understood, than is the case for other types of historic asset such as a building or designed landscape. Nevertheless, archaeological assessment and mitigation procedures have a well-developed modular format which lends itself well to integration within the staged project appraisal process promoted for coastal defence by MAFF. The normal stages in the archaeological assessment procedure are appraisal, desk-based assessment, and field evaluation, following which it is possible to define a mitigation strategy.

Archaeological inputs will be at the levels of the Shoreline Management Plan, Strategy Plan and Scheme Design. These will be considered at greater length in the *Coastal defence and the historic environment: English Heritage guidance* (2003), and are only outlined here.

Shoreline Management Plans and archaeological appraisal

A detailed archaeological appraisal is likely to be the most appropriate response in the process of formulating a Shoreline Management Plan. This will involve consulting all sources of available archaeological information, and should include a broad-brush assessment of the archaeological potential. The work should be carried out in accordance with a brief provided by local authority archaeological officers and by a qualified consultant well versed in the strengths and weaknesses of heritage databases. The appropriate regional team of English Heritage will be able to offer advice.

Strategy Plans and archaeological desk-based assessment

An archaeological desk-based assessment is likely to be the most appropriate response in the process of formulating a Strategy Plan. A desk-based assessment is more detailed than an appraisal, but still essentially a desktop exercise. Assessments may also include field inspection of the study area, but will normally stop short of collecting new data through interventional fieldwork. However, FCDPAG 2 (MAFF 1999a, paragraph 3.4), recognises the difficulty of determining the character or importance of archaeological remains based on existing records, and allows for 'quite detailed archaeological investigations at this stage'. There may be cases where field evaluation is necessary to resolve critical questions, but generally archaeological field evaluation should be related to Scheme Design. Assessments should be carried out in accordance with briefs provided by local authority archaeological curators and in accordance with the standard published by the Institute of Field Archaeologists (IFA 1994a).

Scheme Design, field evaluation and the mitigation strategy

Where it is clear that a scheme will affect historic remains, an archaeological field evaluation will be required. This should include a mitigation scheme which aims to avoid or reduce damage to these remains. A field evaluation involves a structured, systematic and problem oriented programme of site investigation through non-intrusive and/or intrusive fieldwork programmes, designed to determine the presence, extent, character and significance of archaeological remains within a given area. Evaluations should be carried out in accordance with briefs provided by local authority archaeological curators and in accordance with the standard published by the Institute of Field Archaeologists (IFA 1994b).

Adverse impacts of schemes can often be avoided by siting groundworks so as to minimise archaeological impacts. This might not always be possible, in which case recording of significant sites and structures will be necessary. The significance or importance of archaeological sites is best assessed objectively with reference to national priorities and strategies (eg English Heritage 1999) and to regional archaeological research frameworks (eg Brown and Glazebrook 2000; Williams and Brown 1999). Local authority archaeological officers, Inspectors of Ancient Monuments and other English Heritage staff are also available to advise on what level of recording would be reasonable for any given site. The reasonable costs of recording important remains should be considered as part of the cost-benefit analysis for a scheme, and incorporated within the total scheme costs.

The archaeological implications of managed realignment schemes

Managed Coastal Realignment is one of the five generic coastal options likely to be identified in SMPs. All coastal defence options may have impacts on archaeological sites. For example, 'holding the line' by beach recharge can result in damage to sites in offshore areas from which the required aggregate is extracted; or the running of barges onto the foreshore to deliver rock rubble can also damage sites. However, realignment is of special significance in terms of the historic environment, particularly archaeological remains, both in terms of its potential physical impacts and because it is an option likely to be increasingly adopted in the future.

At its simplest, realignment involves breaching sea walls to permit development of new areas of salt marsh on formerly claimed land. However, groundworks are often necessary to facilitate shoreline evolution; and often new defences will be constructed landwards of the existing ones. A number of schemes, around the coast of England have been designed, mainly on the east coast, of which several have been implemented. Over the last three years, some 6ha of land has been given over to the creation of mud flat and salt marsh habitats. English Nature (2000), however, estimates that, in order to achieve Biodiversity Action Plan targets for these habitats, it will be necessary to realign 740ha of coast every year for the next 15 years

The main types of works and the potential archaeological problems or advantages associated with them, are as follows:

Sea wall breaching

Sea walls can be complex earth and timber structures of some antiquity. Where an early sea wall is to be breached, detailed recording and sampling for dendrochronology, palaeoecological analysis, and studies of wood technology should be undertaken. Experience shows that conditions for archaeological recording during machining are poor, but recording can be undertaken from exposed sections after breaching, at low tide. After breaching, there may be a relatively narrow confined channel to drain the area. Resultant scour and erosion may result in exposure and destruction of archaeological wooden structures.

Salt marsh re-creation

Measures to recreate salt marsh further inland by managed realignment are a priority in terms of nature conservation. Such measures can often broadly coincide with archaeological interests. Facilitating sediment accretion leading to salt marsh development should not disturb sites and could result in greater physical protection of underlying deposits. Associated works such as creek, lake and lagoon excavation may, however, cause damage to archaeological sites and structures, and re-wetting dewatered deposits can have detrimental effects on archaeological materials.

Mud flat re-creation

The main coastal habitat under threat is salt marsh, but very occasionally grading to lower the land surface to an appropriate level for development of other habitats, such as mud flats, may be undertaken to comply with the EU Habitats and Birds Directives. This has been done to compensate for loss of mud flat habitat associated with development elsewhere.

Lagoon creation

This involves mechanical excavation of lagoons to provide open water habitats for birds and other species. These excavations are commonly deeper than for mud flat re-creation, and hence are likely to destroy sites of earlier date (Fig 28.5).

Creek creation

To encourage the re-establishment of extinct tidal creeks as active channels, they may be dug out mechanically. These creeks are exactly the locations where there is the highest probability of finding hulks, prehistoric settlement evidence and wooden structures.

Re-wetting

Archaeological sites and structures in sediments beneath land currently protected by sea walls are partly dewatered, though anoxic micro- and macro-sites within the deposits may still survive. It is assumed that land claim generally occurred sufficiently long ago (>100 years) for a new metastable physico-chemical and biological equilibrium to have become established. Introduction of oxygenated water could lead to a burst of microbial activity, with consequent accelerated degradation of wood and other organic materials. Accelerated corrosion and disintegration of metal artefacts and poorly fired ceramics might also occur. Further research on the effects of re-wetting and salinity changes on these and other categories of material is required as a priority.

Fig 28.5 Timber jetty exposed during excavation of a lake, as part of a Managed Realignment Scheme initiated by the Essex Wildlife Trust at Abbot's Hall, Great Wigborough, Essex. The new lake compensated for loss of freshwater habitat resulting from sea wall breaching to create new salt marsh (Hall 2000). Archaeological evaluation had been undertaken (Colchester Archaeological Trust 2000) previously, but this structure was detected during a watching brief. It was fully recorded. Dendrochronological dating proved unsuccessful, and a radiocarbon date is currently awaited. Provisionally, it is thought to be of Roman date

An outline protocol for archaeological evaluation and mitigation during implementation of coastal management schemes

The purpose of evaluation is to characterise the archaeological remains in a given scheme area in order to reduce the likelihood that they will be damaged by the scheme, or to allow a programme of recording to be designed where damage is unavoidable.

Full details of Scheme Design should be discussed with the local authority archaeological officer and, where nationally important remains are affected, with English Heritage. In particular, details of the extent of any surface lowering, 'lagoon' and 'creek' excavation and sea wall breaching are needed. It is imperative that once a mitigation strategy is formulated, the Scheme Design should be implemented on site without amendment or that any variations are discussed and agreed with the local authority archaeologist.

Archaeological evaluation may involve some, or all, of the following techniques.

Desk-based assessment

Examining documents, maps and aerial photographs (APs), to help establish the date of land claim and detect other archaeological features. Experience gained so far from the English Heritage RCZAS indicates that large structures, such as fish weirs or wrecks, are readily visible on APs, but that (as would be expected) ground survey is necessary to detect artefact scatters and small structures.

Fieldwalking

This will detect archaeological sites and deposits at the margins of land claimed, where there was only ever thin sediment cover, and which may have been disturbed by ploughing: especially saltern sites with their conspicuous 'red earth' deposits. Surface artefact scatters that 'disappear' beneath the coastal sediment prism will point to the existence of buried sites. Deeply buried sites will not, however, be picked up by fieldwalking.

Earthwork survey

Earthworks and other landscape features (eg partly infilled palaeochannels) commonly occur on unploughed grazing marsh behind extant sea defences. Survey of these may be necessary, initially for record purposes, but also to guide subsequent evaluation by trial trenching. Conventional surveying techniques will often suffice, but where earthworks are partly buried by later sediments high-resolution methods employing GPS and GIS (to enhance the data) may be helpful (Chapman and Van de Noort 2001).

Boreholes or machine cut trial pits

Where there is recorded evidence for archaeological remains or high archaeological potential, a grid of boreholes and/or machine cut trial pits should be made in all deposits likely to be impacted. However, most sites are likely to be small and easily missed. In some cases magnetic susceptibility determinations on samples from cores may be pertinent. In the Netherlands, samples from palaeosurfaces obtained from boreholes are often analysed for cultural indicators such as bone, charred plant material and ceramics (eg Goudswaard 2000), and it is surprising that this low tech but effective methodology is not more widely applied in the UK. Three-dimensional deposit models based on stratigraphic data from boreholes can be generated using appropriate software. Deposit models enhance understanding of buried topography, thereby highlighting locations with the highest archaeological potential.

Geophysical prospecting

Where evaluation of very large areas must be undertaken, geophysical prospecting techniques may be applicable (English Heritage 1995). Electromagnetic (conductivity) survey has been shown to be effective for the detection of major buried topographic features (palaeochannels, gravel 'islands') where a significant contrast in the conductivity of different geological deposits exists. Bates and Bates (2000) present case studies, ground-truthed by borehole survey. Magnetometer survey, utilising either fluxgate gradiometers or high sensitivity caesium vapour instruments, may also be successful particularly for the detection of intensely magnetised archaeological features such as salterns or other semi-industrial activity (eg Cole and Cottrell 1998; Linford 1999; Martin 1999). Other geophysical techniques, including earth resistance, ground penetrating radar (GPR) and certain shallow marine applications for underwater sites (eg Draper-Ali 1996) may also prove fruitful. However, the success of geophysical techniques in such challenging environments will be highly site specific: for example the presence of conductive clay or salt water will severely attenuate GPR signals and may well restrict the application of this technique.

Evaluation/excavation trenches

Evaluation/excavation trenches should be placed at locations where archaeology has been proved, or may be inferred with reasonable confidence. Ideally, sites detected during evaluation should be preserved *in situ*, usually by redesign of groundworks. Where this is not possible, an appropriate level of excavation and recording will be required. On-site sampling for macrofossils and microfossils should be undertaken, and materials for scientific dating should be collected. All archaeological deposits and sedimentary contacts should be levelled in relation to OD (UK Ordnance Datum).

Watching brief

There should be provision for a subsequent watching brief on contractors' excavations. Wherever possible, cutting through sea walls which may be of some antiquity should be avoided, especially when they may have internal timber structures; though if this proves unavoidable, recording and sampling for dendrochronology, wood technology and palaeoecology should be undertaken from exposed sections at low tide, after breaching.

Archaeological monitoring

Following sea wall breaching, a programme of archaeological monitoring should be put in place, primarily to detect sites subsequently exposed by scour and erosion.

Preservation *in situ*

The protocol outlined above is essentially one of reaction to coastal management schemes unrelated to the historic environment. Nevertheless, archaeological mitigation, including preservation of sites *in situ*, has already been achieved alongside scheme implementation at several sites in England; funding for such measures in the UK is potentially available as part of the wider environmental mitigation associated with schemes. Heritage managers may, however, find themselves confronted with the need to preserve sites in circumstances where such funding cannot be found.

Either way, the key point is that the preservation of coastal archaeological sites must be considered within a broad and long-term context. On eroding unmanaged low coastlines, or along those where realignment, or non-intervention, is the intentionally chosen option, it would be absurd to attempt to preserve a site by means of hard defences. The end result, as the coast rolls back, would be preservation of the site as an offshore island encased in concrete or rock-rubble armour, inaccessible and divorced from its original environment. Moreover, maintenance would call for a substantial, and unending, financial commitment. Where cliff erosion is inevitable, or is deemed necessary to provide sediment supply elsewhere, cliff top sites will be lost. Archaeologists have better things on which to spend their limited funds than concrete and rubble, and it must be accepted that some coastal sites are unsustainable. Assessment of the significance of sites will determine whether they should be allowed to erode with only minimal record, or whether full-scale excavation and recording is necessary.

This is not to say that hard defences are *never* the best option. Plainly, where sites of national significance are threatened, serious consideration must be given to hard defence works. Where sites are subject to localised erosion within a generally stable coastline hard defences are certainly appropriate. Even in this situation, however, to construct hard defences can be to write a blank cheque, in terms of long-term maintenance. Before doing so, managers must be convinced that the site is of sufficient importance to justify such measures.

Soft options will probably prove to be the most sustainable in the long run. Beach nourishment (emplacement of sand or gravel seawards of sites to absorb wave energy) is one. This is most appropriate where a site is eroding due to loss of sediment eroded from locations elsewhere which formerly supplied it: usually where hard defences have been installed further along the coast and, consequently, where cliffs or beaches no longer erode. Beach nourishment does, however, involve a continued commitment. Best of all are options involving schemes designed to trap sediment, usually by construction of polders, groynes or similar structures. In principal, these will require a one-off cost, with only periodic maintenance costs: but heritage managers would be well advised to take advice on the efficacy of such measures from coastal managers.

Erosion is only one aspect of coastal site preservation, however. It should not be assumed that because a site maintains its physical integrity it has been preserved, especially in the case of sites within waterlogged anoxic sediments or sediments that were formerly so. Management schemes which result in dewatering of sites, or their re-watering with oxygenated water, are bound to modify the sedimentary environment. There is very little information on the hydrological, chemical and biological changes that may result. This point is made elsewhere in this volume, in relation to waterlogged freshwater sites, but it is even more apt in relation to sites in saline environments with their more complex chemistry. There is a pressing need for monitoring in order to provide hard data, so as to inform future site management schemes.

Acknowledgements

The authors are grateful to colleagues working within local authorities who have provided information. They include Mark Brennand (Norfolk Archaeological Unit), Nigel Brown (Essex County Council), Steve Catney (Lincolnshire County Council) and Stuart Boulter (Suffolk County Council). We also wish to thank colleagues from DEFRA, English Nature and the Environment Agency for their helpful comments. Thanks are also due to colleagues at the English Heritage Centre for Archaeology, who provided helpful advice, notably on applications of geophysics.

Bibliography

Bates, M R, and Bates, C R, 2000 Multidisciplinary approaches to the geoarchaeological evaluation of deeply stratified sediment sequences: examples from Pleistocene and Holocene deposits in southern England, UK, *J Archaeol Sci* 27(9), 845–58

Brown, N, and Glazebrook, J, 2000 *Research and archaeology: a framework for the eastern counties, 2: research agenda and strategy*, East Anglian Archaeol Occas Pap 8, Norwich

Chapman, H P, and Van de Noort, R, 2001 High resolution wetland prospection using GPS and GIS: landscape studies at Sutton Common (South Yorkshire) and Meare Village East (Somerset), *J Archaeol Sci* 28, 365–75

Colchester Archaeological Trust, 2000 *A fieldwalking evaluation at Abbot's Hall Farm, Great Wigborough, Essex*, Colchester Archaeol Trust Rep 105, Colchester

Cole, M, and Cottrell, P, 1999 Adlingfleet, Lincolnshire; report on the geophysical surveys, December 1996 and September 1998, unpub Engl Heritage Ancient Monuments Laboratory rep 27/99

Crump, R W, 1981 Excavation of a buried wooden structure at Foulness, *Essex Archaeol and Hist* 13, 69–71

DEFRA, 2001 Department for Environment, Food and Rural Affairs, *Shoreline Management Plan: a guide for coastal defence authorities*, London

DEFRA, 2002a Department for Environment, Food and Rural Affairs, *Safeguarding our seas: a strategy for the conservation and sustainable development of our marine environment*, London

DEFRA, 2002b Department for Environment, Food and Rural Affairs, *Seas of change: the Government's consultation paper to help deliver our vision for the marine environment*, London

DETR, 2000 Department of the Environment, Transport and the Regions, *Modern ports: a UK policy*, London

Draper-Ali, S, 1996 *Marine archaeology and geophysical survey: a review of commercial practise and its contribution to archaeological prospection*, Hampshire and Isle of Wight Trust for Maritime Archaeology

English Heritage, 1995 *Geophysical survey in archaeological field evaluation*, London

English Heritage, 1999 *Exploring our past 1998; implementation plan*, London

English Heritage, 2003 *Coastal defence and the historic environment: English Heritage guidance*, Swindon

English Nature, 2000 *Annual report 1st April 1999–31 March 2000*, Peterborough

Flemming, N C, 2002 *The scope of strategic environmental assessment of North Sea areas SEA3 and SEA2 in regard to prehistoric archaeological remains*, Department of Trade and Industry, London

Fulford, M, Champion, T, and Long, A, 1997 *England's coastal heritage; a survey for English Heritage and the RCHME*, Engl Heritage Archaeol Rep15, London

Goudswaard, B, 2000 Archaeological assessment within the Betuweroute Cargo Line Project, in *Archaeological evaluation strategies in Belgium (Flanders and Wallonia), England, France and the Netherlands* (eds K Evans and J Williams), Kent County Council (Planarch), Maidstone

Hall, J, 2000 Abbott's Hall: leading the way on coastal and farmland conservation, *Essex Wildlife* 51, 3

Hulme, M, and Jenkins, G J, 1998 *Climate change scenarios for the UK: scientific report*, UK Climates Impact Programme Technical Rep 1, Norwich

IFA, 1994a Institute of Field Archaeologists, *Standard and guidance for archaeological desk-based assessments*, Reading

IFA, 1994b Institute of Field Archaeologists, *Standard and guidance for archaeological field evaluations*, Reading

Linford, N T, 1999 Whitehall Farm, Kelk, East Riding; report on geophysical survey December 1998, unpub Engl Heritage Ancient Monuments Laboratory rep 20/99

Linford, P, 2000 Greyfriars, Dunwich, Suffolk: report on geophysical survey, March–April 1994, unpub Engl Heritage Ancient Monuments Laboratory rep 43/2000

MAFF, 1999a Ministry of Agriculture, Fisheries and Food, *Flood and coastal defence project appraisal guidance 2: strategic planning and appraisal* (draft, July 2000)

MAFF, 1999b Ministry of Agriculture, Fisheries and Food, *Flood and coastal defence project appraisal guidance 3: economic appraisal*, London

MAFF, 1999c Ministry of Agriculture, Fisheries and Food, *Flood and coastal defence project appraisal guidance 4: approaches to risk*, London

MAFF, 1999d Ministry of Agriculture, Fisheries and Food, *Flood and coastal defence project appraisal guidance 5: environmental appraisal*, London

MAFF, 2000 Ministry of Agriculture, Fisheries and Food, *Shoreline management plans: a guide for coastal defence authorities*, draft for consultation, July 2000

MAFF, 2001a Ministry of Agriculture, Fisheries and Food, *Flood and coastal defence project appraisal guidance 1: overview*, London

MAFF, 2001b Ministry of Agriculture, Fisheries and Food, *Shoreline Management Plans: a guide for coastal defence authorities*, London

Martin, L, 1999 Marshchapel, Lincolnshire; report on geophysical survey December 1999, unpub Engl Heritage Ancient Monuments Laboratory rep 15/2000

Pethick, J S, 1992 Natural change, in *Coastal zone planning and management* (ed M C Barrett), 49–64, London

RCHME, 1996 Roy Comm Hist Monuments Engl, *The national inventory of maritime archaeology for England*, Swindon

Roberts, M B, 1998 Middle Pleistocene sediments and archaeology at the ARC Eartham Quarry, Boxgrove, West Sussex, in *The quaternary of Kent and Sussex, field guide* (eds J B Murton, C A Whiteman, D R Bridgland, A J Long, M B Roberts and M P Waller), 187–213, Durham.

UKCIP, 2002 United Kingdom Climate Impacts Programme, *Climate change scenarios for the United Kingdom; the UKCIP02 briefing and scientific reports*, Tyndall Centre for Climate Change Research, Univ East Anglia

West, S, 1970 The excavation of Dunwich town defences 1970, *Proc Suffolk Inst Archaeol Hist* 32, 25–33

Williams, J, and Brown, N (eds), 1999 *An archaeological research framework for the Greater Thames Estuary*, Chelmsford

Wymer, J, 1999 *The Lower Palaeolithic occupation of Britain* (2 vols), Salisbury

29

The National Trust soil policy

ROB JARMAN

Abstract

In 1996, the Royal Commission on Environmental Pollution (RCEP) published its report *Sustainable use of soil* (RCEP 1996). The National Trust responded by recognising that soil is the fundamental natural resource that underlies all its interests in the natural and cultural environment. It is the key to global and local economic activity; and on it depend most of the needs of modern society: water supply, waste treatment, pollution mediation, air quality, food and fibre production, protection of cultural heritage and biodiversity, etc.

The National Trust has decided that soil deserves equal status in its conservation work with ancient monuments, habitats and species, and outstanding cultural landscapes.

The main recommendation of the RCEP report was that UK Government should prepare a Soil Protection Strategy. Anticipating this, the National Trust prepared its own Soil Protection Strategy in 1998. The Government eventually published a *Draft England soil strategy* early in 2001 and the final version is still awaited (October 2001).

The National Trust is applying its Soil Protection Strategy to its land and building management and also to its work with government, agencies and other institutions. The priority is to educate staff, land managers, planners, accountants, project managers, and also conservationists from a wide range of disciplines, in the functions, values, processes and sensitivities of soil. This paper presents some of this work; and emphasises the need for all conservation organisations to respect soil.

National Trust Soil Protection Strategy

The National Trust considers soil to be a critical natural and cultural resource that requires understanding, respect and stewardship. The National Trust will strive to safeguard the conservation interests and natural processes associated with soils on its properties; to ensure that the soils on its properties can fulfil a diversity of functions (such as the production of food, fibre and materials, or the protection of the environment) without loss of productive potential; to integrate soil protection into the National Trust's policies and practices for the acquisition, and subsequent management, of land and buildings; to promote soil protection in the National Trust's advocacy work and in the National Trust's environmental education programme.

Rob Jarman
Head of Environmental Practices
National Trust
Sheep Street
Cirencester GL14 1JJ, UK

National Trust Soil Protection Strategy implementation plan

Implementation of the Soil Protection Strategy requires the National Trust to develop a soil management strategy which will obtain information on the nature and state of the soil resources on National Trust properties and identify specific conservation interests, natural processes and soil functions which merit protection. The strategy will review the management of National Trust properties to see how soil degradation can be prevented and how any desired soil function can be sustained. It will to prepare guidance on soil protection for estate managers and farm tenants, and arrange training. The state of soils on National Trust sites will be monitored and appropriate quality targets will be set. In consequence the National Trust and tenant responsibilities for soil stewardship will be clarified and formalised in the National Trust's agricultural policy, in its Whole Farm Plans and in its farm tenancy agreements. The National Trust will enable its farm tenants to implement sustainable land uses and to practice soil protection, by giving technical support and incentives. The National Trust's interest in soil conservation will be conveyed to members, the public and to schools – including preparation of a *Soil matters* education resource pack. The remit of the National Trust Committees will be extended to include soil conservation. At the statutory level the National Trust will advocate soil protection measures in reforms of the Common Agricultural Policy, Common Land law, and pollution control laws; and advise the Government on its proposed Soil Protection Policy.

Rationale for a National Trust Soil Protection Strategy

Overview

There is a growing movement across the world for soil protection; soil is crucial to support human existence and yet it is being severely and irreversibly degraded by human activities (such as mining, deforestation, bad farming and pollution). In the UK, soil erosion and degradation are now officially recognised as problems; government departments are trying to incorporate soil protection into agriculture, forestry, water resource management, waste disposal and land use planning. The conservation and heritage agencies now emphasise the importance of soil in sustaining biodiversity and cultural heritage.

The National Trust has not typically regarded soil as a natural or cultural resource requiring preservation, although it has included soil protection in some of its conservation programmes (eg Kinder Scout, Kinver Edge, Long Mynd).

Since 1996, when the Royal Commission on Environmental Pollution published its study on *Sustainable use of soil* (RCEP 1996), National Trust staff have been reassessing the National Trust's responsibilities for soil. Soil protection has begun to permeate the National Trust's work, as in the drafting of the Agricultural Policy (National Trust 2000), the preparation of Whole Farm Plans, and the registration of National Trust farms with the Soil Association Organic Farming Scheme. National Trust woodlands have been registered with the Forestry Stewardship Council (National Trust 2001a). The National Trust has confirmed a peatland protection policy, and developed a range of peat-free substitute manures and composts for soil conditioning. The National Trust's management of the catchment lands of its lakes in historic landscapes has been reviewed to reduce soil movement and siltation problems. The causes of soil erosion on many National Trust properties have been identified and rectified.

During 1998–1999, the Department of the Environment, Transport and Regions (DETR) started to prepare a *Soil strategy for England*, as the Government's response to the Royal Commission's Report. A *Draft soil strategy for England* was published for consultation in February 2001; a final Strategy is still awaited (October 2001).

Soil protection is now a fundamental part of the National Trust's work of promoting the permanent preservation of places of natural beauty and historic interest. This paper describes why, and presents the National Trust's Soil Protection Strategy.

What has soil got to do with us?

Soil erosion and degradation is a real issue

Soil is derived from the breakdown of solid geology (by weathering or glaciation), accumulation of organic matter and the development of a community of living organisms. The UK has a great diversity of soil types, due to its geology, topography, climate, hydrology and historical land use. Soil is the substrate on which terrestrial plant and animal communities depend; and is the main repository of zoological and botanical remains and archaeological artefacts from previous natural and cultural environments. Soil is central to the hydrological cycle; and it determines land use options and practices in agriculture and forestry.

Soil is a living resource (a typical English lowland loamy soil will contain 25 tonnes of micro-organisms per hectare) including 4 tonnes of earthworms, 10 tonnes of fungi, 1 ton of springtails, and spiders, beetles and snails. Yet soil flora and fauna have not yet been considered worthy of inclusion in nature conservation or biodiversity strategies.

In the UK, soils have no specific protection: description and conservation have largely been geared to the productive use of soil for forestry and agriculture. High-grade (Grades 1 and 2) farmland has been protected from 'development' (ie building) for several decades by the Ministry of Agriculture, Fisheries and Food (MAFF, now DEFRA), but the soil and its productive capability have not been protected from degradation by intensive land uses and farming. Some aspects of soil erosion, such as obstruction of highways, and pollution of watercourses by sediment, are subject to legal powers, yet they are rarely enforced.

Soil receives only incidental protection in sites of Special Scientific Interest, Special Protection Areas and Special Areas of Conservation, which are designated exclusively for habitats, species or physiographic processes. The Geological Conservation Review describes and protects only a few sites (for their fossil soils, as do Regionally Important Geological/Geomorphological Sites). Archaeological designations (Scheduled Ancient Monuments or World Heritage Sites) and landscape designations (Areas of Outstanding Natural Beauty, or National Parks) are very weak in their protection of soil.

Soil in the UK is being damaged or irreversibly lost, directly through construction, mining or erosion, and indirectly through pollution and mismanagement, leading to reduction of productive capacity, loss of biodiversity and the historic archive, and 'downstream' impacts of sedimentation (eg smothering of fish-spawning gravels).

Globally, in recent years some 24 billion tonnes of topsoil a year has been eroded from farmland, leading to the projection that 30% of the world's arable land could be depleted within 20 years (Brown 2000). In the UK, 450,000m^3 a year of soil is eroded from farmland (DETR 2001): at the National Trust's Arlington estate in North Devon, the lake has accumulated 18,000m^3 of silt since it was desilted ten years ago.

The RCEP reported on the *Sustainable use of soil* in 1996: it examined the uses of soil in the UK, and the environmental and economic problems arising from mismanagement. It recommended that soil should have equal status with air and water in environmental legislation and requested Government to prepare a Soil Protection Strategy.

The Forestry Authority has published *Forests and soils: conservation guidelines* (Forestry Commission 1998); MAFF has published a *Soil code* (MAFF 1998) and a *Soil erosion manual* (MAFF 1999). The Environment Agency (EA) is presently developing its Soil Policy (Environment Agency 2001). Scottish Environmental Protection Agency (SEPA) published a State of the Environment Soil Quality Report (SEPA 2001). The nature conservation agencies are evaluating soil as a nature conservation resource in its own right. Scottish Natural Heritage (SNH) published an information booklet on soil (SNH 1996). The Council for the Protection of Rural England (CPRE) has published a *Soil charter* (CPRE 1999). British Society of Soil Science (BSSS) has long promoted the values and functions of soil (eg BSSS nd), but has long been ignored by those who should know better.

The National Trust soil resource

The Soil Survey and Land Research Centre has been assisting National Trust staff in their consideration of soil protection matters. It has advised that the National Trust should cherish the soil on its properties, both for the specific functions that the soil can fulfil and for the specific interests that the soil contains. Soils meriting protection include:

Soils for wildlife

Soils support a range of plant and animal communities in woodland, moorland, heathland, montane, wetland, grassland and coastal habitats. Soil is integral to the ecology of such habitats, and soil management requires better understanding, such as how to protect soil.

Undisturbed soils

Some soils may never have been ploughed or disturbed, in sites such as ancient woodlands, historic parklands and old grasslands: they constitute a direct link with the post-glacial evolution of the landscape and ecosystems prior to human intervention.

Soils for archaeology

Soil protects archaeological sites and preserves artefacts, through physical cover and through sustaining specific conditions of humidity and chemistry. Soil contains evidence of past activities, climates and ecosystems, and may be protected beneath and within buildings, earthworks, submerged forests, sand dunes, peatlands, caves, etc.

Soils as landscape

Soil influences physical and cultural patterns in the landscape and contributes to local distinctiveness, through determining land use and farming systems and providing materials for buildings (such as cob, daub, brick etc).

Soils as industrial archaeology

Soils have often been degraded by past land uses and now support specific habitats or remnants of industrial archaeology eg heathlands or woodlands developed on 'worn-out' farmland (dating as far back as the Neolithic); or mineral workings.

Soils for food and fibre

Soils support agriculture and forestry, yet production systems often neglect the inherent capabilities of the soil and cause degradation of structure and fertility. Sustainable land use depends primarily on looking after the soil – how else will continued production of food, fibre, energy and materials be possible?

Soils as waste treatment systems

Soils mitigate environmental pollution. We rely on soils to soak up toxic and polluting elements and store them or convert them to benign elements. Most sewage and waste treatment systems rely on soil for the final cleansing; we designate land as 'buffer' strips, 'nutrient sinks', or as 'soakaways'.

Soils for buildings

Soils support buildings and structures, both physically and also as sources of materials (cob, daub, clay, thatching straws, etc); most historic buildings rely on the indigenous soils for their vernacular character, and their conservation requires continuing availability and use of those soils.

Soils for electrical earthing

Soils provide electrical earthing for buildings and transmission systems. Changes in soil structure, moisture and chemistry affect their conductivity and thus their capabilities for earthing, sometimes with severe consequences.

Threats to National Trust soil resources

The RCEP report summarised the main causes of soil degradation as erosion, compaction, acidification, contamination, biological depletion, loss of organic matter, and extraction and depletion. These all arise on National Trust land, in the following ways.

Agriculture

Intensive arable farming

Continuous arable cropping, autumn sown cereals, cultivation with heavy machinery, ploughing of old pasture, minimal rotations and use of agrochemicals, have caused loss of soil organic matter and soil biodiversity, degradation of soil structure, and destruction of archaeological interests. Damage may be insidious and hard to see, but many National Trust properties have vulnerable soils. Water erosion on arable land in the UK has increased over the last 20 years, and severe erosion does occur on some National Trust farms.

Livestock production

Over-grazing damages vegetation and soil and can lead to serious erosion. Upland mineral and peat soils are particularly vulnerable. Over-stocking is often linked with slurry and silage problems, and bad spreading practices can severely damage soil and water resources.

Land drainage

Land drainage dramatically alters soil chemistry and biology and has serious consequences for wetland ecosystems. Drainage also damages buried archaeological remains, especially organic materials.

Contamination

Eutrophication, acidification and contamination of soils occur from nitrate, phosphate and biocide uses. Soils can filter and buffer pollutants, and prevent their release into watercourses, but inappropriate soil management causes pollution.

Forestry

Conifer plantations in heathlands, peatlands, natural woodlands and sand dunes can have a significant and irreversible impact on soil. Soil erosion and sedimentation arise from logging, harvesting, drainage and road making.

Recreation

Footpaths and car parks cause soil compaction and erosion, and require stabilisation which can further affect soil structure and natural processes.

Atmospheric pollution

Soil acidification is of particular concern for vulnerable upland soils in England and Wales, arising from sulphur and nitrogen emissions, damaging upland and lowland heathlands, meadows, pastures, watercourses and wetlands.

Extraction and depletion

Building work removes or covers soil and can completely destroy the soil resource. Peat cutting has damaged many upland blanket bogs and most lowland raised bogs, of great value for their plant and animal communities and their archaeology and records of past environments. Turf stripping can also seriously damage soil.

Remediation

Ironically, work to stabilise or remedy damaged or contaminated soil can destroy nature conservation and industrial archaeological interests.

Limitations of the National Trust's historic approach

Soil protection by the National Trust has been inadequate

The National Trust has no inventory or evaluation of its soil resources (Biological and Archaeological Surveys only note the soil types present on a property). Farm tenancy agreements prescribe 'good husbandry', but the National Trust's long-term interest in soil and land quality has not been protected. The National Trust Whole Farm Plan does now cover soil protection and restitution of damaged soils. Responsibility for conservation of the National Trust soil resource has been uncoordinated between disciplines, ie nature conservation, forestry, agriculture, archaeology, gardens and environmental practices. Finally National Trust Committees and Council have had no remit for soil conservation.

The National Trust's approach from 2000 onwards

Appropriate land use: a soil driven policy

The National Trust is committed to developing the principles and practices of sustainable land use (National Trust 2001b). In practice, this depends on the assessment of land capability; what can a piece of land be used for without diminishing it and its natural and cultural significances? How can the function(s) of land be perpetuated to benefit future generations? How can present land uses be devised so as to be reversible without detriment to future function?

There is clearly competition between land uses and between soil functions even when those uses are as sustainable as possible. Even organic arable cultivation may still not be 'good' enough to preserve archaeological remains *in situ*; if the land is needed for archaeology, then it cannot be used for arable crops. Choices have to be made between competing land uses but if soil type, capability and range of functions are taken into account, then the best-fit decision can be taken.

The National Trust is advocating a much closer link between conservation and land capability; some 'historic' land uses are no more sustainable than 'modern' uses. Many upland pastures were created by clearance of woodland, which lead to soil erosion and degradation and significant alteration of watercourses and estuaries etc. Even a millennium later, grazing (even under extensive pressure) still causes soil movement and erosion: the 'traditional' land use is not sustainable by the land capability. Yet reversion of these areas to woodland or heather moorland is resisted by conservationists and landscape historians.

We should learn from the past that many civilisations have declined due to their abuse of their soils and loss of productive capacity. Present global soil loss trends indicate the growing importance of fertile and productive soils in the UK to world food production. We are likely to experience increasing pressures for land to be used intensively for farming – with increased competition with environmental land uses. Climate change and pollution are projected to exacerbate these trends in soil degradation (DEFRA 2001) primarily from increased extremes of soil moisture with hotter drier summers and warmer wetter winters. Regional climate change studies, such as in north west England (SNW 1998) and south east England (Wade et al 1999), co-sponsored by The National Trust and coordinated by the UK Climate Impacts Programme, have illustrated how soils may adapt to various climate scenarios and how land use changes might develop.

All conservation organisations from whatever discipline need to be more aware of the importance of soil to their interests. They need to be especially aware of changes in the soil environment that are presently occurring.

Bibliography

Brown, L R (ed), 2000 *State of the world 2000; a Worldwatch Institute report on progress toward a sustainable society*, New York

BSSS, nd British Society of Soil Science, *Soil: the vital natural resource and life-support system*, Aberdeen

CPRE, 1999 Council for the Protection of Rural England, *Soil charter*, London

DEFRA, 2001 Department for Environment, Food and Rural Affairs, *Climate change UK programme*, London

DETR, 2001 Department for Environment, Transport and Regions, *Draft England soil strategy*, London

Environment Agency, 2001 *Best farming practices: profiting from a good environment*, Bristol

Forestry Commission, 1998 *Forests and soil conservation guidelines*, Edinburgh

MAFF, 1998 Ministry of Agriculture Food and Fisheries, *The soil code*, London

MAFF, 1999 Ministry of Agriculture Food and Fisheries, *Controlling soil erosion*, London

National Trust, 2000 *Agriculture – 2000 and beyond: an agricultural policy for the National Trust*, Cirencester

National Trust, 2001a *Woodland management: a forestry policy for the National Trust*, Cirencester

National Trust, 2001b *Farming forward*, London

RCEP, 1996 Roy Comm Environmental Pollution, *Sustainable use of soil*, 19 rep, London

SEPA, 2001 Scottish Environment Protection Agency, *State of the environment: soil quality report*, Stirling

SNH, 1996 Scottish Natural Heritage, *Soils: Scotland's living landscapes*, Perth.

SNW, 1998 Sustainability North West, *Everybody has an impact; climate change in the north west of England*, Manchester

Wade, S, Hossell, J, Hough, M, and Fenn, C, 1999 *The impacts of climate change in the south east*, Epsom

30

A study into the mitigation of construction impact on archaeological remains

MAT DAVIS

Mat Davis
Land Quality Coordinator
Environment Agency
Kingfisher House
Goldhay Way, Orton Goldhay
Peterborough PE2 5ZR, UK

Abstract

This paper presents an overview of an important study commissioned by English Heritage into the mitigation of construction impacts on archaeological remains. The need for the study can be linked to the publication in 1990 of *Planning policy guidance note 16* (PPG16, DoE 1990) and specifically, its promotion for the use of positive management and sympathetic engineering designs to achieve the *in situ* preservation of important archaeological remains. Within a development project the study recognises four stages of construction activity that have the potential to cause an impact on a site's archaeological remains. During each of the four construction stages the key management approach (or mitigation measure) to avoid or lessen the construction impact are discussed. The incorporation of the various mitigation measures into an overall mitigation strategy for a site, and the role which an archaeologist and engineer will play in development of the strategy is then given. Finally mention is made of the importance which monitoring can have in giving confidence to the archaeological and developer community, that the adoption of a specific mitigation strategy has successfully enabled the development to proceed without a resulting loss of the site's *in situ* archaeological remains.

Introduction

Many engineering operations employed during the course of a construction project have the potential to seriously impact in a negative way on important archaeological remains. The impact, which may be physical, chemical or biological, can damage the archaeological remains directly and indirectly. Indirect impacts occur if the construction impact is on the site's burial environment, which is promoting the continuing *in situ* preservation of the archaeological remains. Impacts from construction have long been recognised and hence archaeological considerations have been incorporated into engineering programmes for at least two decades, though usually with the emphasis on removal of the archaeology through excavation and preservation by record. For a variety of economic, scientific and social reasons this approach to archaeology under went a significant shift 11 years ago with the publication of PPG16 (DoE 1990).

This guidance document places great emphasis on the physical preservation *in situ* of important archaeological remains by the use of positive management and sympathetic engineering designs. However, in following this policy approach it has become increasingly apparent that the overall impacts of construction on archaeological remains are not fully understood. As a consequence, decision-making to agree acceptable engineering designs to mitigate against impact on the archaeology is seen as difficult, costly and time consuming. The urgency to address this gap in our knowledge is illustrated by the Monument at Risk Survey which revealed that road building and construction (property development and urban expansion) accounted for 36% of observed cases of wholesale archaeological monument destruction (Darvill and Fulton nd). This figure does not take into account demolition and building alterations, which account for a further 20% of all monuments destroyed. As a step towards addressing the knowledge gap, English Heritage commissioned in 1998 a study of the mitigation of construction impact on archaeological remains (Davis et al 2003). This paper presents an overview of the study, together with a summary of the suggested key good practice approaches necessary to achieve the successful mitigation of construction impacts on *in situ* archaeological remains.

Drivers for change

In 1992 the revised European Convention on the Protection of the Archaeological Heritage was signed by representatives of 20 member states of the Council of Europe. It was accompanied by the publication of *Charter for the protection and management of the archaeological heritage* (ICAHM 1995), which sets out the principles and guidelines to preserve *in situ* or by record archaeological remains found either in research or development projects. This Charter and its convention endorsed in the remit of PPG16 which, coupled with *The management of archaeological projects* (English Heritage 1990; its revision in 1991 now commonly known as MAP 2), brought about a revolution in the concept, planning and management of archaeological projects of all kinds. The Charter was directed at local and central government planning authorities and developers, with the aim of producing common links and routes to achieve a successful balance between economic development or scientific research work and the preservation of the archaeological heritage for future generations. The archaeological profession therefore helped drive the change because, with the adoption of the Charter and the publication of PPG16, the process for preserving archaeological remains *in situ* as part of a mitigation strategy for a development became a necessary part of the management repertoire of archaeological curators in England.

Prior to the publication of PPG16, developers were already very familiar with the drive by policy makers and others for them to contribute to the sustainable management of finite and non-renewable resources. For example, because land is a finite resource the application and then enforcement of planning guidance now strongly encourages developers to use brownfield (ie previously developed land) rather than greenfield sites (ie agricultural land or similar). Therefore, the concept assertion in PPG16 that 'archaeological remains should be seen as a finite and non-renewable resource' is already understood by developers. However, in seeking to conform to the requirements of PPG16, developers are often confounded at the outset of their projects by the current paucity of detailed archaeological information in public databases that could assist in delimiting areas of significant or known archaeology, and initiate the early use of avoidance measures. Therefore, developers and their consultants have also pushed for a change in the approach to archaeological site management. Their reasoning can perhaps be most directly linked to the financial implications of encountering important archaeological remains on their site. That said, economic considerations should also unite all professionals involved because the cost of change increases greatly during a project, such that the unplanned incorporation of an engineered mitigation strategy into the construction phase of a project may be financially disastrous (as demonstrated in Fig 30.1). For example, massive cost overruns occurred on the Rose Theatre site in London, where the developer spent an estimated £11 million funding both a six-month excavation and subsequent alterations to their building design (Ove Arup 1991).

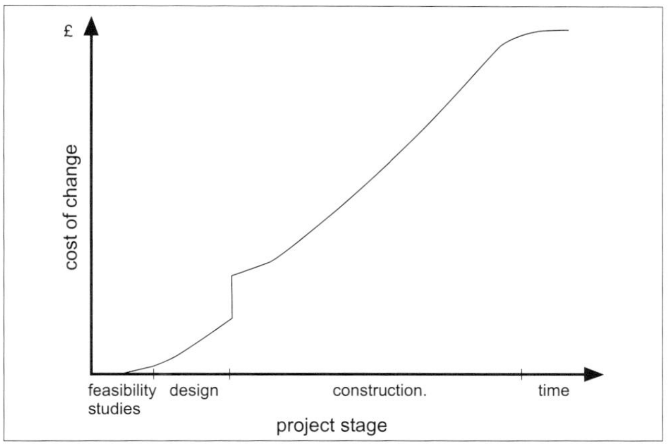

Fig 30.1 Cost of change against project stage (after Institute of Civil Engineers 1996)

Overview of the four stages of a construction project

The development of a site generally comprises four stages of construction activity, each of which will involve a wide variety of potentially ground disturbing (archaeologically damaging) engineering operations. Though construction impacts may be physical, chemical or biological, this paper gives particular attention to disturbances that result in physical movement (excavation, displacement, compaction, heave and stresses etc) and water regime changes (retention or exclusion, flow patterns and rates, quality and temperature etc). This is because it is these two parameters which are thought to most significantly change the character of the burial environment, as well as impact directly on specific archaeological remains. For example, ground movement can physically damage archaeological remains, physically move the remains and so alter their site context, and structurally alter the deposits in which the remains are preserved. Impact on archaeological remains by the absence or presence of water can be physical (eg abrasion, solution and volume changes), chemical (eg universal catalyst and an electrolyte), and/or biological (eg favours anaerobic or aerobic microbial populations, Thompson 1997). Finally, though regularly reported, it is worth repeating that in England the presence on a site of waterlogged and anaerobic burial conditions often gives the greatest preservation of organic artefacts and environmental remains.

Stage 1: pre-construction ground investigation

The pre-construction ground investigation generally commences with a desktop study in which documented information relating to the site chosen for development and its immediate vicinity is complied and evaluated. Because the information will influence the early stage of a project (ie feasibility studies and design stage shown in Fig 30.1) it is vital for archaeological input to the project team by this time. To ensure the input is effective there must also be good communication between the developer, engineer, archaeologist and the main contractor team.

Following the desktop study a variety of ground intrusive geochemical, geotechnical and archaeological investigations may be required. Each investigation is undertaken by the relevant specialists who will follow their own professional design specification. For example, the geotechnical engineer may conduct the investigation in accordance with British Standard Institutions *Code of practice for site investigations* (BS5930 1999), a geochemical specialist may design their investigation with reference to the British Standard Institutions *Investigations of potentially contaminated land* (BS10175 2000), and an archaeologist may refer to the Institute of Field Archaeologists *Standard and guidance for archaeological field evaluations* (1994).

Each specialist is therefore selecting the investigation technique and sampling protocol that will provide the information they require for their own specific project needs. This can range from information on the depth through 'made ground' (eg fill material) to solid rock for foundation design, the chemical composition of buried industrial waste materials or, the extent of waterlogged archaeological remains. Unfortunately, all the investigations are likely to employ their own specific engineering operations that will be ground intrusive and therefore potentially destructive to the *in situ* archaeology. Mitigation of the impact will require communication and joint planning between the different specialists, perhaps to the extent that the use of boreholes and trials pits is undertaken jointly and at the same location (ie geotechnical borehole at the base of an archaeological/geochemical trial pit).

Stage 2: pre-construction activities

Before construction of a new development can commence a certain amount of site preparation is usually required. Conducted as part of the main construction contract, or let as a separate 'enabling works' contract, the engineering operations used may range from simply clearing away areas of surface vegetation to a major operation involving soil stripping and *in situ* remediation of land contamination. The site preparation activities also involve mobilisation of personnel and equipment (plant) to be employed on the third and main stage of a construction project.

Provided the good practice suggested for the first stage of a development project has been followed, the archaeological sensitivity of a development site should largely be known. The importance of early archaeological input and the establishment of good communication between all members of the project team is again vital because it can be possible to mitigate impact of the engineering operations used during this phase of a project by careful zoning of the site. Zoning involves reaching agreement on the types of engineering operations permitted in each part of the site in order for construction impact to avoid areas of archaeological sensitivity, eg the careful routing of access roads across the site and, the above ground placement of service lines (power, communication, drainage) to offices on the site. Zoning need not be purely 'place specific' as it can also involve the creation of operating windows for certain engineering operations. For example, the traffic of heavy machinery on parts of the site can be prohibited during the wet winter months when soil conditions are more susceptible to damage by surface rutting and compaction.

Stage 3: construction activities

The operations employed during the main phase of construction activity generally have one of two purposes: those connected with earthwork construction and, those concerned with the construction of structures and buildings.

Earthworks are most likely to be carried out on highways contracts, reservoir contracts and, in a minor way, landscaping contracts for other developments. The term earthworks describes the excavation of soil to form cuttings, usually with side slopes but occasionally within retaining walls, and the placement of compacted soil in layers to form embankments. The potential impact on *in situ* archaeology can be severe and therefore the design of a mitigation strategy will not only require the close liaison between all parties in the project team, it may also require the knowledge of various specialists (eg conservators and soil mechanic experts). There is frequently a need for innovation and an acceptance by the developer that any mitigation strategy will be very site specific. It may be necessary to draw upon experience from other countries and from areas of knowledge beyond the construction industry. For example, a number of construction loading experiments are reported from projects in America (Matthewson and Morris 1995).

During the construction of structures and buildings, the engineering operations that can potentially create the greatest impact on archaeological remains are those connected with the construction of foundations and services. Foundations can be defined as that part of a structure which directly transmits load to the ground, and they can be either shallow or deep (Cole 1988; Tomlinson 1995). As a guide, shallow foundations often extend to depths of less than 2.0m but may be as much as 5.0m (excluding the special case of deep basements), and deep foundations can be taken as being more than 5.0m deep. If a basement is to be constructed then all soil to the full depth of the basement will be removed by excavation, and additional foundation elements may then be installed below the basement's base.

Shallow foundations are those which generally transfer loads from a building to the near-surface soil. Under normal soil conditions, shallow foundations (ie strip footings, pads and rafts) will yield greater settlement and lower load carrying capacity, and cost less than deep (ie piled) foundations. Therefore shallow foundations tend to be used on lower cost projects, where foundation loads are low in comparison to the allowable bearing capacity of the soil, or where settlement (and differential settlement) criteria are not too onerous. Deep foundations tend to be used on projects where foundation loads are high in relation to the allowable bearing capacity of the surface soil (settlement criteria are stringent) or construction factors (such as a high water table) make shallow foundations less economic due to the difficulties in making the necessary excavations.

By the construction stage of a project the overall mitigation strategy to avoid, reduce or remove the impact of engineering operations being employed by the developer should already form part of the approved engineering design and contractor's method statement. The mitigation strategy may range from specific measures to limit the impact from a single engineering operation (eg low pressure tyres and load spreading plates on an item of heavy machinery) to a fully engineered solution (eg sand backfill and geomembrane covering system below a raft foundation). In practice a mitigation strategy will need to incorporate a range of individual and possibly very detailed mitigation measures that together achieve an agreed and acceptable level of construction impact reduction. The specific mitigation measures may include the enforcement of good working practices (eg trained operators using appropriate equipment), protection of ground surfaces (eg placement of load, spreading geotextiles on soft ground), selection of low impact engineering operations (eg mini-piles), and modification of techniques (eg reuse of existing piles).

Broadly speaking a site specific mitigation strategy can be viewed as following one or both of the following approaches.

Avoidance mitigation strategies

An avoidance mitigation strategy may be adopted in which all archaeologically damaging engineering operations are excluded from an area of archaeological sensitivity, ie avoidance of ground disturbance and therefore removal of the threat of construction impact on *in situ* remains.

A total avoidance strategy is most often achieved by zoning a site to create 'open' areas in which no (or restricted) construction activity is permitted. Each designated area will contain the archaeological remains and most probably a surrounding buffering zone. The size of buffer zone depends on, among other factors, the fragility of the remains, character of burial environment, and the type of construction impact affecting them.

On a large site, avoidance mitigation strategy may be viewed as the most cost effective mitigation strategy; for example, the design of mineral extraction sites may permit avoidance (ie sterilisation) of an area of mineral in order to achieve preservation *in situ* of archaeological remains. However, avoidance as a mitigation strategy has two disadvantages, for the developer it is financially undesirable to exclude areas of the site from the development proposals, and secondly, it may not be possible to effectively exclude all construction impacts from the *in situ* archaeological deposits. This second point particularly applies when a complex or fragile burial environment surrounds the archaeological remains. For example, saturated deposits in which organic remains are preserved within an anaerobic environment are very sensitive to remote changes in the water regime of the site, as would be caused by dewatering operations used during ground improvement and stabilisation. On such a site, an avoidance strategy would not be sufficient to remove the construction impact of dewatering and consequently, an engineered mitigation strategy may also be needed as part of the overall management plan for the site.

The redesign or relocation of an element from a construction project can also be viewed as an avoidance mitigation strategy. For example, the relocation of below ground petrol interceptors away from areas of archaeological sensitivity or, the movement of car parking areas from a below ground basement onto the roof top of a new development.

Engineered mitigation strategies

Engineered mitigation strategies generally incorporate mitigation measures, which reduce the impact of specific engineering operations on the ground containing the *in situ* archaeological remains. The measures generally involve modification of the engineering operation to limit the surface extent or depth of its impact on the ground (eg load spreading devices and modified piling rigs).

Alternatively, engineered mitigation strategies may adopt measures to isolate the *in situ* archaeology from ground disturbance caused by engineering operations. The measures will generally require either a containment system or a covering system. A containment system involves the installation of a bund, membrane or similar barrier around the undisturbed ground containing the archaeological remains. On appropriate sites the archaeological remains and their surrounding burial environment are preserved within the enclosed ground despite disruption of the surrounding ground during site development.

However, the isolation of undisturbed ground by a

containment system may remove it from the natural soil processes which are responsible for its specific burial environment (eg the input of water to maintain saturated conditions). Therefore, an alternative approach to isolate the archaeological remains from the construction project is to contain them within or below an engineered covering regime. The aim of this approach is to actively maintain the burial environment conditions thought to be responsible for the *in situ* preservation of the remains, whilst still permitting development at ground level. Though a variety of different covering systems have been used on sites in England perhaps the most experience has been gained with the use of sands in site reburial (Canti and Davis 1999). The use of covering systems can involve the reintroduction of burial conditions that are thought to promote archaeological preservation. This may be necessary if construction activities have caused an unavoidable or accidental disturbance of the ground, such as a loss of surface cover or lowering of the water table. Though the reintroduction rather than maintaining of a burial environment is on the whole undesirable, it can be a valid objective of an engineered mitigation strategy.

Stage 4: remedial and maintenance activities

In addition to producing new developments, the construction industry is responsible for maintaining, improving and adapting the existing stock of buildings, roads, etc. The engineering operations associated with these maintenance activities are often very similar to those used during the first three stages of a construction project. Therefore, the good practice suggested earlier in this paper should also apply to this stage of a project. Hence, the early involvement of an archaeologist in a maintenance project team and, the agreement of mitigation measures to form an overall mitigation strategy should ensure construction impacts on *in situ* archaeological remains are minimised.

An extra consideration when undertaking maintenance activities is that an earlier archaeological mitigation strategy may have been developed for the original construction project. Any maintenance operations must not therefore compromise the integrity of the earlier mitigation strategy. For example, open areas retained on a site may form part of an avoidance strategy and therefore, any remedial drainage or landscaping works on these areas may need to be modified to ensure the archaeology continues to remain undisturbed. Equally, if a covering or containment mitigation system exists on a site, any underpinning works on a building's foundation will need to ensure the membrane or other device used in the mitigation strategy is not damaged.

Therefore, the importance of good record keeping to document the mitigation strategy implemented on a construction project cannot be stressed enough. Finally, because the remedial and maintenance activities do not always require planning permission, the role of management agreements should be considered as part of the mitigation strategy. The agreement may need to include a programme of monitoring to verify the continuing integrity of the mitigation strategy once the construction project has been completed.

Conclusions

A mitigation strategy usually comprises a coordinated sequence of measures aimed at avoiding or minimising the impact of construction activities on the *in situ* archaeological remains at a site. When the measures are managed together the overall mitigation strategy will be based on either an avoidance or engineered approach, or possibly a combination of both.

The generally smaller size of development projects, high land prices and the possibility of external disturbances occurring during the redevelopment of surrounding sites puts a bias on the adoption of engineered mitigation strategies in urban locations. It is also the urban locations that have seen the greatest adoption of innovative mitigation strategies, for example the use of novel piling solutions (Chapman et al 2001). The development by specialists, and then adoption by developers and decision makers, of innovative and novel mitigation strategies is important because it is contributing to a reduction in conflicts that can exist between 'the need to preserve nationally important archaeological remains and the need to allow our towns to thrive and develop' (Wainwright 1993, 418).

Conflicts between different parties has been one of the drivers for English Heritage commissioning the study into the mitigation of construction impacts on archaeological remains. It is without a doubt that, the evaluation of archaeological 'problems' during the early stages of a development project results in a more cost-effective construction programme in which archaeologically sympathetic redesigns or modifications enable a more direct and trouble-free path to development completion than might otherwise be the case. Therefore, the successful design, installation and maintenance of an archaeological mitigation strategy is achieved through early and continuous communication between the developer, engineer, archaeologist and the main contractor team.

Whichever mitigation strategy is adopted, it should have a design life that will ensure its effectiveness for the life span of the new development. To create confidence in the successful adoption of a mitigation strategy it may be necessary to incorporate a programme of monitoring before, during, and after the construction project. The monitoring requirements will be both site and project specific and as a minimum will need to focus on ensuring the correct installation of the mitigation strategy, and its protection during the construction phase of the project. In some cases the monitoring programme may need to be extended to verify the continuing operating performance of the mitigation strategy for a period after completion of the construction project.

Bibliography

BS10175, 2000 British Standards Institution, *Investigation of potentially contaminated land – code of practice*, London

BS5930, 1999 British Standards Institution, *Code of practice for site investigations*, London

Canti, M G and Davis, M J, 1999 Tests and guidelines for the suitability of sands to be used in site reburial, *J Archaeol Sci* 26, 775–81

Chapman, T, March, B, and Foster, A, 2001 Foundations for the future, *Civil Engineering* 144, 36–41

Cole, K W, 1988 *ICE works construction guides: foundations*, London

Darvill, T, and Fulton, A K, nd [1998] *MARS: the monuments at risk survey in England, 1995 main report*, Poole

Davis, M, Gdaniec, K L A, Brice, M, and White, L, in prep *Study of the mitigation of construction impact on archaeological remains*

DoE, 1990 Department of the Environment, *Planning policy guidance note 16: archaeology and planning*, London

English Heritage, 1991 *Management of archaeological projects*, London

ICAHM, [1995] International Committee on Archaeological Heritage Management, *Charter for the protection and management of the archaeological heritage*, [London]

Institution of Civil Engineers, 1996 Planning and control of construction, *Civil engineering procedure*, 5 edn, London

Institute of Field Archaeologists, 1994 *Standard and guidance for archaeological field evaluations*, Birmingham

Mathewson, C C, and Morris, L E, 1995 Intentional burial of two archaeological sites below a highway in Montagu County, Texas: analysis of dynamic loading during construction, *Proc symposium on engineering geology and geotechnical engineering* 31, 260–75

Ove Arup, 1991 *The York development and archaeology study*, Manchester

Thompson, G, 1997 The impact of engineering on buried archaeological artefacts in the UK, unpub MSc thesis, Imperial College, Univ London

Tomlinson, M J, 1995 *Foundation design and construction*, 6 edn, Harlow

Wainwright, G J, 1993 The management of change: archaeology and planning, *Antiquity* 67, 416–21

31

Turning made ground into monuments: preservation of archaeological remains in Old Cairo, Egypt

PETER SHEEHAN

Peter Sheehan
Sub-project director
Archaeological Monitoring of the USAID Groundwater Control Project in Old Cairo Egyptian Antiquities Project
American Research Centre in Egypt
2 Midan Simon Bolivar, Garden City
Cairo, Egypt

Abstract

This paper considers the impact on *in situ* archaeological remains of the groundwater control projects currently being carried out in Old Cairo. The site contains a number of historic churches and a synagogue located within the Roman fortress of Babylon. Rising groundwater levels over the past 20 years have accelerated the rate of decay of these designated 'monuments' and have led to the design of schemes aimed at reducing groundwater levels in targeted areas around them. Archaeological information for the area has previously been limited, so the direct impact of these schemes is being monitored to provide data for the creation of a deposit model which may enable its preservation in the future. The need for a re-evaluation of the relative significance of 'monuments' and intact archaeological deposits is suggested. Discussion of the various engineering methodologies leads to a consideration of the acceptable level of impact caused by engineering measures on both monuments and archaeological deposits. Directions for future research are discussed, as are strategic ways forward for sites similarly affected by groundwater problems. A conceptual advance is envisaged by which those engaged in future dewatering projects in the city might view the preservation of intact archaeological deposits as worthy of a monumental effort.

Introduction

Begun in early 2000, the ongoing Cairo Wastewater Organization (CWO)/United States Agency for International Development (USAID) funded groundwater control project in Old Cairo aims to lower groundwater levels in a number of targeted historic monuments in this historic area of the city. Archaeological monitoring of this project has been provided by the Egyptian Antiquities Project (EAP) of the American Research Center in Egypt (ARCE), in collaboration with the Supreme Council of Antiquities (SCA).

In the relative absence of previous scientific archaeological information from the site the aim of the archaeological monitoring project has been to maximise the opportunity afforded by the groundwater control operations in Old Cairo to record and evaluate the archaeological resource, whilst at the same time providing appropriate mitigation strategies in collaboration with the consulting engineers, C C Johnson and Malhotra (CCJM), USA, and the contractors, The Arab Contractors, Bridges and Special Structures Division (AC-BSSD).

The long term strategic issue raised by this project is the degree of acceptable impact on intact archaeological deposits and subsurface structures caused by measures designed for the preservation of defined monuments. At the curatorial level archaeological information gained during the project should allow for a greater understanding of this impact via reassessment of the relative significance of listed monuments and the *in situ* archaeological remains below and around them. Rising groundwater levels in historic Cairo and indeed at archaeological sites and monuments throughout Egypt indicate that approaches developed during this project may be widely applicable in the near future.

The project has demonstrated the importance of liaison between the archaeological and engineering communities, both to broaden the basis of observational data and to refine for their mutual benefit the rather generalised preconceptions of 'made ground' and its cultural and structural/geotechnical significance. Monitoring of the impact of engineering groundworks and the proposed dewatering regime on both the targeted monuments and the burial environment is most likely to be achieved as part of longer term conservation planning linked to a monument maintenance regime.

Historical background

The core project area is defined by the *c* 4ha enclosed by the known walls of the Roman fortress of Babylon, built by the Emperor Diocletian around AD 300, but certainly enclosing earlier structures located at this vital strategic junction where the Nile valley opens into the broad floodplain of the delta (Grossmann et al 1998). The fall of the fortress marked a pivotal moment in the Arab Conquest of Egypt in AD 641 and this location was chosen, in preference to Alexandria, as the new capital of Egypt which would dominate the medieval Arab world and ultimately become the teeming 16 million megalopolis of today.

In the early medieval period the former fortress became an enclave for the Christian and Jewish minorities of Cairo, and each of the seven churches and the one remaining synagogue within its ruined walls can claim, to a greater or lesser extent, to be grounded in antiquity. The proximity of this enclave to the heart of the Islamic city and its monuments has evoked an added significance for Old Cairo as a place where the conservation of monuments of the three religions has taken on a political and symbolic aspect in recent times (Lambert 1994).

A gradual historical movement northwards of the centre of administration left Babylon isolated amidst the abandoned ruin heaps of the early medieval city, and only during the last 100 years with the explosion of Egypt's population and the use of more efficient means to transport it has this part of the city been comprehensively resettled. The increase in population in this area, much of it informal settlement without infrastructure, has resulted in considerable additions to the groundwater table. At the same time since the construction of the Aswan High Dam in 1971 the River Nile no longer acts as a drain for this groundwater during the winter months. The agricultural and economic possibilities offered by the Dam have in fact had serious negative consequences for archaeological sites and monuments throughout the Nile valley (D'Monte 2001). The first appearance of water within the historic monuments of Old Cairo targeted by the present project was in 1979, since when the groundwater level has increased by 2.0m (Hefny 1996; Fig 31.1). In that same year the Roman fortress was specifically included in the nomination submitted by Egypt for the inclusion of 'Islamic Cairo' on the World Heritage List (UNESCO 1979).

Fig 31.1 Current groundwater level within the church of St George

Conditions for preservation

Natural environment

Egypt is characterised by a dense concentration of archaeological sites so rich in structural and artefactual remains that the country is often cited as a living example of archaeological preservation *in situ*. For these very reasons it is perhaps not surprising that there has been a structural bias in the perception of what constitutes *in situ* archaeology, nor that deposits have been viewed as merely collections of objects. A negligible rainfall and a hot arid climate can provide optimal conditions for preservation of both buildings and organic material. Conditions (as in Old Cairo) around the former floodplain area subjected to annual inundation and a fluctuating groundwater level are not conducive to good organic preservation. On the other hand the rising ground level over time and the use of resilient building materials and forms have contributed to a very high level of structural preservation. The chemical and mineral composition of the raised modern groundwater level is clearly affecting both standing and subsurface structures, particularly the former, where damage caused by capillary action and the movement of hygroscopic salts is clearly visible throughout the area.

The curatorial environment: initial definitions of significance

Ironically the preservation practiced over the last 120 years, a period which has witnessed a great awakening in and interest in Egyptian antiquity and antiquities, has contributed greatly to the destruction of the archaeological sequence preserved in structures and deposits. The choice, given such good preservation conditions, of what was significant to retain and worthy of presentation continued to have a heavy impact on the upper levels of the archaeological sequence until relatively recently. In the early years of the last century not even the Roman and Hellenistic levels at some Pharaonic temple sites escaped clearance.

In Cairo the mechanisms for defining and protecting significant components of the built heritage date back to the formation of the Comité pour la Conservation des Monuments de l'Art Arabe (CCMAA) in 1881, a body given the twin tasks of drawing up an inventory of Arab monuments and recommending the means for their preservation (El-Habashi and Warner 1998). By 1910 the Coptic churches and the Roman fortress of Old Cairo had been added to this original list. Part of the reason for this was that in the intervening 30 years the 'monumental' status of the Roman fortress had become clear as a rather inevitable result of excavation in such a building-friendly environment, while the churches were beginning the process of shedding their accumulated modifications and reverting to an 'original form' which could be appreciated as such.

Problems of being a significant monument

Subsequent to this initial 'dégagement' almost all of the buildings brought to light and defined as monuments have required or received repeated interventions which have been achieved at the expense (in more than one case total) of the historic structure. To these 'preservation impacts' can of course be added losses caused by the effects of pollution and rising damp common to buildings throughout the area. Indeed the degradation suffered by archaeological sites and monuments, coupled with the lack of resources to conserve and store the sheer volume of artefacts preserved within archaeological layers, remains the strongest argument for archaeological preservation *in situ*.

A natural desire of clergy and communities for upkeep and renovation has been exacerbated in Old Cairo by the proximity of the monuments to each other. The contagious effect of individual projects has tended to produce cycles of restoration in a confined space like that of the former fortress. During these various projects the largely period-based and visual definition of significant monuments has created a framework which militates against the preservation of later modifications to a building as well as of its subsurface features.

Even the original *de facto* statement of significance made by the decision of the CCMAA to display and conserve parts of the fortress in the 1890s was not universally shared, as Alfred Butler's comments of 1902 reveal,

> ...with the British occupation came a sense of security which has led to the most deplorable destruction. The need of a fortified enclosure having vanished, Copts, Greeks and Jews vied together in demolishing the walls ... It is the simple truth that in the last eighteen years more havoc has been wrought upon the Roman fortress than in the previous eighteen centuries (Butler 1978, 238).

These factors have combined, over the relatively short period in which preservation has existed as a strategic issue, to produce a situation in which the recognition and definition of Old Cairo as an area of historic significance has been accompanied by an accelerated rate of decay and loss of archaeological information within its standing structures.

On the other hand the remaining *in situ* buried archaeology, protected to some extent in the past by its sheer depth (at least 10m throughout the whole area of the fortress), has until recently largely survived the difficulties experienced by those monuments unfortunate enough to be caught out in the open air of Cairo.

However the rise in groundwater and other threats to the monuments have over the past 10 to 15 years led to an increasing tendency to employ geotechnical solutions involving interventions with both short and long term effects on buried archaeological deposits. Fears caused by 1992 earthquake damage to the listed 'Hanging Church' built over the surviving South Gate of the Roman fortress has, for example, led to extensive grouting of the buried archaeological deposits under this gateway. Another formerly listed building, the Coptic Church of the Virgin was demolished in 2001 and is currently being rebuilt using deep piled foundations.

Contract 102 operations

Aims

The geotechnical part of the USAID funded Greater Cairo Wastewater Project (Contract 102; sewers, house connections and geotechnical work) is targeted at reducing the current groundwater levels within the churches of St George (Greek Orthodox) and St Sergius (Coptic Orthodox), the Ben Ezra Synagogue and the large round tower of the Roman fortress which lies within the grounds of the Coptic Museum (CCJM 1999). Other groundworks originally scheduled for the Coptic Church of the Virgin were obviated by the permission given by the SCA in 2001 to demolish and rebuild the church.

Design

Essentially the design is based on a gravity flow system employing a linear arrangement of reinforced concrete caisson shafts (Fig 31.2). Close to the targeted monuments

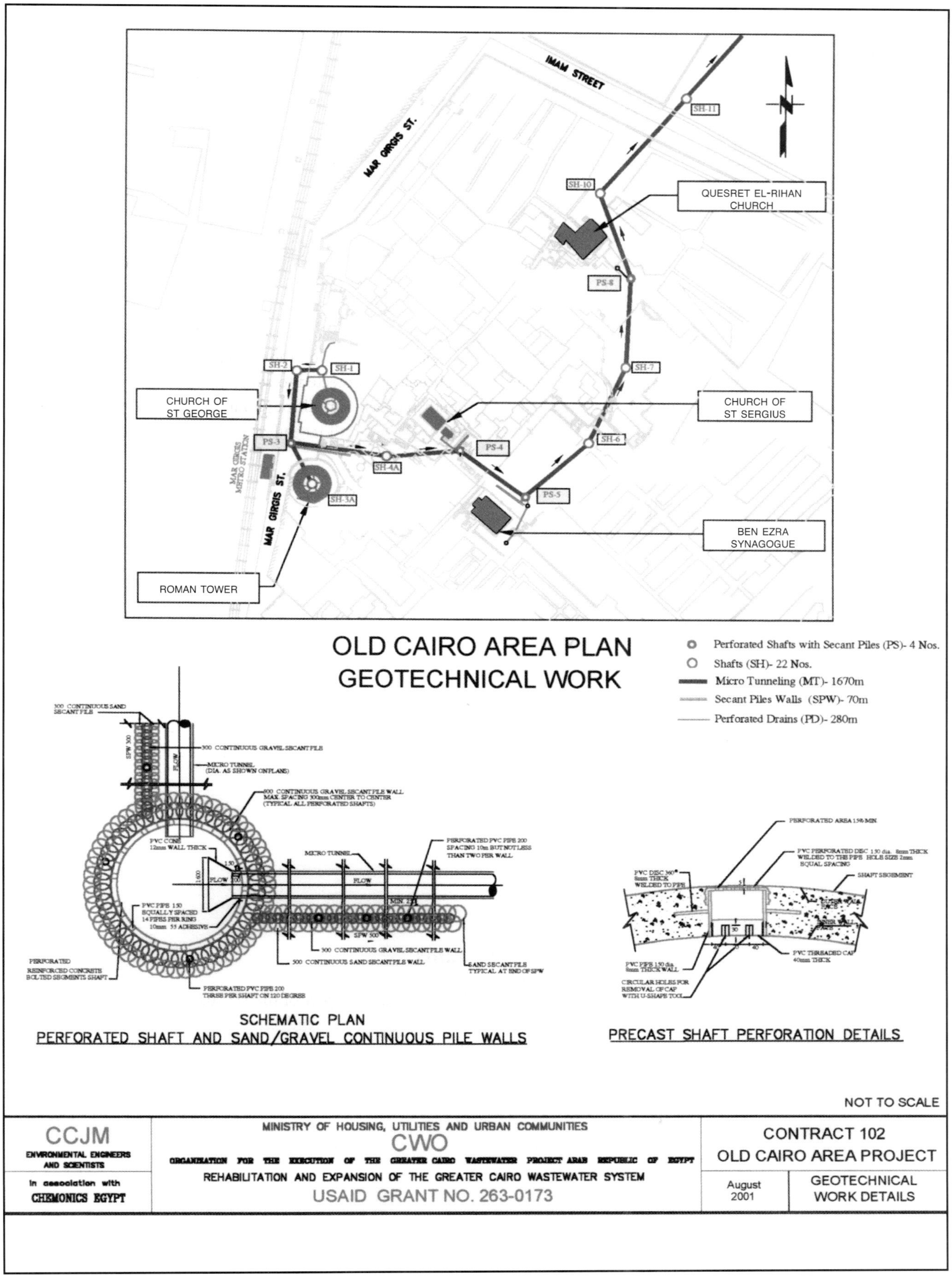

Fig 31.2 Old Cairo area geotechnical work, plan and constructional details (courtesy CCJM)

three groundwater collection shafts 3.66m in diameter and 12–15m deep have been sunk. These have precast prefabricated perforations in the segments of the shaft wall, allowing controllable collection of water draining into the shaft either from 70–100m of 1.0m wide vertical sand/gravel filters (mechanically drilled as a series of overlapping individual 300mm piles, forming a sand and gravel secant pile wall) or 280m of 340mm perforated pipes laid in traditional open-cut trenches. Water is ultimately conveyed out of the area by 1670m of 600/700mm laser-guided microtunnels linking the perforated shafts via an additional 22 unperforated 3.66m/2.59m diameter caisson shafts. The gravity flow avoids the movement of fine material inherent in pumping, and with it the consequent dangers of soil consolidation and differential settlement. The microtunnel route has been largely dictated by the contingency of having to excavate from above in the case of the microtunnel drilling head becoming stuck. The excavated volume of archaeological material can be estimated at around 3000m^3 or less than 1% of the total potential archaeological deposits within the core area of the project.

Monitoring as part of the construction works

The objective of monitoring and instrumentation requirements specified in the contract documentation was 'to monitor movements of the ground and buildings around each working front during construction to aid in evaluation of the Contractors construction procedures' (CCJM 1999, 02015-1). In practice this monitoring consists of visual observations, crack monitors (telltales) and regular checking, calibration and analysis of elevation reference points (ERPs). The aim of this monitoring is to provide indications of soil consolidation and consequent differential settlement caused by construction operations. However since dewatering has not yet occurred (other than localised and small scale surface pumping) this monitoring programme will need to be extended for a considerable time after the system becomes active in order to provide data on the amount of consolidation likely to be produced.

Archaeological monitoring and mitigation

All the shaft excavations and other groundworks have been monitored by the EAP project and each has produced significant new information on the nature and condition of archaeology on the site. The 'secant pile' work is perhaps an exception to this, since material encountered during drilling for the 300mm piles comes to the surface as a bentonite slurry. Nevertheless voids and obstructions encountered during this drilling have provided additional topographical information. Two of the shafts were excavated archaeologically to groundwater level to provide a control for the other shafts. Monitoring of the perforated drain excavations was facilitated by temporary dewatering, which increased the opportunity for archaeological recording. A high level of cooperation between the archaeological monitoring project and all the elements of the construction team – from geotechnical consultants to the divers removing obstructions from beneath the caisson shafts – has yielded a wide range of mutually useful data. Information from these groundworks is being combined with borehole and microtunnel logs to create a deposit model for the area related to the national grid. Mapping and deposit modelling of the site will allow intrusions like temporary wells to be reused without further impact on the archaeological resource.

Mitigation of groundworks has sought to reduce the impact only on structures revealed in open trenches (Fig 31.3). The shaft locations however were dictated from the outset by the confines of the site and the necessity to allow access from above to the microtunnelling drilling head in the event of it becoming stuck. Total destruction of archaeological deposits and structures at these locations was therefore predicated from the outset. Changing the location in response to archaeological features revealed by the groundworks would have required value judgements or an assessment of relative significance which the current lack of data precluded (Fig 31.4). Additionally since most shaft excavation took place below groundwater level the traditional period-based/visual conditions necessary for ascribing monumental significance to structures or deposits did not arise.

Fig 31.3 Mitigation of impact; perforated pipes being tunnelled through foundations of a Roman wall at 5.0m below modern ground level

Fig 31.4 Predicated destruction; top of a nearby Roman wall (similar to Fig 31.3) encountered at 2.0m below ground surface in shaft 5

In the case of the monuments targeted for dewatering however this rationale of 'necessary destruction' clearly should not equally apply (Fig 31.5). These locations have demonstrated the need for an *a priori* definition within geotechnical design parameters of the degree of acceptable impact upon subsurface elements of the monument, as well as adequate archaeological investigations of these elements, well in advance of the exigencies of a construction schedule.

Fig 31.5 Shaft inserted within the Roman tower, part of a listed monument; perforated pipes to dewater the radial chambers are being laid through the thickness of the tower walls

Future directions

A project conceived for the preservation of particular monuments has thus, a hundred years after these monuments were evaluated and given their present significance, in turn evaluated the spaces between them. The groundworks have *de facto* produced a representative sample for preservation by record. The synthesis of these records should now allow definition of the reasonable limits for the representative sample to be preserved *in situ*.

The information derived from the monitoring project represents the key factor affecting future preservation. The first step is simply being there to record it, and the significance of USAID's desire for an archaeological dimension to accompany, if not to underpin, the construction project should not be underestimated. Similarly influential in the context of conservation in Cairo is ARCE/EAP's recognition of the literally fundamental relationship between archaeology and conservation projects. Conservation projects in the historic city have been dogged in the past by partial or inadequate 'solutions' to the groundwater problem, to such an extent that it has now become almost a mantra that this problem must be addressed before any other works can proceed. Since the current project is seeking to provide this essential prerequisite to conservation for not one but four listed monuments, its archaeological component may come to be viewed as similarly essential.

Communicating the archaeological record in a way which will aid subsequent preservation hinges on format, language and 'usability' to curatorial authorities whose duties are to the monuments and who may have difficulty in seeing 'soft and eminently undisplayable deposits' (Nixon nd, 39) in the same light as the more familiar large masonry structures.

Another essential step is the provision of a mechanism by which the nature and significance of the monuments' archaeological context can be communicated as a design parameter to engineers designing and implementing preservation projects. USAID has recognised this by including an archaeologist on its proposed committee to review the progress of Contract 102. The role of this archaeologist is to create 'dossiers' on each monument demonstrating their relationship with the archaeological context and the potential benefits of this information to design issues. The conservation plans which these dossiers could and should subsequently become are therefore well placed to provide for the preservation of hitherto unconsidered aspects of the monument such as its burial environment.

Future research

The archaeological monitoring project has already produced baseline data on the burial environment and underlying topography to inform all subsequent studies. This research can expand the current project based pragmatic aim of monitoring building movement into a much wider and more significant study of the effects of localised groundwater reduction on archaeological features and deposits throughout the area. The combination of former, current and planned post-project hydrology of the site provides unique opportunities for research. Before the construction of the High Dam, the position of the site in the Nile floodplain produced periodic immersion followed by a gradual fall throughout the rest of the year. The importance of the level of the Nile flood to the very life of the country means that precise data on the flood level each year is available (with some lacunae) from AD 622 until the present day (Toussoun 1925). In the last 20 years these periodically immersed strata have been saturated in an additional 'layer' of groundwater. Since the amount of soil consolidation to be expected as a result of dewatering is related to historic water levels this research has practical implications for the monitoring and maintenance of the installed systems once they become active.

Strategic ways forward for archaeological preservation *in situ* in Cairo and other comparable historic cities affected by rising groundwater levels

1) Linking with powerful engineers' groups to take opportunities for refinement to existing deposit models afforded by the largely untapped vein of engineering work in the city. Contract 102 is part of a major project aimed at overhauling the wastewater system in Greater Cairo over the past 10 years, much of which has taken place without archaeological monitoring. There has therefore been no real tradition of meaningful interface between infrastructure projects and the archaeological curators, a situation which can be modified by demonstrating how much information these projects can provide. Agencies funding large infrastructure projects have the financial and political power to ensure that effective monitoring takes place and that their obligations towards the conservation of cultural heritage in accordance with international charters and national legislation are fulfilled.
2) A definition of what constitutes an acceptable impact of 'preservation engineering' on intact archaeology, either below ground or within standing structures. Inclusion of defined archaeological layers or deposits as a constraint within design parameters. In particular the application of tile drain schemes to the internal spaces of historic structures needs to be clearly related to the level of surviving archaeological deposits. Decisions on the level to which dewatering should take place should reflect intact archaeology and long term building conservation issues as much as the desire to return to arbitrary and/or relatively recently submerged floor levels. The size, scope and sequence of engineering projects designed to dewater groups of monuments should also be assessed so that experience and information gained early in the process can actively contribute to the ongoing design. Equally more preconstruction archaeological investigation, supported by adequate temporary dewatering facilities would reduce the necessity for *ad hoc* mitigation solutions arrived at during construction.
3) Archaeological preservation by record during the project followed by adequate publication and preservation of these records in an accessible curator-friendly form can lead to future preservation *in situ*. Web-based archiving would in addition widen access to this information and enable issues concerned with archaeological or architectural preservation to be addressed by the wider community.
4) Documentation of extent and depth of engineering intrusions for future reuse, for example temporary wells used around monuments for dewatering during perforated pipe work can be located and utilised during further works or maintenance.
5) Re-evaluation of interpretation of World Heritage status and other voluntary or statutory definitions to clarify the relationship between monuments and their archaeological context and the value of each. The continued survival and conservation of intact archaeological deposits is another layer of value for historic areas which complements their religious and tourist/economic significance.
6) Detailed follow-up assessment of the effect of temporary and permanent groundwater control on monuments and archaeological deposits. Consequent characterisation of stable and non-stable environments. Monitoring methods used must be regularly analysed to clarify these effects rather than merely producing a mass of impenetrable data.

Bibliography

Butler, A J, 1978 (1902) *The Arab conquest of Egypt and the last thirty years of the Roman dominion*, 2 edn, Oxford

CCJM, 1999 C C Johnson and Malhotra, P C, Arab Republic of Egypt; Ministry of Housing, Utilities and Urban Communities; Organisation for the Execution of the Greater Cairo Wastewater Project, Old Cairo Area Project; Contract 102, Sewers, House Connections and Geotechnical Work, unpub contract documents vol 2 of 3, Specifications, Cairo

D'Monte, D, 2001, The Aswan Dam is destroying the Temple of Karnak, *The Art Newspaper* 110, January 2001, 28

Grossmann, P, Jones, M, Le Quesne, C, Noeske, H-C, and Sheehan P, 1998 Zweiter Bericht über die Britisch-Deutschen Grabungen in der Römischen Festung von Babylon – Alt-Kairo, *Archäologischer Anzeiger*, 173–207, Berlin

El-Habashi, A, and Warner, N, 1998, Recording the monuments of Cairo: an introduction and overview, *Annales Islamologiques* 32, 81–99, Cairo

Hefny, K, 1996, Feasibility study of water problems affecting the Coptic area at Cairo, unpub American Research Center in Egypt rep

Lambert, P (ed), 1994 *Fortifications and the synagogue; the fortress of Babylon and the Ben Ezra synagogue, Cairo*, London

Nixon, T, nd [1998]Practically preserved: observations on the impact of construction on urban archaeological deposits, in *Preserving archaeological remains* in situ: *proceedings of the conference of 1st–3rd April 1996* (eds M Corfield, P Hinton, T Nixon and M Pollard), 39–46, London

Toussoun, Prince Omar, 1925 *Memoires de l'Institut d'Egypte: Vol 2, L'histoire du Nil* 8, 361–411, Cairo

UNESCO, 1979 United Nations Educational Scientific and Cultural Organisation, Convention concerning the Protection of the World Cultural and Natural Heritage: World Heritage List: Nomination submitted by Egypt: Islamic Cairo, unpub rep

32

Wetland archaeology and peat wastage

RICHARD BRUNNING

Abstract

To provide background information for a Water Level Management Plan the Environment Agency commissioned Somerset County Council Heritage Group to conduct a desktop study into the effect of peat wastage on waterlogged archaeological remains in the Somerset moors (Brunning 2001).

Rates of peat wastage

The rates of peat wastage under arable and pasture regimes were derived from three sources: ground anchors; measurements carried out elsewhere and historical and archaeological evidence.

Ground anchors

Topographic information is available from a series of six permanent ground anchor points installed by the Environment Agency in two areas of Somerset peatlands (ADAS 1996).

If the average of the first three measurements (December 1987 to April 1988) are averaged and compared to the average of the last three measurements (September 1996 to August 1997), peat wastage losses of between 40mm and 71mm are suggested equivalent to wastage rates of 0.44m to 0.79m of peat over an 100 year period.

Measurements carried out elsewhere in the UK and Europe

In arable fields peat wastage rates varying between 10–18mm per year (Richardson and Smith 1977), 20–30mm per year (French and Pryor 1993) and 39mm per year (Hutchinson 1980) have been recorded.

In grassland peat soil wastage by oxidation has been measured at rates of between 10–20mm per year in central and western Europe and a Polish study suggested losses under grassland were about half those under arable cultivation (quoted in Armstrong 1996, 8).

Historical/archaeological evidence

Historical coring records (Anon 1881; Bulleid 1933, 21), the nature of past and present discoveries of archaeological sites (eg Fig 32.1 and Brunning 1998; Stradling 1849, 52) and recent palaeoenvironmental studies (Housley et al 2000, 16) all provide evidence of significant peat wastage occurring in Somerset over the last 200 years with a loss in some places of over 4.0m of peat.

Richard Brunning
Somerset Levels and Moors Archaeologist
Somerset County Council
County Hall, Taunton
Somerset TA1 4DY, UK

Fig 32.1 *The Late Bronze Age pile alignment at Harter's Hill, Somerset, showing how peat wastage has brought the ground surface down to the top of the structure*

What are the implications for archaeological remains?

The density of archaeological sites in the Somerset peatlands has been estimated in a recent study at 3.4 sites per km^2 (Van de Noort et al 2001). In Somerset peatlands 53 waterlogged archaeological sites are known to survive wholly or partially *in situ*, 14 of which are wholly or partially designated as Scheduled Ancient Monuments. All of these sites exist so close to the surface that their waterlogged components will be completely destroyed by peat wastage within the current century if the grassland wastage rates shown above are correct. The palaeoenvironmental potential of the upper peat levels will also be completely destroyed. Climate change is likely to hasten this destruction (May 1997; Bird 1993).

Fig 32.2 *Trench to assess the condition of the Neolithic Sweet Track in Shapwick Heath National Nature Reserve where a pumping system maintains high water levels*

What are the solutions?

1) Assess the condition and hydrological regime of the most important sites.
2) Hydrologically protect important sites at risk (eg as in Fig 32.2 and Brunning et al 2000) or rescue excavation of sites that are impossible to protect.
3) Change the Environmentally Sensitive Area scheme operating in Somerset peatlands to enable it to help prevent further wastage by following the recommendations of relevant reports (Spoor et al 1999).
4) Create more sustainable wetland landscapes in some areas of the floodplain (eg Rasmussen 1999)

Bibliography

ADAS, 1996 Monitoring of raised water levels in the Somerset Levels and Moors ESA 1989–1995, unpub ADAS rep

Anon, 1881 Section of strata near Shapwick, *Proc Somerset Archaeol Natur Hist Soc* 27

Armstrong, A, 1996 The conservation of peat soils on the Somerset Levels and Moors, unpub ADAS rep

Bird, E C F, 1993 *Submerging coasts: the effects of a rising sea level on coastal environments*, Chichester

Brunning, R, 1998 Two Bronze Age wooden structures in the Somerset Moors, *Archaeol Severn Estuary* 8, 5–8

Brunning, R, Hogan, D, Jones, J, Jones, M, Maltby, E, Robinson, M. and Straker, V, 2000 Saving the Sweet Track; the *in situ* preservation of a Neolithic wooden trackway, Somerset, UK, *Conservation Management Archaeol Sites* 4, 3–20

Brunning, R, 2001 *Archaeology and peat wastage on the Somerset Moors*, Taunton

Bulleid, A, 1933 Ancient trackways in Meare Heath, Somerset, *Proc Somerset Archaeol Natur Hist Soc* 79, 19–29

French, C A I, and Pryor, F M M, 1993 *The south-west fen dyke survey project, 1982–86*, East Anglian Archaeol Rep 59, Peterborough

Housley, R, Straker, V, and Cope, D W, 2000 [1999] The Holocene peat and alluvial stratigraphy of the upper Brue valley in the Somerset Levels based on the soil survey data of the 1980s, *Archaeol Severn Estuary* 10, 11–23

Hutchinson, J N, 1980 The record of peat wastage in the East Anglian Fenlands at Holme Post, AD 1848–1978, *J Ecology* 68, 229–49

May, R, 1997 *Climate change; a note by the UK Chief Scientific Adviser*, London

Rasmussen, J B, 1999 *Denmark's largest nature restoration project: the Skjern River restoration project*, Copenhagen

Richardson, S J, and Smith, J, 1977 Peat wastage in the East Anglian Fens, *J Soil Sci* 28, 485–89

Spoor, G, Gilbert, J, and Gowing, D, 1999 *Conservation of peat soils on the Somerset Levels and Moors, part 4: safeguarding peat soils*, Silsoe College, Cranfield University

Stradling, W, 1849 The turbaries between Glaston and the Sea, *Proc Somerset Archaeol Natur Hist Soc* 1(2), 48–62

Van de Noort, R, Fletcher, W, Thomas, G, Carstairs, I, and Patrick, D, 2001 *Monuments at risk in England's wetlands*, draft v 2, Univ Exeter

33

'Cleaning up': protecting deposits on brownfield sites under environmental impact assessment

BRIAN DURHAM

Brian Durham
Archaeology Officer
Oxford City Council
PO Box 10
Oxford OX1 1EN, UK

Abstract

Five recent construction projects in Oxford have thrown up a pattern where preservation of archaeological remains *in situ* is overridden by statute. Migration of hydrocarbon contaminants (including methane) under an archaeological site can contaminate it by association, leading to extraction of clean archaeological deposits. Case studies include a Cistercian abbey Scheduled Monument, a medieval suburb and a Roman kiln site.

Should archaeologists complain? How does the historic environment compare in significance with clean water and clean soil? Under environmental impact assessment (EIA), we can ask how many people they each matter to, and how much they matter to them: difficult and potentially emotive questions.

Introduction: brownfield heritage issues and PARIS

A series of construction projects in Oxford over the last four years have thrown up a pattern where preservation of archaeological remains *in situ* provided by planning and policy guidance (PPG16) is overridden by statute, and not always reasonably.

St Thomas in Oxford is a medieval suburb on the Thames floodplain just west of the castle. Over the last twenty years there have been a series of excavations, each adding significantly to an unusual story of artificial reclamation of the floodplain since the early 13th century. One affected site is a scheduled Cistercian abbey. The chain of discovery proves to have a flaw in the form of hydrocarbon contamination which can move on the groundwater, spreading underneath clean sites.

Oxford like everywhere in south-east England is subject to development pressure on land freed up by a previous development cycle, ie brownfield. Much of this land will be on river floodplains, and in historic towns it will include ancient domestic sites which have evolved into successful trade/industrial sites with consequent contamination.

Hydrocarbon spillage on a floodplain will tend to percolate to the surface of the groundwater and then move gradually with the flow, forming a 'plume'. Standard ground- and water-quality targets require that this contamination be removed on the polluter pays principle, so it becomes part of the land transfer requirement, ultimately achieved by a condition on planning consent. The problem is that the 'plume' can extend under an innocent archaeological site. The standard practice has been to remove clean overburden to get at the plume which may itself be only inches thick. Conventionally this is costed on the basis that the superficial deposits are merely fill, made ground, with no intrinsic value. Clearly this is not so in a medieval suburb, but if a developer regards the cleaning up of someone else's pollution as an imposition, how much worse if he is asked to comply with PPG16 before they start?

There are green alternatives to bulk extraction, as this paper will demonstrate, mainly involving microbial attenuation, but they need time to plan and time to implement. They must therefore be programmed in at the earliest stage in a development. This brings to the fore the concept of environmental impact assessment, whether under the statutory process required for larger developments, or under the same intellectual concept as it is applied by planners across the full range of economic development. Foremost is a need to assess whether it is worth it: how does the protection of the historic environment compare in significance with clean water and clean soil; how many people does it matter to; and how much does it matter to them if estimated in cash terms? These are difficult questions. The ideal would be if both ground quality and historic environment can be mitigated in an integrated programme, in which the archaeology is evaluated early enough so that a decision can be made in favour of archaeologically 'green' mitigation.

Brownfield heritage: the conflict

Polluter pays has been adopted universally in western planning legislation and seems to be here to stay. The polluters in the present four case studies are identified thus: pollution of the cultural heritage is deemed to be caused by the would-be developer, if their proposals would cause significant loss; pollution of the ground and groundwater is deemed to be the responsibility of the occupier/owner who failed to take sufficient care of contaminants. It will be the job of the would-be developer to manage the resource of land and groundwater quality, normally under a financial arrangement with the vendor; they will similarly need to manage the cultural heritage resource by minimising/mitigating the impact of their development, either before or after they acquire rights over the land.

Relevant planning guidance

The optimising of land and groundwater quality can be achieved by several alternative technologies, and we will see that the range is expanding as pressure of space leads increasingly to the reuse of brownfield sites; optimising the cultural heritage resource of a site can be achieved by changes in design which tend to preserve deposits *in situ*, or by creating a permanent record. Both resources are overseen by the planning system. However a significant difference is that land quality is a resource where the planning authority is prevented from overriding the authority of the body with expertise and responsibility, ie the Environment Agency (PPG23, DETR 1999, para 42). The authority of English Heritage over cultural heritage matters is not prescribed in the same way, except possibly in respect of Scheduled Ancient Monuments.

Four Oxford case studies

Our four case studies will show how conflicting remedies may be presented to the competent authority at a point in the development programme when no effective compromise between them is possible. The relevant background to each case is tabulated in Table 33.1 and they may be characterised briefly as follows.

Site 1: 64–66 St Thomas Street, Oxford

Laing Homes redevelopment of redundant GPO depot on the site of a medieval tenement, with significant hydrocarbon groundwater pollution (Cook 1999).

Site 2: Rewley Road, Oxford

Persimmon Homes redevelopment of the former railway yard for residential use, affecting the site of the medieval Rewley Abbey, Scheduled Ancient Monument, with significant heavy metal ground contamination.

Site 3: Rewley Road, Oxford

Said School of Management Studies redevelopment of former railway yard, affecting the site of the medieval Rewley Abbey, Scheduled Ancient Monument, with significant hydrocarbon groundwater contamination.

Site 4: Guelder Road, Frys Hill, Oxford

Oxford City Council remediation of significant methane ground contamination arising from recent landfill in connection with new housing.

The planning status of the case study developments

It should be understood that only Site 3 was subject to formal environmental impact assessment (Waterman Environmental 1998). Site 1 was too small, Sites 2 and 4 are parts of larger projects which might have qualified for environmental impact on grounds of size and traffic generation for instance, had the process of such assessment been more common locally in the mid 1980s when these developments were conceived. In the event they saw no formal environmental impact assessment; moreover for the 24ha Frys Hill site (Site 4) there was no pre-determination assessment of the archaeology (not illustrated). For three of the sites a compromise mitigation strategy was negotiated through the planning process, the exception being Site 1 where there was merely a passive watching brief. The four are presented here as examples of a conflict which emerged in the planning process and which could arise again under environmental impact assessment, with a view to pre-empting it. Because of the concentration of significant archaeological deposits in urban areas, this is

Table 33.1 Informal case studies of potential mitigation conflict between hard and soft environmental impacts

Site name	St Thomas Street former GPO depot	Rewley Road Persimmon Homes development	Rewley Road Said Business School	Guelder Road Frys Hill
Site area	0.1ha	1.8ha (part of 4ha)	2.2ha (part of 4ha)	2.5ha (part of 24ha)
Local Plan status	Primarily residential	Residential (outline consent)	Business or hotel (outline consent)	Recreation
Development	Residential	Residential	Higher education institution	Remediation of methane venting
Archaeological constraint	Buried floors and walls of medieval and later tenements	Scheduled remains of Cistercian abbey including enclosing ditches and farm	Scheduled remains of Cistercian abbey including enclosing ditches and farm	Roman pottery production site, identified by geophysical and trenching in 1995
Contamination	Hydrocarbons, VOCs and SVOCs in groundwater (LTG June 1998)	Mercury, lead, copper PAHs Hydrocarbons (Environment Agency letter June 1997)	Petroleum Hydrocarbons in groundwater; ash, coal and PAH in made ground; Ammonia in groundwater (LTG risk assessment April 1999)	Methane and carbon dioxide, risk to new housing
Quantification of contaminant	Post-treatment levels elevated w.r.t. Dutch Target Level, but within Dutch Intervention Level	Not available	Not quantified in risk assessment April 1999	30–40% methane v/v in samples (BS requires <1.5%)
Likely source	Fuel tanks related to use as GPO transport depot	Former use as railway goods yard. Bunded fuel tanks mapped c 1900, centrally within site	Railhead oil depot to east in 1900 (site of present fire station). Rail to road coal distribution businesses, one still operating in early 1990s	Potentially from old peat deposits along Northfield Brook, trapped by landfill related to residential development.
Potential remedies	Martin and Bardos 1996 list containment, pumping, bioventing, excavation	Extraction to landfill; capping with clay	Pumping/venting Vacuum extraction	
Chosen remedy	Total excavation of 270 sq m to 0.9m depth, further deepened over 120 sq m to below water table (>2m depth)	As set out in Broughton Beatty fax 12 and 17 June 1998, approved by LA, plus excavation of one additional TPH hot spot	Excavation of c 900 sq m to groundwater and removal to licensed landfill. Pulse pumping of free product hydrocarbons Monitoring of remainder	Install collector pipe and vent tubes
Archaeological effect	Eastern half had been open car park, where development impacts mitigated by evaluation trenching and then by open-area excavation. This showed development of tenement from initial reclamation to foundations of building which survived to 1940s. Western half of development had been covered by depot building, and because of perceived petroleum hazard had not been evaluated in the significant frontage area. Passive watching brief on the remediation excavations reported garden soils on the western half of area (plausible), but layered deposits were visible in side of excavation towards medieval tenement, and there may have been significant loss here. A ditch-like feature at a deep level might also have been significant. Informal on-site discussion with the hydrogeologist suggested that the medieval garden levels had not been contaminated, and were only removed because of the Environment Agency's direction that hydrocarbons at groundwater level be cleansed, which in the chosen remedy meant excavation	Residential development constituted c 40% of large brownfield site, part of which was designated as SAM (arbitrary catch-all shape). The vendor Railtrack commissioned archaeological survey which enabled identification of heritage sensitive areas. On this basis the least sensitive northern area sold to Persimmon Homes for residential development. Proceeded with implementation of archaeological mitigation but ahead of agreement on groundwater remediation, which was then interpolated, alongside various other issues. No reason to suspect that damaging deep extraction was done	This site was residue of the railway yard after Persimmon Homes site had been taken out. University needed to dismantle a listed station structure before proceeding, and in May 1999 the above remediation procedure was carried out for one of the areas of lower sensitivity within the scheduled area. Informal discussion on site suggested this was less than optimal area for extraction; confirmed by Environment Agency (but not LA); also (as with St Thomas) non-intrusive technologies available, but needed longer lead-in time, which was not possible given the change in consulting engineer from Arup to Witby and Bird. Passive watching brief reported no identifiable archaeological deposits as viewed from atop of excavation.	Following advice of archaeological (OAAS), most sensitive 50m of the collector trench stripped archaeologically in advance of construction, and fully recorded archaeologically. Complex multi-phase Roman deposits with much pottery, but (fortunately) no kiln or other structure
Sustainable resolution	LA to overlay Land quality GIS on Archaeological GIS to identify potential conflicts; Sustainable solutions to be explored with Environment Agency at initial scoping/liaison stage	No change	Establish extent and velocity of plume distance over which it is detectable in relation to distances to any abstraction points or other sensitive receptors. Revise risk assessment, and remediate by pulse pumping, bioventing and or natural attenuation	No change

now potentially one of the most significant uncontrolled threats to the cultural heritage in the UK.

Ground/water quality issues in the case studies

All four cases are brownfield sites, in shades of brown: a small redundant transport depot, an ex-railway goods yard, and a large area of mildly contaminated ex-sewage farmland. The issue for the archaeologist is that extraction of contaminated soil to licensed landfill in three cases amounts to ground disturbance of a scale which, in a planning proposal, would attract a condition for archaeological mitigation on the basis PPG16 and local environmental policies. If the realistic costs of such archaeological mitigation were built into the budget for remediation of contamination, it would make non-intrusive remediation significantly more attractive on a purely cost basis. Informal discussion with consulting hydrogeologists on Sites 1 and 3 indicated that non-intrusive remediation systems were available, but had been excluded by their client in each case not because of expense, but because they would take longer to implement. Programming of mitigation to avoid conflict is a fundamental management function of environmental impact assessment.

Archaeology and contaminated sites

The archaeological literature does not seem to have registered this conflict as such. This may be because of a perception that the quality of land and water takes automatic precedence of over the cultural heritage. The issue is nevertheless unlikely to go away, because government policy is presently directing 60% of new residential development towards brownfield sites, and these sites may have a relatively high incidence of archaeological deposits simply because of their geographical location as areas amenable to human settlement, whether modern or ancient. Even if partly disturbed, hectare for hectare it is more likely that industrial contamination will coincide with sites with cultural heritage value than with sites with no cultural heritage value.

Local authority obligations on contaminated land

Furthermore in three cases we are dealing with hydrocarbon contamination of the groundwater arising (presumably) from leakage from relatively minor road freight operations during the last seventy years or so. These are not sites which have been identified in any sort of comprehensive register (a national frameworks survey in the 1990s set out to plot them from land-use histories, but was discontinued because it appeared to be leading to land blight (Crabtree 1997, 5)). Oxford City Council operates a land quality geographical information system (GIS) prepared by their land quality consultant, recording sites which have been investigated intrusively, one of the more advanced Local authority systems countrywide (J Juneman, pers comm).

National standards in land/water quality

We have seen that for reasons of government policy and historical geography the issue of invasive remediation versus archaeological protection is likely to be a recurring one. In three of the case studies the perceived conflict arose because the preferred system of remediation was the surgical removal of the contaminated soil to a licensed landfill. The driving force here is to comply with national standards and guidance on groundwater quality; thus PPG23 poses the question; 'does the site threaten surface or groundwater?'

In all three projects the answer was (clearly) affirmative. We have seen that a requirement to defer to Environment Agency expertise and responsibility is expressed in PPG23, and it is reiterated in Circular 2/99 para 42 (see also Glasson et al 1999, 272–7). Thus the quality standard will continue to be that of the Agency, even if the method of mitigation is for the developer to select and seek approval. The two Laing Technology Group (LTG) reports compare site groundwater quality with Dutch Intervention Levels (DIL), which are surpassed in the post treatment monitoring at Site 1, and with the higher standard of Dutch Target Levels (DTL), which Site 1 had yet to achieve. The Environment Agency's approach to target levels is based on site-specific risk management of the 'source-pathway-receptor' model of waterborne pollution (Tim Johns, pers comm).

There is an extensive literature on procedures for remediation, which appears to show growing confidence in relatively green biological remedies, in terms of value-for-money and sustainability (Crabtree 1997, 67). These are discussed below, but programming issues, which appear to be the major difficulty with implementing these greener technologies, will be considered first in case they present insurmountable difficulties.

Programming of information gathering

Planning to mitigate contamination is the subject of PPG23. Planning to mitigate cultural heritage impacts is the subject of PPG16. A significant difference between these two instruments for present purposes is the timescale needed to evaluate the resource before devising mitigation. For archaeology the identification of impacts is often a staged process, in which screening, scoping, desktop study, field collection, geophysical and intrusive field evaluation may all be necessary precursors to devising a scheme of mitigation. The same position may be reached in land/water quality studies by fewer stages.

The temptation to delay finding out

Thus the developer has two reasons for leaving the archaeological investigation as late as possible in the development process, ie the time taken by the staged evaluation process and his need for value-for-money from his capital. In parallel with this process he will be taking advice on likely contamination, but this too may be subject to cash-flow consideration, and he will be looking for expressions of commitment from the deciding authority before spending money. In the hiatus, the developer's archaeological consultant will need to qualify his ballpark proposals with the presumption that the preferred method of archaeological mitigation (ie preservation *in situ*) may not be feasible in the case of a site which is contaminated; the developer's soil engineer will similarly need to recognise that the economics of total excavation and removal off site will be distorted if there is intrinsic cultural value in the soil, ie archaeology.

The perils of late-breaking information

In the case of our Site 3, it should be noted that the Environmental Statement dated May 1998 had the benefit of full staged archaeological evaluation over a period of eleven years, which confirmed the national importance of Rewley Abbey. Three potential impacts were listed, but remediation was not included, perhaps because the contamination report had identified only 'slightly elevated levels' of contaminants (Waterman Environmental 1998, 33). Further ground investigation was proposed for later that year however, and it was presumably this which led to a requirement for extensive ground extraction within the scheduled area (J Juneman, pers comm).

Balancing the importance of hydrological and archaeological resources

So we can conclude that the timescale is not easy, and there will be a tendency to defer these relatively expensive investigations. An assessment of the importance of the two resources might help to judge the extent to which such data collection ought to be prioritised. Water quality relative to heritage is not an easy equation to solve, but it is possible to put economic and monetary valuations on such resources, if only by directly consulting the sensitive receptors. Winpenny (1991) showed how a crude estimate can be made by economic cost-benefit analysis, tailored to the requirements of the receptors, in our case the users of the aquifer and the people of Oxford whose cultural heritage is affected. Their requirements can often be valued in monetary terms, ie what would they pay to retain a given resource? Here the techniques of market research can be employed (Tunstall and Coker 1992).

There are however well documented concerns about the objectivity of public opinion (Rubin 1994), so directions in management of pollution should not rely exclusively on popular sentiment, and a rational approach is safest at this early stage in the process, springing from a tried and tested bureaucracy. In the UK this consists of a well-established planning code, advised where necessary by the respective agency. The Environment Agency has responsibility for control of groundwater quality, and can direct the enquiring developer to extensive technical documentation, although they do not publish their own guidance. English Heritage has responsibility for cultural heritage matters, and is presently preparing a circular on urban regeneration and sustainable development of brownfield sites. Their own scale of significance is traditionally based on a checklist used in the identification of nationally important sites for scheduling purposes.

Land/water quality remediation technologies

In order to decide where it is feasible to make judgements between the relative values of water quality and cultural heritage, it is important briefly to review the rapidly developing technologies for remediating pollution, which could provide room for manoeuvre. Crabtree investigated the feasibility of sustainable development on brownfield sites, and concluded that older remediation systems were relatively unsustainable in global terms, but this was changing (1997, 67). Many of the innovative systems are documented in detail by Martin and Bardos (1996). They consider five main treatment technologies: biological; chemical; physical; solidification/stabilisation and thermal, and they provide a range of costs per cubic metre of soil for the various treatments. They indicate which type of contamination can be successfully dealt with by each technology, and they are positive about biological clean-up of organic compounds. Thus they report the effective *in situ* remediation of ground soaked in creosote and other wood preservatives in Canada, *in situ* degradation of petroleum hydrocarbons in Germany and Holland, and three successful bio-venting remediations (Martin and Bardos 1996).

Sometimes remediation of organics contamination can be done *in situ* biologically or by 'bioventing'; but in some cases is likely to require cultivation of the surface levels and is hence destructive or at least difficult to control. Other biological remediations such as bio-reactors and windrow farming require temporary extraction of the deposit and are therefore as destructive of archaeology as incineration or removal to licensed landfill.

The remediation of hydrocarbon contamination has recently posed difficulties on the archaeologically rich Thames floodplain in Oxford. The more mobile contaminants percolate down to the water table, at a depth of 2–3m, where they rise and fall with changes in groundwater level, and move with the groundwater flow to form a plume. Larger molecules, including hazardous polyphenyls, may tend to be adsorbed onto soil components, and stay closer as most of the volatile and mobile fractions will have moved away. It is worth noting that certain aromatics can, as a result of microbial attack, be converted to something even nastier, and this needs to be watched; and may justify a need to alter the cultural conditions.

In situ bioremediation of hydrocarbons could also bioremediate a lot of organic deposits related to the monument in question if the water table is lowered significantly for any length of time. Pulse pumping will be important, also avoiding injecting oxygen and nutrients at a level below permanent groundwater.

The above comments relate mainly to a potential source of groundwater contamination, not 'contaminated land' as such. The latter may have different archaeological implications depending on the proposed end land use. For groundwater, heavy metals contamination can be more serious in acid soils (inf J Juneman), where heavy metal ions are less likely to be absorbed onto soil elements.

Environmentally friendly bio-technologies

On two of the sites under discussion it is the total petroleum hydrocarbons which are the main problem, and it is important to consider these technologies in terms of how damaging to archaeological deposits they may be. The results of an informal review are tabulated as Table 33.2. The remediation data are mainly from Martin and Bardos (1996), the opinions mainly those of the present writer except where stated. It should be borne in mind however that published results are not necessarily transferable between sites, and each would need to be reviewed by risk assessment on a site-by-site basis. The circumstances assumed are the Upper Thames floodplain which is rich in both archaeology and brownfield sites (including Sites 1–3), but the findings may be applicable with caution to a range of temperate lowland valley bottoms with alkaline geology.

Crabtree (1997) prefers bioremediation technology for most organic pollution, for the following reasons: it is often as effective as the other systems (of which few can be relied upon to achieve 100% removal of the contaminant); it is one of the most environmentally sustainable of the remediation technologies; it can be half the price. However the incentive to use it remains less because: developers (and others) still question its effectiveness; the cost saving may not benefit the developer directly (ie the benefit may be to the vendor of the land as the 'polluter', or to the ultimate occupier); it may take longer to complete; *ex situ* technologies may require additional space to implement.

So in the present legal situation the developer has less concern about removal of the contaminated soil to landfill. He may be accused of exporting the problem at someone else's cost, but he is seen as having done something physical to the

Table 33.2 Informal review of available technologies with respect to archaeology

Remediation technology	Remediated *ex-situ* (whether on-site or off-site); includes solidification, vitrification, incineration, soil washing; bio-reactor	Remediate *in situ* but with cultivation	Remediate by isolating using coffer dam or bund	Remediate *in situ* using point interventions, eg vacuum extraction, bioventing, pulse-pumping	Naturally attenuation (do-nothing option)
Pollutant to which the technology may be appropriate on an Upper Thames floodplain site	Heavy metals, mineral contaminants	NVOC, SVOC	All non-volatile contaminants, contaminants, but not a permanent solution	Free product petroleum hydrocarbons VOC	Petroleum hydrocarbons
Cost range per cu m (Martin and Bardos 1996)	£14–150	Possibly less than £30–£40	No quote, engineering costs and monitoring	£17–85	?monitoring costs only
Negative archaeological impact	Total destruction of all deposits to depth of extraction impact	Intensive ploughing will destroy all archaeological deposits to depth of cultivation. Added nutrients may affect artefacts	Physical severance of deposits by coffer dam	Possible degradation of topmost organic remains (ie treat surface of groundwater only, see note 3)	None, subject to monitoring
Positive archaeological impact	Potential for controlled removal of soil to expose archaeological deposits for investigation and recording	Results of mitigation archaeological work may be significant	Potential enhanced organic preservation arising from management of water level	Preservation *in situ*	Preservation *in situ*
Potential for archaeological mitigation	Full archaeological record	None?	Investigate and record line of coffer dam		
Resourcing for archaeological mitigation	Evaluate, excavate, record in advance under CDM regulations, to depth of proposed extraction impact	Evaluate, excavate, record in advance under CDM regulations, to depth of proposed cultivation	Evaluate, excavate and record line of bund; monitor	Evaluate and monitor	Evaluate and monitor only

site by replacing contaminated with uncontaminated material, minimising the risk of residual contamination, and has done it quickly so as not to tie up precious investment capital. To use bioremediation would take relatively longer and therefore require different financing.

Archaeologically friendly technologies

Crabtree clearly prefers this technology for sound environmental reasons (1997, 67), but as Table 33.2 shows, it is only *in situ* bioremediation which is truly green in terms of being archaeology friendly as well as environmentally sustainable. This *in situ* technology has an additional handicap in the development world however, in that its effectiveness can only be judged by post-treatment sampling, which requires another leap of faith on the part of the end user. The net effect is that any but the most sophisticated developer will see it in his own interest to go for the quick fix type of remediation. Crabtree finds therefore that there is a need for financial incentives to be green, ie subsidy (1997, 67). Until it can be confidently shown that there is no significant risk in biological techniques, and the procedures therefore begin to be commonplace, archaeologically friendly technologies may be rarity.

The ultimate archaeologically friendly technology: natural attenuation

Given the uncertainties about the technical remedies, there remains the final option of 'do nothing' or 'natural attenuation', as raised by J Juneman (pers comm). It is clear that the assessment process needs to model the route followed by pollutants, from source through pathway to a target receptor, the latter being (say) a watercourse or an extraction borehole. Present integrated methodology uses four-tier predictive modelling as set out in Environment Agency documents P12 and P13; the Agency uses a scale of sensitivity related to the importance of the water resource for abstraction purposes, and operates to a code of practice DD 175 1988 as revised (T Johns, pers comm). With hydrocarbon pollution no doubt this will demonstrate the pattern of natural attenuation in the aquifer, and for the 'low sensitivity' Thames floodplain this may mean some indication of the distance down gradient of a hot spot at which these contaminants became undetectable as a result of microbial action.

Spin-off from archaeology to the hydrogeologist: 'reservoir' effect

There may be feedback from archaeology to the hydrogeologist. It is clear from LTG's report on Site 3 that there was less of a plume effect than had been anticipated, and there may be an archaeological explanation for this. The hydrocarbons being less dense than water tend to accumulate at the surface of the groundwater, where they are likely to be carried with the gradient. However on a former habitation site like Rewley Abbey there is a network of enclosing ditches which in their period of use will have been dug to a level well below the modern water table. These now exist as large silt-filled channels, blocking the water flow in the surface levels of the groundwater although presumably having no effect on the deeper flow. So while soluble pollutants (and perhaps some emulsified insolubles) will form a plume down gradient, the bulk of the hydrocarbons from a hot spot could find themselves penned up in areas edged by these old watercourses as if constrained by impermeable clay bunds.

This reservoir effect need not be confined to settlements, it will arise from the reticulated regime of natural channels which is characteristic of floodplains like the Thames at Oxford. These again will tend to control the shape of hydrocarbon plumes, making them accessible for pumping out. Study of plumes to track them to their historical source, estimate how long they have been there, and what threat they represent, is important to the risk assessment. If for instance this is old pollution on a low sensitivity aquifer and say 1000m up gradient of any receptor, it may prove that remediation of a hot spot will have an insignificant effect on water quality; in which case we should be pumping out whatever free product exists, and leaving the residue for soil processes to attenuate.

Viewed in this perspective, it may be that the onerous task of making a value judgement between groundwater quality and protection of cultural heritage resources can be deferred until such time as the impact on aquifer quality is confirmed and quantified. Such a rational approach to natural attenuation could lead to major savings in the development sector, in parallel with reduced impact on the cultural heritage resource, and will make the present bulk extraction something of the past.

How can environmental impact assessment help with brownfield heritage?

A brief history of environmental impact assessment in the USA and Europe

The history of environmental management in post-war UK includes major environmental management schemes such as Green Belt legislation and the Clean Air Act; post-war England already had some of the safest water supply in Europe. However the first comprehensive approach to management was the USA's *National Environmental Protection Act* (NEPA) of 1969. The Act is seen as a 'fluke', worded in general terms and relying on the courts for its developing strength, but also gaining strength from amendments which were really intended to weaken it (Glasson et al 1999, 29).

NEPA heralded a more systematic approach to environmental management worldwide, and during the years

1970–1987 in the UK some 250 environmental impact assessment type reports were prepared. In 1988 UK's Department of the Environment issued regulations and an accompanying circular (DoE 1988). These provided a mechanism for integrating the 1985 EEC directive (85/337/EEC) into the existing UK planning code, which some (eg Weston 1999, 19) believe to have left the working of the native code largely untouched.

The effect of the UK 1988 regulations was to formalise the presentation of information needed for safe and authoritative decision-making on development proposals, and to identify systematically the people (and other receptors) who are likely to be affected. Since their implementation there has been, in one decade, a ten-fold increase in environmental impact assessments over the total for the previous fifteen years. Interestingly this reverses the trend in US, where environmental impact assessments peaked in the first ten years of NEPA; in that case the decline is attributed to a diffusion of the spirit of environmental impact assessments so that formal assessments were becoming less important, and also a recognition of the importance of statements of 'no significant impact' (FONSIs, see Glasson et al 1999).

Screening proposals for their significance

The process of environmental impact assessment starts with screening all development proposals which arise in a rapidly evolving economy, to identify at an early stage those which are likely to have significant impact. Under the 1988 UK regulations, environmental impact assessment immediately became mandatory for certain types of project, which are specified under Schedule 1; less significant projects under Schedule 2 ranged from those which are more likely to need environmental impact assessment to those which are less likely to, with a grey area in between where a decision is based on indicative thresholds, including size, materials throughput or sensitivity of the receiving landscape.

Of the UK statements produced from 1988–1998 a useful tabulation exists for those available to Oxford Brookes School of Planning; it illustrates the range of development types to which the regulations have been applied, and provides some statistical data on the effectiveness of the system (Wood and Bellenger 1999). For Schedule 2 with its grey area there is a perception of foot-dragging on the part of developers. This could be construed as a protest at having to do it in the European way, but alternatively developers may have been seeking to establish ground rules in the grey part of the spectrum; this uncertainty may be partly addressed by the 1999 Regulations and circular (SI 1999; DETR 1999), which put emphasis on significance in respect of size and nature of the project in relation to its location (see also below).

Agreeing the scope of an assessment

Once it has been demonstrated by screening that a project requires environmental impact assessment (and perhaps partly in parallel with it) there is need to establish the scope of the investigation, ie identify those impacts that will need data collection and assessment because they are potentially significant. Scoping needs considerable perception given that it is done before there has been any new data collection, and this is the point where a developer will ask specialist consultants to identify any known 'scheme stopper', ie an impact which is not reasonably open to mitigation. Projected size, visibility, traffic growth, waste generation, effects on groundwater and air quality, destruction of habitat, damage to cultural heritage, and risk are all material considerations in this equation. Also to be considered are positive impacts such as new employment, regeneration of derelict land, and (as demonstrated in the abortive search for a third London Airport) socio-economic factors; and there needs to be a consideration of alternative means of achieving the objective, not just alternative sites but alternative processes, one of the areas reinforced by the 1999 legislation.

Some form of tabulation is usually employed in the scoping process, either presence/absence or weighted. The equation must take account of Government directives and the terms of Structure and Local plans, because in most cases the inquiries behind these documents amount to strategic environmental assessments, which may pre-empt some issues for the incoming developer. Therivel (1995) has discussed the appraisal of local authority Structure and Local Plans. And it should be noted that the most recent (1999) UK regulations allow developers to request the local planning authority to provide a 'scoping opinion' in a form which can be registered and publicised (see below).

The process of assessment

Any process of assessment of widely differing factors needs to be treated systematically in order that nothing is overlooked, and so that the relative and absolute significance of each is properly evaluated against established standards and thresholds, or against policy and guidance. Planning consultants use matrices for this, and mechanistic scoring systems have been developed to provide some valuation within the topic, or against the same topic on other developments (on a scale of 1 to 7, from major negative impact to major positive impact). For lesser impact they may develop thresholds ('scale it matters'), and use a quantitative valuation for assessing cumulative effect of multiple impacts (DETR 1998). However it is increasingly clear that no individual professional can reliably balance the full range of impacts, and at least the final stage needs to be done by a panel, known as a Delphic assessment. Prediction of effects is based on 'reasonable foreseeability' or 'worst case', and uses statistical models and geographical information systems.

Many attempts have been made at a definition of the environmental impact assessment process outlined above, one of the simplest and most general being Wathern (1990), 'EIA is a process ... providing decision makers with an indication of the likely consequences of their actions'.

These likely consequences are eventually embodied in an Environmental Statement (ES), whose value and effectiveness will be judged on the basis of how it proposes to mitigate impacts. The final stage of the assessment process needs therefore to consider the range of options for mitigation. It is at this point, as we have seen with respect to brownfield heritage, that conflicts can arise if proposed mitigation of one impact undermines or even nullifies the mitigation of another. Thus the overall mitigation needs to be worked up sufficiently to be able to identify conflicts, so that in the worst case the decision-making body is presented with clear issues.

There has been a move to provide a quality control yardstick for environmental impact assessments by academic departments specialising in the field, and Oxford Brookes University has accumulated a library of environmental impact assessments. Quality control of Environmental Statements is no more straightforward than any other part of the process however, for a rational reason discussed by Beattie (1995). He notes a widely acknowledged uncertainty about the objectivity of these documents; while they contain scientific data, the data is not used scientifically, because they are 'advocacy documents'. Beattie's perceptive paper (1995) argues that they are valuable none the less.

Discussion: can the environmental impact assessment process balance the scales?

In this adversarial situation, how is ground/water quality to be balanced against other parameters of a development? Are we not likely to find that the soft environment and the regenerative environment lose out every time? Issues of this sort are conventionally balanced by two concepts, the 'best practicable environmental option' (BPEO) and a fall back position, the 'best available techniques not entailing excessive cost' (BATNEEC). Glasson et al (1999, 275–7) give examples of their application in the area of conflict arising from waste disposal proposals.

This paper has reviewed informally a current example of conflict in mitigation strategies for very different parts of the spectrum of significant impacts. We have also reviewed the established and evolving mechanism of environmental impact assessment which may be able to broker a compromise in such a situation. We have seen that this could be achieved in a relatively direct way if the project came with pre-packaged data on each impact; but we have also sympathised with the developer in his reluctance to spend good money on collecting this data themselves until other factors are in place, ie until they have some good news about the acceptability of their proposals generally.

No doubt a resolution is achievable, and it is likely to be in the hands of those carrying out the provisional scoping, and the updating of the scope as the investigation proceeds. If these personnel, whether private consultants or planning control officers, are aware of an area of conflict called provisionally 'brownfield heritage', then a marker can be put in either to prioritise data collection, or to ensure that ballpark resources allow strategically for the increased risk of expense in this area. Environmental impact assessment is an evolving discipline with the potential to accommodate these issues, and the main challenge will be to ensure that it is flagged before the arrival of a crunch situation.

Effects of the 1999 regulations

How will the new regulations affect this? As commented in the professional press (Weston 1999) at the appearance of the circular DETR 02/99 (1999), there will have to be more transparency and accountability on both sides of the adversarial divide, ie local planning authorities will need to register and publicise their screening and scoping opinions, and publicise updates to the proposals; developers will need to ensure that when they have considered and rejected an alternative location, process, method of construction etc, that this is recorded in their Environmental Statement for public dissemination, even if it may lead to planning blight by reflection. The new regulations clarify Schedule 1 with a list of development types. They also bring important clarity to Schedule 2: developments which qualify are clearly predicated on 'significance of impact, under size, nature and location'. Paragraph 44 lays emphasis on 'location'. Annex A lists indicative thresholds and criteria for Schedule 2 projects, 5ha being a threshold for area; Annex B gives selection criteria.

Conclusion: how can environmental impact assessment contribute to preservation of archaeological remains *in situ*?

On a straw poll of Oxford floodplain sites, including a Scheduled Ancient Monument, it has been demonstrated that preservation of archaeological remains in situ can become a major challenge, once the juggernaut of ground and water quality remediation has started rolling.

The reasonable solution is to make allowance for possible conflict at the earliest stage, before budgets and programmes are set. This is the discipline of environmental assessment. As an intellectual exercise it can be followed independently by a formal environmental impact assessment. It can thereby ensure that all sides are in a position to make whatever tactical decisions are needed at the stage when the relative significance of threatened resources becomes clear.

The ultimate decision-making process is thereby based on a constructive and sustainable strategy, identifying a 'best practicable environmental option', potentially at a stage when there is still time to implement relatively 'green' and sustainable (if time-consuming) mitigation technologies.

A second and more subtle but longer term benefit is that environmental impact assessment offers a professional structure in which quieter voices can be heard, which has to benefit the historic environment. If a hydrocarbon plume is only say 30m in length, and if the nearest sensitive receptor is 300m down gradient, then natural biological attenuation is probably doing its job, and the 'do nothing' option is self-evident, the ultimate in environmental friendliness.

Acknowledgements

The present paper arises from a certificate course in Environmental Impact Assessment at School of Planning, Oxford Brookes University in 1999; the remediation prices quoted in Table 33.2 are those available at that date. Thanks are owed to the following for comment and advice: J Barber, M Corfield, G Lambrick, A Miller, R Thomas and G Wait. The writer was employed by Oxford Archaeological Unit throughout, mainly on tasks related to planning control for Oxford City Council. Responsibility for opinions expressed resides solely with the writer.

Bibliography

Beattie, R B, 1995, Everything you already know about EIA (but don't often admit), *Environmental Impact Assessment Rev*, 15, no. 2, 109–114

Cook, S, 1999 Archaeological investigations at 64–66 St Thomas St, Oxford, *Oxoniensia* 64, 285–97

Crabtree, R C, 1997 Contaminated land policy: a sustainability appraisal, unpub Oxford Brookes University school of planning pap 173

DETR, 1998 Department of Environment, Transport and the Regions, *A new deal for trunk roads in England* (2 vols), London

DETR, 1999 Department of Environment Transport and the Regions, *Environmental impact assessment*, Circular 02/99, London

DoE, 1988 Department of Environment, *Town and country planning (assessment of environmental effects) regulations* 1988, Circular 15/88, London

Glasson, J, Therivel, R, and Chadwick, A, 1999 *Introduction to environmental impact assessment*, London

Martin, I, and Bardos, P, 1996 *A review of full-scale treatment technologies for the remediation of contaminated soil: final report*, Roy Comm Environmental Pollution Rep, Surrey

Rubin, C T, 1994 *The green crusade: rethinking the roots of environmentalism*, New York

SI, 1999 Statutory Instrument 293, *Town and country planning act, environmental impact assessment (England and Wales) regulations* 1999, London

Therivel, R (ed), 1995 *Environmental appraisal of development plans 2; 1992–1995*, Oxford

Tunstall, S M, and Coker, A, 1992 Survey-based valuation methods, in *Valuing the environment: economic approaches to environmental evaluation* (eds A Coker and C Richards), 104–129, London

Waterman Environmental 1998 The Said Business School, Oxford: environmental statement unpub Waterman Environmental rep

Wathern, P (ed), 1990 *Environmental impact assessment: theory and practice*, London

Weston, J, 1999 Environmental rules in council impact, *Planning* 1317

Winpenny, J T, 1991 *Values for the environment, a guide to economic appraisal* London

Wood, G, and Bellenger, C, 1999 *Directory of environmental impact statements*, Oxford

34

A damn'd bad place, Sheffield: challenges and opportunities of the 'urban renaissance'

ANNA BADCOCK, ANDREW LINES and JAMES SYMONDS

Anna Badcock
Assistant Director

Andrew Lines
Project Archaeologist

James Symonds
Executive Director

Archaeological Research and Consultancy at the University of Sheffield (ARCUS)
Graduate School of Archaeology
West Court, 2 Mappin St
Sheffield S1 4DT, UK

Abstract

This paper explores the challenges and opportunities facing archaeological research on brownfield sites in the north of England. There is increasing governmental pressure to develop brownfield sites, yet the historical development of such sites is often poorly understood. Crucially, it must be recognised that brownfield sites are not just containers of useful evidence. They are themselves the consequences of the histories that we are interested in and need to be recognised as such.

We argue that the archaeological investigation of such sites, although technically daunting, can offer valuable insights into everyday life and working practices in the early industrial age. This paper draws upon recent Archaeological Research and Consultancy at the University of Sheffield (ARCUS) project work on 18th- and 19th-century steel-making sites in Sheffield.

Introduction

The 'urban renaissance' has a very clear expression in Sheffield. Investment in inner city housing and commercial building stock has increased dramatically in the last few years. Like many other industrial towns or cities, some of the major sources of land, which often includes prime city centre locations, are brownfield sites.

The government aims to provide 4 million additional homes over the next 25 years, with a target figure of 60% of these new homes being built on brownfield land. Government figures suggest that by 2030, 45,000ha of previously developed land will have been reused for new development (DETR 2000).

This kind of land has often been used for industrial purposes in the past, and in many cases may be partially or completely contaminated and require special treatment. Other problems mean that brownfield sites may be far less economically viable for the investor than their greenfield counterparts; financial assistance may be required to make such sites attractive to developers. These factors can have implications for archaeology that goes beyond the scope of the current planning environment of Planning and Policy Guidance PPG15 and PPG16.

Archaeological work on brownfield sites can be difficult, but very exciting. These sites require a flexibility of approach that is not always provided by the standard, desktop, evaluation and excavation procedure that has become second nature, and we have had to change our ways of thinking and working in order to do the justice to the archaeology.

Crucially, the investigation of these sites should be seen as a unique opportunity to further our understanding of the changing character of everyday life in the industrial age. If we are to understand what it was like to work and live in industrial cities, we need to interact with the smallest of material remains, as well as the legacy of buildings and town plans that is more often identified as our 'industrial heritage'.

Flexible approaches

In 1996 we started work at Riverside Exchange, a 4ha brownfield site in the heart of Sheffield. The site is now being developed to accommodate office blocks, leisure facilities, a hotel and flats, most of it targeted at the middle class market.

No trace of historic land use survived above ground before work started, but it was clear that the deposits would be very mixed in terms of their character and survival. While desk-based studies provided useful information, the usual technique of evaluation needed to be modified, as small trial trenches in these kinds of deposits were found to be less than helpful. Bold action and flexibility was needed, and we were assisted greatly by a cooperative developer and curatorial archaeologist. In this case, mechanical excavators were used to examine the deposits over large areas, to enable us to make a rapid assessment of areas suitable for further investigation.

Given that development is actively encouraged on brownfield sites, archaeological investigations may end up being undertaken as a condition of planning permission, and may, conceptually at least, play a slightly different role in the planning procedure. This does *not* mean that preservation by record becomes the only viable option, as both these case studies will demonstrate. During our further excavations on the Riverside site, deposits were characterised on a continual basis. Appropriate specialist input at all stages, and on-site interpretation enabled development to continue without interruption, hand-in-hand with the archaeological work.

At the Jessops site, in the original heartland of steel-production, archaeological intervention took place after part of the site had been remediated. The remediation involved the excavation of up to 2m depth of ground deposits being fed through a crusher and recompacted back into the ground to form a uniform base. The central part of the site was examined archaeologically, and important remains were encountered and recorded. The *in situ* preservation of some of the remains at both Jessops and Riverside Exchange was achieved.

In these kinds of conditions our techniques may require adjustment. The scale and speed of landscape change during the industrial and modern period is remarkable in these settings. This is something that we often underestimate and is itself an important dimension to the experience that people living and working in Sheffield would have had. Where large areas of land were built up or razed in a matter of hours or days, it is not particularly helpful to insert a narrow slot trench in the hope of understanding the site. Likewise, well-formulated sampling strategies and collection policies can become useless in the face of tons of waste products from large-scale operations that may have been shifted around a site countless times.

More often than not, the remains that are present comprise structural foundations and redeposited material, with survival of occupation levels being much rarer. Broad and rapid characterisation of deposits and areas becomes an important part of understanding the potential of the site, while finer details are investigated through targeted excavation.

Riverside Exchange

Excavations at the Riverside Exchange, in the centre of Sheffield, have shed light on the development of the steel industry. The known history of this part of Sheffield goes back to the 12th century, when a corn mill was constructed on the River Don (Belford 1996). The presence of the river was a key element of the development of the site through time, as corn milling gave way to an increasing amount of forging and turning of edge tools, all using water power.

The site boomed in the 1760s, with the founding of John Marshall's Millsands steelworks. The Marshalls site was remarkable for being what may have been the first integrated steel works in the world; housing two vital processes under one roof. Here, wrought iron was converted into steel in cementation furnaces, a process that had been developing since the later part of the 17th century. Secondly, the converted (or blister) steel was re-melted and refined in crucibles, a technique perfected by Huntsman around 1740, and one which revolutionised steel-making. For over a hundred years, the crucible process was the only means by which an ingot of steel could be cast (Hey 1998).

Archaeologically, these two processes can be easily detected, but with one or two exceptions, such sites have never been excavated before now. Currently, we believe there are only four upstanding remains of cementation furnaces in Britain, and three of these are in Sheffield. At Riverside Exchange, the buried remains of a further three furnaces were uncovered, one of which appears to be a very early prototype which may date to the later part of the 18th century (Fig 34.1). As far as we know this single chest furnace is unique. It seems that this design was not particularly successful, as no evidence for its continued development and use has been found, although reference to this design was made in 1774 by Jars in his *Voyages métallurgiques* (translated in Barraclough 1984, appendices 3, 4, 22, 26 and in text).

From the late 18th century then, the Riverside Exchange site became a focal point for iron and steelworking in the city. Several cutlery workshops were established to make use of the specially converted cementation (or 'blister') steel. In the early 19th century William Vickers built his first water-powered rolling mill on the site, heralding the start of something which would become much larger; Naylor Vickers became pioneers of large steel castings, gaining worldwide recognition for their production of armaments.

But what of the people that worked here, and the smaller

Fig 34.1 *Early cementation furnaces at Riverside Exchange*

industries that grew up alongside the steel makers? The post-excavation work for Riverside Exchange is still being undertaken, but even in these early stages we can begin to ask questions about the nature of this work. Wheel pits and water channels provided the infrastructure of power generation for the various workshops which grew up on the site (Fig 34.2). It is from these features that many of the most interesting items have been recovered, including knives, forks, files, tools worked bone for handles, pins and wire (Symonds and Lines 2001). Many of these items were unfinished, giving us vital information about the various stages of production and insights into how these were organised. The museums may be full of the best examples of the products of industry in Sheffield, but until now we have had few chances to investigate the processes behind these products in archaeological settings. It is in these contexts that we can get a sense of the conditions that attended living and working; the mistakes, discards, experiments and local solutions to more general problems. These insights are of great importance when trying to understand the development of craft skills and industry in the region.

Fig 34.2 *Cutlers wheel goit at Riverside Exchange*

Likewise, the huge numbers of ceramics and clay pipes recovered will contribute to our understanding of working class life in the later 18th and 19th centuries. There was a surprising amount of domestic pottery and alcohol bottles from Riverside Exchange, most of it cheap, mass-produced and widely used. While some of this material is likely to have been brought in as landfill, much of it relates to activity on the site itself. Oyster shells were found in some quantity; some were the result of consumption by workers whereas others, judging by the size and quality of the shells, the result of dumping from restaurants and hotels in the district (Symonds and Lines 2001). These and other deposits have encouraged us to reconsider our modern ideas of industrialisation, and to examine how domestic, craft and industrial arenas may have been much more closely linked (Lines 2000). While owners and entrepreneurs lived up and away in what are still the leafy suburbs, much of their workforce lived according to a very different pattern. Our notions of distinct spheres of home and workplace may not be doing justice to patterns of dwelling for many at the time.

Many of the clay pipes found had what we might call 'classy' designs such as flutes, stags heads, Masonic motifs, intertwined grape vines and elaborate floral motifs. Some of the broken stems had multi-faceted edges, worn away by abrasion with a harder surface; perhaps this is our first evidence for industrial doodling? Or possibly, as elsewhere, the broken stems were being used like chalk, to mark the goods themselves or to keep track of productivity (Lines 2000).

Even though such objects may be mass-produced, certain styles were obviously considered to be highly fashionable within specific social circles. Some of the designs are common to those found on many other sites around the country, but others were more popular locally. The display and use of these objects in the workplace was deliberate and considered and gives us a material insight into the subtle ways in which people recognised themselves as members of particular communities.

Jessops Brightside

Investigations at the Jessops Brightside site were similar in many ways to those at Riverside, and although post-excavation again is ongoing, the results will complement the work undertaken at Riverside.

The Jessops site also developed from medieval beginnings, being based around the site of a corn mill founded on the River Don in at least the 14th century (Badcock 2000). Again, water power was the key to the early development of this site; the corn mill became a cutlers wheel and workshop, which evolved through the 18th century into a water-powered forge. Steam power was introduced in the mid 19th century, and under William Jessop, the site boomed, becoming one of the major crucible steel producers for the American market (Hey 1998, 151).

One of the most surprising aspects of this site was the

degree of archaeological preservation. Over 200 timbers, some of them enormous structural elements from the tilt hammer forge, had survived in waterlogged conditions. In other areas, the brick and stone-built features such as culverts, furnace bases, hearths, crane footings and fire-pits had survived relatively undamaged underneath the cement floorings of the later factories that had simply been built over the top (Fig 34.3; Badcock 2001).

Fig 34.3 Furnace floors, fire pits and a possible casting pit at Jessops

The amount of material culture recovered from the Jessops site was a tiny proportion of that from Riverside, and perhaps more like the assemblages we would expect of this kind of industrial archaeology. But there are still things to be learnt of the working practices, and the way in which technological developments can maintain or change those practices. One of the most poignant finds from the Jessops site was the rusted remains of a clocking-in machine – a symbol of regulated working hours, clock time and the increased surveillance and control of the workforce.

Here too we found the remains of four cementation furnaces (Fig 34.4), and ash-pits, furnace bricks, clay linings, fire grates and metal doors were all intact. We are using scanning electron microscopes to assess whether the samples we have recovered are the raw iron or the converted 'blister' steel – quite salutary then to realise that a skilled worker could tell by eye whether or not the metal had been properly converted and the furnace could be allowed to cool.

Fig 34.4 Cementation furnace foundations at Jessops

Hidden histories

So why do we do this kind of archaeology? When archaeologists are 'intervening' on a site, and may be perceived as nothing but a nuisance, it can be hard to convey the fact that even very recent archaeology forms a vital part of our cultural heritage. But the pressure of working within the construction industry, and in these kinds of situations may be good for us as a discipline, and as individuals, in forcing us to be self-critical and focused.

We think we know about steel and cutlery in Sheffield because the industries are well documented, but all too often the smaller voices are lost from these accounts, in favour of the 'big men' and the larger picture. In February of this year, as Corus announced job cuts in the UK, a report on the Today Programme on Radio 4 commented that 'modern steelworkers are skilled ... not the leather-aproned men who scampered in wooden shoes around molten metal' (BBC Radio 4, 14 February 2001).

This can be seen as evidence of a continued misunderstanding and undervaluing of technology and skilled labour in the industrial past. In 1892, Vickers at Riverside produced the largest steel casting ever to have been made England. This single casting required the contents of 176 crucibles, *each* weighing 90 pounds, to be poured by hand into the mould in 11 minutes flat (Pawson and Brailsford 1862). This was a masterly combination of strength, teamwork, knowledge and coordination. We have to challenge the preconceived ideas of 'unsophisticated' or 'quaint' small working practices of the pre-industrial age. Similarly, we have to show that industrialisation was not a relentless and seamless progression of improvement and change, but a complex integration of established skills and knowledge with new forms of power and production.

The histories are relevant on a number of different scales, from the individual, local and regional to the national and international arenas which these industries were a part of. The material remains help us understand how people worked in these industries, the material objects they used, and sheds light upon the changing concepts of the home, the workplace and the nature of work itself. The histories of class, gender, and ethnicity are present in this archaeology, but have too frequently been omitted from the textbook stories.

Yes, you could say that sie matters! The steel industry was big business, and the men at the forefront of its 19th century expansion knew it – the classical architecture of the steelworks, and their names such as Atlas, Cyclops, Aetna and Hecla presented a fascinating combination of power, strength and myth. But now we are beginning to engage with the men, women and children that made it all possible, and think about the routine and the working practices of the every day.

Towards sustainability and social inclusion

So if this is social inclusion of the past, what about the present? Our work is meaningless if we can not adequately bring it into the public arena, and allow people to use the material to form their own ideas about the past. People quite rightly get angry and frustrated if they feel excluded from their past or the material remains of their heritage. Graffiti on a partially demolished cutlery and edge tool workshop in central Sheffield is heartfelt (Fig 34.5). The loss of such buildings or archaeological remains is certainly seen by some as an erosion of local identity and that loss is compounded when the results of our work remain trapped in the grey literature.

Developers, regional development agencies and the like have a duty to address the issues that archaeology raises on brownfield sites. Sustainability is a term we hear more and more with reference to new development, be it for housing or commercial investment. Sustainability surely requires development to have as active an engagement with the past as it does with the present and future. This not about making buildings look like their older predecessors, but about understanding the histories of particular areas, and what has given them their character. This means following the histories of brownfield sites right through to the present. At Riverside Exchange we took time to record post-industrial activities on what had become waste ground; the temporary shelters and residual traces of those who currently live on the margins. It also means a commitment to the idea that the histories we are beginning to explore do not remain the sole preserve of specialists or professionals. We have a responsibility to communicate, and where possible, to involve people in the exploration of their own relationship to the local, regional, national and international histories that surround them.

In conclusion, the challenge of brownfield archaeology is a complex but exciting one, brought to us recently by the urban renaissance. The challenge is both physical and mental, and requires us to be flexible and innovative if we are to be able to work on these sites effectively. We must rise to this challenge, however, to do right by the archaeology, the public, and the sustainability of future communities.

Fig 34.5 *Graffiti on demolition hoardings, Sheffield*

Bibliography

Badcock, A, 2000 Archaeological appraisal of land at Jessops Riverside, Brightside, Sheffield, unpub ARCUS rep 595

Badcock, A, 2001 Jessops Riverside archaeological field evaluation fieldwork summary report, unpub ARCUS rep 595b

Barraclough, K C, 1984 *Steelmaking before Bessemer* (2 vols), London

Belford, P, 1996 Desk-based assessment of Riverside Exchange, Sheffield, unpub ARCUS rep 240

DETR, 2000 Department of the Environment, Transport and the Regions, *Towards an urban renaissance*, http://www.regeneration.detr.gov.uk/utf/renais/.htm

Hey, D, 1998 *A history of Sheffield*, Lancaster

Lines, A, 2000 The making of Riverside, unpub ARCUS paper

Pawson, H, and Brailsford, J, 1862 *The illustrated guide to Sheffield*, Sheffield

Symonds, J, and Lines, A (eds), 2001 Riverside Exchange: interim evaluation report, unpub ARCUS rep 240d.1

35

Putting remains back into history: creating a sustainable historic environment

JOHN CARMAN

John Carman
Affiliated Lecturer
Department of Archaeology
University of Cambridge
Downing Street
Cambridge CB2 3DZ, UK

Abstract

Archaeology is a historical discipline or it is nothing. The purpose of preserving remains from the past is therefore to provide a basis upon which that past can be understood. Current policy imperatives require that we create a historic environment, but the concept of sustainability has been introduced from outside the discipline, from development economics and biology. To make the link between scientific practice and policy, this paper will present an understanding of sustainability that goes beyond the concern with the mere physical survival of ancient remains to treating them as part of a fully sustainable historic environment.

Archaeological remains as record, resource and heritage

A concern with the relationship between the management of the archaeological resource and archaeology as a research discipline also involves a concern with the three different phenomena that these fields represent. These are: the archaeological *record* which is the focus of research; the archaeological *resource* which is the object of preservation; and the archaeological *heritage* which is what we make available to the wider public. In doing so, it is always important to grasp that the material constituting the record, the resource and the heritage is identical – it is the full range of material with which all archaeologists deal. The differences between the concepts can be seen as a series of transformations along various 'dimensions' which are applicable to all of them (Fig 35.1).

These three phenomena are of concern to different groups of people. As 'record' or 'resource', archaeological material is primarily of concern to field archaeologists researching into the past. As 'heritage', however, it is primarily considered to be of concern to the public (McGimsey 1972, 5; Fowler 1984; Merriman 1991). The diagnostic attribute of the record is that of variability: for Binford it takes the form of variability in assemblages of artefacts, in form, in function and in style (Binford 1983); for Hodder it takes the form of temporal, spatial, depositional and typological variability (Hodder 1986, 125). The diagnostic attributes of the resource, however, are that it is finite and consequently non-renewable (Darvill 1987, 1; McGimsey 1972, 24). As heritage, the key attribute it must possess is representativeness of a period of the past or a class of material (Cleere 1984, 127). The use to which the record will be put is always that of research, whereas the resource and the heritage will be preserved, the one for future research, the other for public access. In fulfilling its research role, the record and the resource both constitute evidence to be used in understanding the past, the one immediately and the other in the future (Barrett 1987, 6). By contrast, the heritage constitutes a resource for creating and sustaining a

Criterion	The archaeological...		
	Record	Resource	Heritage
Of concern to	archaeologists	archaeologists	the public
Key attribute	variability	finite/non-renewable	representativeness
Used for	research	preservation for research	preservation for access
Creation of	archaeology theory/method	regulation	regulation
Considered as	evidence	resource	source
Identified by	*survey*	*categorisation*	*categorisation*
Selected for	relevance	significance	importance
Concerns (time)	past	present	present

Direction of travel of material from 'record' to 'heritage' ⟶

But this is also *realm of transformation* from one type of archaeological object to another, because all these categories are made of the same material ⟷

Fig 35.1 *Archaeological objects: measures of difference between the archaeological record, the archaeological resource and the archaeological heritage*

sense of cultural identity in the present (Lipe 1984, 2; Fowler 1992; Darvill 1987, 4). In seeking the record, various techniques of survey are applied (Cleere 1984, 126) while the resource and heritage are subject to characterisation studies and processes of categorisation (Darvill 1993; Carman 1996). Once identified, components of the record are selected for use on the basis of their relevance to specific research projects (Binford 1977, 2), whereas components of the resource and heritage are selected on the basis of their ascribed significance or importance (Cleere 1984, 127; Schaafsma 1989; Wester 1990). Overall, the record relates in general to the past. By contrast, the resource and the heritage are *always* a phenomenon of the present.

As indicated here, these three phenomena represent merely different 'versions' of the same body of material. Accordingly, they are capable of transformation from one to another, back and forth. Indeed, the creation of the heritage as a category is little more than a process of the designation and changes in accepted treatment of particular kinds of material. This is frequently represented in archaeology by the sequence record – resource – heritage, which also represents the normal 'direction of travel' of a body of archaeological material from discovery onwards (cf Schiffer 1972; Carman 1990). On first discovery, archaeological remains are part of the archaeological record and are considered and treated as such. Once discovered, however, they may also become part of the archaeological resource, to be kept for the future benefit of archaeology as a discipline. At a subsequent stage this material may be considered as appropriate for a more public audience, and will become part of the heritage that is of concern to the community as a whole. This trajectory from record through resource to heritage is normal because rarely – if indeed ever – will something be classed as 'heritage' on immediate discovery and subsequently transformed to 'mere' record although the theoretical possibility remains.

The act of preserving archaeological remains *in situ* transforms them from components of the *record* to components of the *resource*. Their subsequent inclusion in a 'sustainable historic environment' will make one more transformation: from *resource* to *heritage*.

Sustainability and archaeological remains

The near universal adoption of Agenda 21 (a comprehensive plan of action to be taken globally, nationally and locally by organisations of the United Nations, governments, and major groups in every area in which human impacts on the environment) and the commitment to creating sustainability in all things has led to the creation of the concept of the 'sustainable historic environment' in which archaeological remains should find their place. But there has been very little discussion of what 'sustainability' means for archaeology; the term is very often used as if it is inherently meaningful but in fact we need to *make* a meaning for it. The meanings applied in other fields do not fit the circumstances of archaeology at all well, since the concept of sustainability has its origins in two fields entirely unrelated to archaeology: biological ecology and development economics.

As a concept originally deriving from biology (cf Adams 1990), sustainability represents the attempt to create or maintain self-sustaining biological systems that will persist through time. Plants and animals can be introduced into a space and allowed to interact as they would if left entirely alone: the result will be a stable ecology where cycles of reproduction and predation will result in the continuing survival of species. Alternatively, existing ecologies may be managed in such a way that the cycle of reproduction and predation already established will be allowed to continue and thus also ensure the survival of species. As a concept adopted by development economists and development agencies, sustainability means the creation of types of new economic activity that – by allowing the creation of new wealth – will allow a human community and its established way of life to continue. 'Development' alone has come to mean the introduction of new opportunities for the creation of wealth regardless of cultural factors, that may lead either to rejection of the new wealth-creating opportunities or their adoption but with significant effects upon the community whose well-being they were meant to ensure. 'Sustainable development' means the introduction of new wealth-generating opportunities which are grounded in local ideology and existing practices, allowing the established way of life to continue.

What biology and economics have in common is a propensity for self-generation. Plants and animals reproduce themselves: as individuals die, they leave behind offspring; as each generation dies, it leaves descendants to carry on the cycle of reproduction and predation. Economic activity generates wealth and this wealth can be used to finance further economic activity of the same kind: economic activity therefore leads to more economic activity. It is this capacity for ongoing and continuing activity – both biological and economic – that gives the idea of sustainability its meaning in these fields. Sustainability here does not mean mere survival of discrete entities but the creation of systems of activity, and it is the *activity* which is sustained, the *system*, rather than the individual components. The problem here for us is of course that archaeological remains – whether *in situ* or not – lack this propensity for self-generation. Archaeological remains do not breed, they do not renew themselves, they do not create further archaeological remains. Instead they are held to be finite and non-renewable (Darvill 1987, 1; McGimsey 1972, 24), even though some commentators (Carman 1996; Holtorf 1998; Holtorf 2001) have argued that the archaeological resource is in fact – at least to some degree – 'renewable' and non-finite. For archaeology, therefore, the concept of sustainability appears to collapse into a concern with merely ensuring the physical survival of discrete objects.

But the concept of sustainability has spilled out beyond biology and economics into other realms. MacFarlane (2000, 152; after Rannikko 1999) lists the 'dimensions' of sustainable development as: ecological sustainability; economic sustainability; social sustainability; and cultural sustainability (Fig 35.2). The common theme here is that of working in harmony with – rather than against – the object of sustainability, whether it is a natural ecology, economic wealth creation or a human community. Where one or more objects are concerned – such as in creating an economic system that will conserve biological systems and existing lifeways – the task becomes one of combining several different imperatives so that they do not conflict (cf Sustainable Communities Task Force 1993).

ECOLOGICAL SUSTAINABILITY

Development in harmony with ecological processes
Biodiversity
Conservation and protection of natural resources

ECONOMIC SUSTAINABILITY	SOCIAL SUSTAINABILITY	CULTURAL SUSTAINABILITY
Economic efficiency Takes into account needs of future generations	Reinforces control over our own lives Equitable distribution of resources	In harmony with cultural concepts of the community

Fig 35.2 Dimensions of sustainable development (after Macfarlane 2000, 152)

At first glance, it would seem that a 'sustainable *historic* environment' would be most similar to the idea of *ecological* sustainability. This would be built around the idea of development in harmony with existing historical and cultural processes; ensuring a diversity of historical resources; and the conservation and protection of historical resources. In other words, something like the current planning control system in England and Wales! This is not however the understanding adopted by agencies such as English Heritage who, from their first thinking about how to make the concept of sustainability apply to archaeology, have opted for an idea more akin to that of social and cultural sustainability (cf English Heritage 1996). This places archaeological and historic structures in the context of the construction of a 'place' which has social and cultural importance to its inhabitants (English Heritage 2000). This style of thinking collapses the idea of a sustainable historic environment into that of the sustainable community, and effectively therefore to that of the 'adaptive re-use' of individual monuments and buildings.

In other words, in current schemes of thought, the concept of the sustainable historic environment – comprising archaeological remains and other components – is subject to reduction to one of two very limited options (Fig 35.3). Either it is conceived as the preservation of individual sites and monuments as something separate from their surroundings; or it is conceived as the adaptive re-use of sites and monuments to fit with the economic revival of a particular place and its community. Neither of these is to be condemned. There is nothing intrinsically wrong with preserving archaeological remains and monuments for their intrinsic historic value, and indeed this is a guiding principle of archaeological resource management (Lipe 1984; Darvill 1993). Similarly, there is nothing inherently wrong in placing archaeology at the service of other worthwhile social and political agendas (McDavid and Babson 1997). It is, however, important to be clear what we are talking about when we consider the notion of sustainability in relation to archaeology. Neither of these limited understandings derives from the idea of archaeology as a research field, and it is with this that this paper is concerned.

OPTION ONE	OPTION TWO
Preservation of discrete objects for their inherent historical value	Adaptive re-use of monuments and buildings as part of a programme of economic and social regeneration

Fig 35.3 Current options for archaeological remains under a regime of 'sustainability'

Putting remains back into history

The following are three examples of possible ways forward in terms of a sustainable historic environment drawing on or involving archaeological research. All of them are either ongoing or past projects which also contain a 'public archaeology' element, and it is this component that makes the connection between archaeology as a research field and the move towards a sustainable historic environment.

'Places' as ongoing history: St Albans, Hertfordshire

The purpose of the Bloody Meadows Project (Carman 1999a; Carman 1999b) is to investigate historic battlefields of all periods from a broadly 'phenomenological' perspective (cf Tilley 1994). The aim is not so much to attempt to recreate what the battlefield was like on the day of battle, but rather to explore the historicity of particular kinds of places through the experience of the physical remains of their pasts. This approach means walking through the space with a keen eye to the different periods of history, and different human uses of the space, represented by buildings, monuments, street plans, different kinds of land use, and different shapes of ground.

In 1455 armies had been raised by the rebel Duke of York and by the King and they came together at St Albans. The King gathered his forces in the centre of the town, where the wide street was suitable for the mustering of an army. The Duke of York sent emissaries to negotiate with the King and the King stood his army down while they talked. The soldiers took off their armour and went into bars and inns to eat and drink. After waiting for a response that did not come, York's army launched an attack that travelled up the narrow streets towards the centre of the town. Barricades along the old town ditch – a line still traceable as modern roads – were thrown down and the defenders retreated towards the town centre. York's soldiers followed. Some of the buildings to be seen today are those that were standing on the day of battle. Suddenly the attackers found themselves in the centre of the town, among the King's soldiers who were simply not ready to fight. Passing up these streets today, you still enter the town centre suddenly, going from quiet residential side-streets into the bustling market area, gaining some sense of what it may have been like for 15th-century soldiers. A confused half-hour after arriving, York's troops stopped their killing. By then, the King was a prisoner, his chief minister was dead, and other great nobles dead, wounded or prisoners. The rest of the royal army had fled. The event is marked today by two small blue plaques at different locations: one on the site where the King's chief minister was killed; the other on the site of the house (now gone) where the abortive negotiations took place. As you pass up these streets today, memorials to the dead of more recent wars remind you what war and battle is all about: in one stand the almshouses dedicated by the Duchess of Marlborough for former soldiers of the early 18th century; and St Albans is one of only two towns in Britain where the dead of the First World War are commemorated in the individual streets from which they came.

The result of this style of encounter is a kind of 'time travel' – not a one-way trip into a singular and particular past and back – but a real journey through various times, where different pasts and an immediate present are met in juxtaposition. Places have histories that are evident in the experiences of them, and it is in experiencing them as places that the histories become evident. The place has meaning because it has a history and that history is manifested in the material evidences of its past which testify to interesting and different pasts. These material things create the drama of the place which is the experience of its history in the present. It is this historicity that such a 'phenomenological' approach to historic battlefields can produce. In taking such an approach, and in being deliberately aware of both past and present in a particular place, the line that lies between the past and the present is walked, where neither dominates the other. Instead, they interact in interesting and challenging ways. It is not a search for an experience of being in the past, but rather an experience in the present which simultaneously reflects and derives from the contribution of history to a particular place. In the case of a historic battlefield, it is not an experience of ancient slaughter, but an experience of a particular place in the present as read through its history as manifested in material form. This history inevitably includes the event of the battle that was fought there, but not exclusively. It is that experience of being in the place that is captured.

Critical theory in the USA: Annapolis, Maryland

Taking inspiration from the post-Marxist critical theory of the Frankfurt School, the aim of the 'Archaeology in Annapolis' project was to reveal how 'many aspects of contemporary ... life [in a capitalist economy] that are taken for granted are neither natural nor inevitable' (Potter 1997, 36). 'Archaeology in Annapolis' was accordingly and expressly an attempt to create 'a critical archaeology [that could] inspire enlightenment [and] lead to emancipatory social action' (Potter 1997, 36). The project focused on two areas: the archaeological evidence for the roots of modern everyday life, and especially the current dominant ideology, in a capitalist economy; and how versions of the past that serve the narrow interests of dominant social groups can be passed off – indeed accepted by subordinate groups – as 'objective' and 'universal'.

Such an approach focused very much on the role of history in identity formation. In Annapolis, that identity had traditionally been forged by a number of elements: the presence of the US Naval Academy; the city's place as the State capital; as a port of call for pleasure boaters; as a tourist town; as an expressly 'historical' place; and as a 'small' town (bearing in mind that the USA generally sees itself as a nation made up not of its major cities but of lots of small communities) (Potter 1994, 46–56). Current economic and social dynamics of the Annapolis community could be seen to revolve around the relationships between the town, the State of which it is the capital and the (Federal) Naval Academy, together with issues of gentrification and the rise of public housing, and issues of tourism and development (Potter 1994, 57–68). In turn, these themes were reflected in aspects of Annapolis' 'standard' history: a vision of the gentility of its population and a consequently ambiguous urban identity; and the siting of the city by the water which encourages a measure of transience among its people (Potter 1994, 82–8). These in turn lead to a general acceptance of certain 'inherent' historical qualities to which Annapolis and its people could lay claim, an emphasis on an imagined colonial 'Golden Age', and the fragmentation of various fields of study so that historical elements which did not fit this vision could be conveniently excluded (Potter 1994, 104–115). In Annapolis, for Potter (1994, 117–21), the past was most generally used as a diversion from problems of the present, as a means for establishing precedents for current policies, and as a means of establishing and maintaining degrees of social precedence among people. Accordingly, by emphasising the imperative of preserving the remains of the past – to serve the tourist industry, to keep the 'character' of the city, and as a reminder to its people of who 'they' are – the means by which the identity of Annapolis could be created and maintained became that very preservation of a particular kind of past (Potter 1994, 123–31).

'Archaeology in Annapolis' aimed to overcome the particular dominant ideology which held Annapolis in its sway, by revealing the alternative stories buried in its archaeology and especially the mechanisms by which the dominant ideology establishes itself (cf Leone 1984). It also sought to reveal the different voices that the dominant ideology kept silent – 'women, children, foot soldiers and sailors, slaves, freed slaves, Native Americans ..., the insane, the gaoled, as well as anybody else who has ever used a dish, a chamber-pot, room, privy, or medicine-bottle' (Leone et al 1995, 110). In particular, an attempt to broaden the scope to reveal the voices of the hitherto silent black population of Annapolis led to a joint project involving the use of artefacts from excavations at sites occupied by African Americans as informed by oral testimony (Leone et al 1995). Overall, 'Archaeology in Annapolis' was a process of opening up dialogue about issues in the present, such as tourism (Potter and Chabot 1997, 46–8) and the place of the past in the present generally:

> Instead of seeing an archaeological site as some kind of 'cache' of truths about the past just waiting to be liberated from the ground, we prefer to see on-site archaeological interpretation as an important way of using sites as an environment in which a relatively wide range of truths about the past may be identified, developed, discussed and negotiated. Initiating dialogues can ... be a far less tidy undertaking than simply providing the fact about a particular piece of the past – but ... we are willing to give up having the final word in exchange for the benefits of being able to hear all kinds of interesting voices other than our own (Potter and Chabot 1997, 53).

'Archaeology from below': Sedgeford, Norfolk

The shift towards an even more democratic approach in archaeology is also evident in certain circles in the UK. At Sedgeford, Norfolk, Neil Faulkner advocates such a democratic approach as a reaction against what he sees as the increasingly 'bureaucratic-professional tendency' in British archaeology (Faulkner 2000, 22–5). He accuses 'rescue' archaeology, which derives from development control policies, of applying the same standards to all projects, regardless of their diversity. He also points out that a presumption in favour of preservation (a principle underlying much archaeological conservation work) means that the historical information represented by much of the archaeological resource remains largely unknown (Faulkner 2000, 26–7). At the same time, rescue excavation is carried out for the purpose of retrieving what information is available in a short time, not because the site has been chosen to answer certain research questions, but merely because it is in danger of destruction, and the result is often the application of standardised recording systems. However, he argues, in good archaeology the material encountered, the methods employed, and the interpretations produced all interact and indeed are dependent upon one another. Faulkner's argument is therefore that while 'rescue' archaeology – concerned exclusively with the production of 'facts' divorced from interpretation and indeed understanding – is positivist and empiricist in approach, 'research' archaeology is much more dialectical in its process (Faulkner 2000, 26–8).

As an alternative to an undemocratic, 'top-down' archaeology he promotes 'archaeology from below' which involves communities in the archaeology of their area. The principles represented by the Sedgeford project assert that sites should not be 'preserved' but should be investigated for what they can tell us about the past – the entire past, not just certain periods – of the locality. Using local volunteers, voluntary involvement by professionals and relying on contributions in kind to finance the project, and by adopting a non-hierarchical structure to encourage initiative among the team, the project seeks to present an alternative to archaeology by professional contracting units supported by state bureaucracies. It also asserts the message that 'there is no single correct method in research', but that the greater the experience of an archaeologist in encountering the archaeological record, 'the more likely it is that critical engagement in the complex relationship between material, method and meaning will produce new ideas' (Faulkner 2000, 32). In challenging the more bureaucratic 'rescue' style of archaeological heritage management, this style of 'public archaeology' seeks to reassert the connection between archaeology as a public endeavour and archaeology as a research process. In this way, the identity between the archaeological *record* and the archaeological *resource* and *heritage* is re-established.

Conclusion

The concept of 'sustainability' has not received the attention by the archaeological community it really deserves. It requires much more thinking on behalf of us all to work out what it may mean for the discipline.

In looking at archaeology as a research discipline, but also as one that reaches out to the wider community, other ways of thinking about the concept of sustainability become possible. The Bloody Meadows Project sees preserved remains not as economic resources nor as isolated 'islands' of history, but as part of a complex story in which past, present and possible futures all flow together to make the experience of history. In Annapolis, archaeology provided the source material for the forgotten and ignored past to reassert itself at the service of understanding how history itself is made in the present. At Sedgeford, archaeology as a practice serves the creation of a sense of 'sustainable' community by creating that community's history.

Here I have placed the emphasis in understanding a 'sustainable historic environment' on the historic rather than the environmental component of the concept. To treat archaeological remains as individual components of an ecosystem is to reduce the role of the archaeologist to a mere preserver of dead material: the trick may be to reunite conservation archaeology with archaeology as a research and – more importantly – an interpretive activity. That way the distinction between the record, the resource and the heritage need no longer concern us.

Bibliography

Adams, W M, 1990 *Green development: environment and sustainability in the third world*, London

Barrett, J, 1987 Fields of discourse: reconstituting a social archaeology, *Critique Anthropology* 7(3), 5–16

Binford, L (ed), 1977 *For theory building in archaeology: essays on faunal remains, aquatic resources, spatial analysis and systematic modelling*, Stud Archaeol, New York

Binford, L, 1983 *Working at archaeology*, Stud Archaeol, San Diego

Carman, J, 1990 Commodities, rubbish and treasure: valuing archaeological objects, *Archaeol Rev Cambridge* 9(2), 195–207

Carman, J, 1996 *Valuing ancient things: archaeology and law*, London

Carman, J, 1999a Beyond the Western way of war: ancient battlefields in comparative perspective, in *Ancient warfare: archaeological perspectives* (eds J Carman and A Harding), 39–55, Stroud

Carman, J, 1999b Bloody Meadows: the places of battle, in *The familiar past? Archaeologies of later historical Britain* (eds S Tarlow and S West), 233–45, London

Cleere, H F, 1984 World cultural resource management: problems and perspectives, in *Approaches to the archaeological heritage* (ed H F Cleere), 125–31, Cambridge

Darvill, T, 1987 *Ancient monuments in the countryside: an archaeological management review*, Engl Heritage Archaeol Rep 5, London

Darvill, T, 1993 *Valuing Britain's archaeological resource*, Bournemouth Univ inaugural lecture, Bournemouth

English Heritage, 1996 *Sustaining the historic environment: new perspectives on the future*, London

English Heritage, 2000 *The power of place: the future of the historic environment*, London

Faulkner, N, 2000 Archaeology from below, *Public Archaeol* 1, 21–33

Fowler, D D, 1984 Ethics in contract archaeology, in *Ethics and values in archaeology* (ed E L Green), 108–116, New York

Fowler, P J, 1992 *The past in contemporary society: then, now*, London

Hodder, I, 1986 *Reading the past: current approaches to interpretation in archaeology*, Cambridge

Holtorf, C, 1998, The past as a renewable resource, *Arkeologen* 5–6, 4–12

Holtorf, C, 2001 Is the past a non-renewable resource? in *The destruction and conservation of cultural property* (eds R Layton, P G Stone and J Thomas), 286–97, London

Leone, M, 1984, Interpreting archaeology in historical archaeology: using the rules of persepctive in the William Paca garden, in Annapolis, Maryland, in *Ideology, power and prehistory* (eds D Miller and C Tilley), 25–38, Cambridge

Leone, M P, Mullins, P R, Creveling, C, Hurst, L, Jackson-Nash, B, Jones, L D, Kaiser, H J, Logan G C, and Warner, M S, 1995 Can an African-American historical archaeology be an alternative voice? in *Interpreting archaeology: finding meaning in the past* (eds I Hodder, M Shanks, A Alexandri, V Buchli, J Carman, J Last, and G Lucas), 110–24, London

Lipe, W D, 1984 Value and meaning in cultural resources, in *Approaches to the archaeological heritage* (ed H F Cleere), 1–11, Cambridge

MacFarlane, R, 2000, Managing whole landscapes in the post-productive rural environment, in *Landscape and sustainability* (eds J F Benson and M H Roe), 129–56, London

McDavid, C, and Babson, D (eds) 1997 *In the realm of politics: prospects for public participation in African-American archaeology*, California, Soc Hist Archaeol (= *Hist Archaeol* 31(3))

McGimsey, C R, 1972 *Public archaeology*, New York

Merriman, N, 1991 *Beyond the glass case: the public, museums and heritage in Britain*, London

Potter, P B, 1994 *Public archaeology in Annapolis: a critical approach to history in Maryland's ancient city*, Washington

Potter, P B, 1997 The archaeological site as an interpretive environment, in *Presenting archaeology to the public: digging for truths* (ed J H Jameson), 35–44 Walnut Creek California

Potter, P B and Chabot, N J, 1997, Locating truths on archaeological sites, in *Presenting archaeology to the public: digging for truths* (ed J H Jameson), 45–53, Walnut Creek California

Rannikko, P, 1999 Combining social and ecological sustainability in the Nordic forest periphery, *Sociologia Ruralis* 39(3), 394–410

Schaafsma, C F, 1989 Significant until proven otherwise: problems versus representative samples, in *Archaeological heritage management in the modern world* (ed H L F Cleere), 38–51, London

Schiffer, M B, 1972 Systemic context and archaeological context, *Amer Antiquity* 37, 156–65

Sustainable Communities Task Force 1993, *Strategic directions for community sustainability, a discussion paper for the citizens of British Columbia*, Victoria, British Columbia

Tilley, C, 1994 *A phenomenology of landscape*, London

Wester, K W, 1990 The current state of cultural resource management in the United States, in *Cultural resource management: an African dimension* (ed B W Andah), *West African J Archaeol* 20, 80–8

36

The value of archaeological remains in urban regeneration

SARAH GIBSON

Sarah Gibson
Senior Archaeology Officer
Southwark Planning and Regeneration
Council Offices, Chiltern House
Portland Street
London SE17 2ES, UK

Abstract

Recent cases in north Southwark demonstrate how the need to regenerate a former industrial area can be reconciled with the requirement to preserve nationally important archaeological remains *in situ*. Assigning a 'value' to the historic environment and its preservation in the context of sustainable development and regeneration is a challenge for the archaeological community. To demonstrate this further, the paper will also seek to illustrate how the process of regeneration and archaeological risk management and investigation can be integrated into community involvement and education initiatives.

Introduction

The London Borough of Southwark has experienced considerable regeneration recently, particularly in the north of the borough, where the decline of the docks and other associated industries has given the opportunity for the renaissance of the area.

Southwark is centrally located on the south bank of the Thames, opposite the City of London and easily accessible for the City, West End and major travel termini. Southwark Council's policy to encourage regeneration is enshrined in its Unitary Development Plan (Southwark Council 1995) and has been supported through Single Regeneration Budget (SRB) funded initiatives, particularly in the north of the borough. The overall intention is by acting as a wick effect, the regeneration and subsequent wealth creation is drawn further to the south and east of the borough.

The need to conserve, enhance, preserve and record the historic environment is also a policy of the Council. Aside from the built historic environment, the north of the borough boasts a rich and varied archaeological resource. The principle area of regeneration coincides with the historic core of this borough, and incorporated into the principal archaeological priority zone (Fig 36.1). The Roman settlement, a suburb of the city of Londinium, was located on the southern bridgehead of the only river crossing at the time, close to the modern London Bridge. The principle medieval settlement was located in broadly the same area, stretching from London Bridge south along Borough High Street to St George the Martyr church, and from Tooley Street in the east and Bankside to the west.

During the course of archaeological investigations in this area significant and indeed nationally important archaeological monuments are found from time to time. Recently, for example, the remains of the Hope Theatre, a 17th-century dual purpose playhouse and bear baiting arena was discovered on Bankside and is now preserved *in situ* beneath a new development.

Conflict between the two policy aspirations can occur.

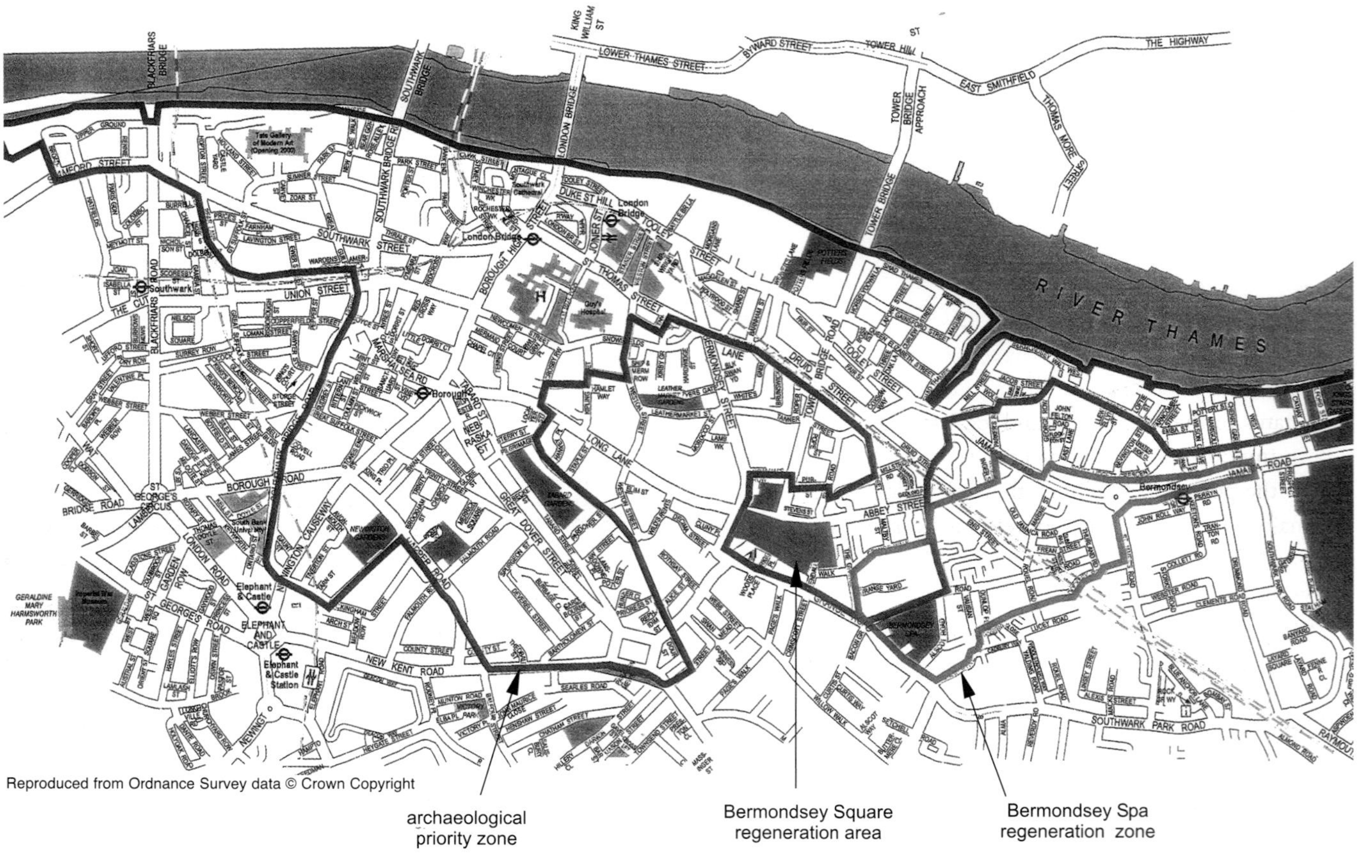

Fig 36.1 Map of north Southwark showing the Southwark archaeological priority area and the two regeneration areas at Bermondsey Square and Bermondsey Spa

Local authority archaeologists, in their advisory capacity to planning authorities and the development industry, must strike a balance between the need for regeneration and the desire to preserve archaeological remains. There are many factors to be considered in these deliberations, not least the significance of the archaeology and the practicality of preservation. The question of value is also a component in these considerations and not just the value which archaeologists attribute to the archaeological resource of the site or area, but the value that other stakeholders give it.

One of the major stakeholders is clearly the community which local authorities serve. The general view of archaeology, as part of the historic environment, is indicated in the Mori poll of 3000 people commissioned by English Heritage and published in their *Power of place* document in December 2000; a response to Government's request to review heritage policies. The results of the poll indicate both appreciation of and support for the preservation of the historic environment. The key point for consideration in this paper is that 85% of those questioned think that the historic environment is important in promoting regeneration in towns and cities. These views are of course, value judgements, and the question of how to quantify these remains.

To the south-east of the key regeneration area lies Bermondsey, a former industrial area, where tanning and the dock trades provided much employment along with food processing activities. The decline of all of these industries has led to an increase in unemployment with resulting socio-economic deprivation. There has long been a strong sense of local identity in Bermondsey, many families having lived in the area for many generations. This has resulted in some tension with those seen as newcomers, with resulting social inclusion issues.

Regeneration initiatives have been local authority led. In particular, two major regeneration projects are developing in the area. The first is on the site of Bermondsey Square, owned by Southwark Council and long identified as a key regeneration site. Still vacant after the destruction by bombing in the Second World War, the site is largely unused except on Fridays when it becomes the site of New Caledonian Antiques market, important for the local wealth creation it provides as well as a focus for the local community.

The road to the regeneration of this site has been an extremely long one and has included considerable public consultation. Indeed it was the public who selected the preferred design of the development which is currently being worked up into a planning application. The site is also the location of a Scheduled Ancient Monument, that of the 12th-century Bermondsey Abbey. In addition, it lies adjacent to a Conservation Area and near to several Listed Buildings.

Acutely aware of its responsibilities for the historic environment, the Council commissioned an archaeological desk-based assessment (MoLAS 1997) and evaluation (PreConstruct Archaeology 1998) to establish the archaeological risk on the site. The results of the evaluation showed that in some areas of the development area, a very high level of survival of the Abbey's remains could be expected. Elsewhere, later buildings on the site had caused severe disturbance.

As a scheduled monument, preservation of the Abbey remains *in situ* is paramount and should the development go ahead, would be achieved by engineering solutions through an ongoing process of archaeological investigation informing design. This approach, quite apart from being in line with the legal status of the monument and local authority policy, is also an important factor in effective sustainable development.

For the potential developer, the archaeology could be seen as a major constraint to its redevelopment, more so as any works affecting the Scheduled Ancient Monument must have consent from the Department of Culture Media and Sport (DCMS) as a legal requirement. In their view, however, it also offers an opportunity; to enhance a project by incorporating physical remains into a development or reinforcing an historic townscape through design. To redevelop a site which has been of such historic significance, where a variety of buildings have stood and fallen over the last 1200 years, demonstrates a continuity of use and a depth of understanding of the unique history of Bermondsey Square. In Bermondsey, where there is a strong sense of local identity, this contributes to telling the archaeological story of the site, providing further value to the potential development as the communication of results offers a way of engaging the local community into the redevelopment and regeneration process.

The process of engagement and consultation is also a key issue for Southwark Council on many issues, including archaeology. The Bermondsey Spa regeneration area is located further to the east of Bermondsey Square and incorporates vacant and former industrial sites. An initial council-commissioned desk-based assessment of the regeneration area demonstrated the archaeological potential for some of the sites, establishing for the Council that there is archaeological risk to development, albeit largely unquantified at this stage (Compass Archaeology 2001). As a contrast to the north of the borough and Bermondsey Square, little archaeological work has been undertaken on these more marginal areas of Bermondsey, a reflection on the small amount of redevelopment which has taken place here in the recent past.

The project is in its early development stages; although some of the sites are being marketed the key activity at present is to consult the local community. A project including the Council's Education Business Alliance, Museums Service, Property Division and Archaeology Officer, in partnership with Birkbeck College was developed. The aim was to engage with primary school children from the area talking about the archaeology of Bermondsey in relation to its historical development up until the recent past, the past that their grandparents can recall. The initiative included an historical walk showing them important local historic buildings, the site of the Abbey and talking about archaeology and how archaeologists do their jobs. This was followed up by visiting the local studies library, looking at old maps of Bermondsey and showing them the changes in the area over time, and then a session with an artist in their own classes creating a future map of Bermondsey. Bringing together the idea of the past, continuity into the future, a future that they can play a part in reshaping through regeneration and the consultation process. The six-week programme centred around visiting the Birkbeck College urban training excavation which was located on one of the Spa regeneration sites (Figs 36.2–36.3).

Fig 36.2 Students from Birkbeck College on the Southwark Council sponsored archaeological evaluation at Old Jamaica Road

Fig 36.3 Local school pupils visiting the Old Jamaica Road site as part of an educational programme run by Southwark Regeneration and Culture and Heritage departments

The project was warmly received and both teachers and children found it useful; the children especially seemed to learn a great deal from the sessions. It told the story of the area where they live and go to school through the historic walk, ancient maps and archaeological excavation. In evaluating the project, one of the teachers stated that '... many of the children had little knowledge of their immediate area and we believe that this project helped them to identify and feel a sense of belonging'. Another teacher called it a thoroughly enriching experience.

From the Council's point of view, relating the archaeology and history of the locality has helped to engage a section of the community which typically receives little attention, drawing them into the consultation and regeneration process and through them, their parents and friends. It helps to encourage sense of ownership of the area by all ethnic groups, demonstrating its past ancient history and linking it with the recent past. It also allowed the archaeological risk associated with one of the regeneration sites to be established; an important factor for both the Council and the potential developer.

In summary, archaeology gives value to urban regeneration by demonstrating continuity, engaging the local community into the regeneration process by telling the story of their local history and archaeology and by contributing to sustainable development. These values are largely shared amongst the stakeholders: archaeologists, developers and local communities.

Local authority archaeologists are in a unique position to influence sustainable development and regeneration policies and maximise the opportunities that the historic environment can bring to the consultation process. However, it is imperative that the wider archaeology community translates the value of archaeology to all end-users and in particular those for whom access to the historic environment has been traditionally more limited in the past.

Bibliography

Compass Archaeology, 2001 Bermondsey Spa Regeneration sites; an archaeological desk-based assessment, unpub Compass Archaeology rep

English Heritage, 2000 *Power of place; the future of the historic environment*, London

MoLAS, 1997 Bermondsey Square, an archaeological desktop assessment, unpub MoLAS rep

PreConstruct Archaeology, 1998 An archaeological evaluation at Bermondsey Square, London Borough of Southwark, unpub PCA rep

Southwark Council, 1995 *Unitary development plan*, London